Fault Detection and State Estimation in Automatic Control

Fault Detection and State Estimation in Automatic Control

Editors

Sheng Du
Wei Wang
Hao Fu
Xiongbo Wan

MDPI

Basel • Beijing • Wuhan • Barcelona • Belgrade • Novi Sad • Cluj • Manchester

Editors

Sheng Du
China University of Geosciences
Wuhan, China

Wei Wang
Zhongnan University of Economics and Law
Wuhan, China

Hao Fu
Wuhan University of Science and Technology
Wuhan, China

Xiongbo Wan
China University of Geosciences
Wuhan, China

Editorial Office
MDPI
St. Alban-Anlage 66
4052 Basel, Switzerland

This is a reprint of articles from the Special Issue published online in the open access journal *Applied Sciences* (ISSN 2076-3417) (available at: https://www.mdpi.com/journal/applsci/special_issues/07C34Z522J).

For citation purposes, cite each article independently as indicated on the article page online and as indicated below:

Lastname, A.A.; Lastname, B.B. Article Title. *Journal Name* **Year**, *Volume Number*, Page Range.

ISBN 978-3-0365-9756-0 (Hbk)
ISBN 978-3-0365-9757-7 (PDF)
doi.org/10.3390/books978-3-0365-9757-7

Contents

About the Editors

Sheng Du

Sheng Du received a B.S. degree in Measurement and Control Technology and Instrument and a Ph.D. degree in Control Science and Engineering from the China University of Geosciences, Wuhan, China, in 2016 and 2021, respectively. He was a joint Ph.D. student at the Department of Electrical and Computer Engineering, University of Alberta, Edmonton, Canada, from 2019 to 2021. He is currently a Professor and Postdoctoral Fellow with the School of Automation, China University of Geosciences, Wuhan, China. His research interests include process control, intelligent control, intelligent optimization, computational intelligence, and artificial intelligence. Dr. Du is a Senior Member of the Chinese Association of Automation (CAA) and a winner of the Outstanding Doctoral Thesis Nomination Award of the CAA. He is an Editorial Board Member of *PLoS ONE, Journal of Sensors, International Transactions on Electrical Energy Systems*, etc.

Wei Wang

Wei Wang received his B.S. degree in automation and Ph.D. degree in control science and engineering from the China University of Geosciences, Wuhan, China, in 2013 and 2019, respectively. He is currently a lecturer at the School of Information Engineering, Zhongnan University of Economics and Law, Wuhan, China. His current research interests include reinforcement learning, adaptive dynamic programming, and multi-agent systems.

Hao Fu

Hao Fu received his Ph.D. degree in control science and engineering from the China University of Geosciences, Wuhan, China, in 2020. He is currently a lecturer with the School of Computer Science and Technology, Wuhan University of Science and Technology, Wuhan, China. His current research interests include multi-robot systems, reinforcement learning and neural networks.

Xiongbo Wan

Xiongbo Wan received his B.S. degree in information and computing science from Huazhong Agricultural University, Wuhan, China, in 2006, his M.S. degree in probability theory and mathematical statistics, and his Ph.D. degree in control theory and engineering from the Huazhong University of Science and Technology, Wuhan, in 2008 and 2011, respectively. From December 2012 to June 2013, he was a Visiting Scholar at Akita Prefectural University, Akita, Japan. From July 2011 to December 2014, he was a lecturer and then an Associate Professor at the College of Engineering, Huazhong Agricultural University. From December 2016 to November 2017, he was a Visiting Scholar with the Department of Computer Science, Brunel University London, Uxbridge, U.K. Since 2015, he has been with the China University of Geosciences, Wuhan, where he is currently a Professor at the School of Automation. His research interests include fault diagnosis, state estimation and model predictive control of complex networked systems.

applied sciences

MDPI

Editorial

Fault Detection and State Estimation in Automatic Control

Sheng Du [1,2,3,*], Wei Wang [4,*], Hao Fu [5,*] and Xiongbo Wan [1,2,3,*]

1 School of Automation, China University of Geosciences, Wuhan 430074, China
2 Hubei Key Laboratory of Advanced Control and Intelligent Automation for Complex Systems, Wuhan 430074, China
3 Engineering Research Center of Intelligent Technology for Geo-Exploration, Ministry of Education, Wuhan 430074, China
4 School of Information and Security Engineering, Zhongnan University of Economics and Law, Wuhan 430073, China
5 School of Computer Science and Technology, Wuhan University of Science and Technology, Wuhan 430081, China
* Correspondence: dusheng@cug.edu.cn (S.D.); imagef@zuel.edu.cn (W.W.); fuhao@wust.edu.cn (H.F.); xbwan23@cug.edu.cn (X.W.)

Citation: Du, S.; Wang, W.; Fu, H.; Wan, X. Fault Detection and State Estimation in Automatic Control. *Appl. Sci.* **2023**, *13*, 12936. https://doi.org/10.3390/app132312936

Received: 19 November 2023
Accepted: 1 December 2023
Published: 4 December 2023

1. Introduction

Fault detection and state estimation play pivotal roles in ensuring the reliability, safety, and performance of automatic control systems. Recently, the integration of advanced methodologies with cutting-edge technologies has significantly impacted the fields of fault detection and state estimation. Particularly with advancements in artificial intelligence, the fusion of deep learning and ensemble methods, such as K -nearest neighbors, random forest regressors, and support vector regression, has garnered considerable attention. These robust, artificial-intelligence-driven approaches have been developed for intricate action recognition, predictive models for fault identification, and fault-tolerant control in complex, multi-sensor systems.

Amidst the challenges presented by packet drops, delays, and the complexities of large-scale networked systems, the pursuit of fault detection and state estimation has ventured into innovative domains. From empowering swarm robots with multitarget search capabilities to developing resilient prediction models capable of navigating uncertainties, this field stands at the forefront of innovation. The integration of these methodologies not only enhances the system's resilience but also ensures its adaptability to unforeseen disturbances.

These studies transcend traditional boundaries, immersing themselves in the domain of Swarm Robots and Multitarget Searches within intricate, interconnected environments. Furthermore, they unveil the pivotal role of fault detection and state estimation in guaranteeing the functionality of automated systems across various industries and sectors.

Fault detection and state estimation stand as imperative tasks for ensuring the reliability, safety, and performance of automatic control systems. Nevertheless, these endeavors encounter numerous challenges, including nonlinear dynamics, uncertain disturbances, incomplete information, sensor faults, and computational complexity. Hence, there is a pressing requirement for novel methods and algorithms capable of surmounting these obstacles and delivering precise, robust solutions.

One promising avenue involves delving into the potential of machine learning and artificial intelligence techniques for fault detection and state estimation [1,2], with a specific emphasis on reinforcement learning (RL) [3]. These techniques possess the capacity to learn from data and adapt to dynamic environments, thereby enhancing the fault diagnosis and state estimation capabilities of automatic control systems [4]. For instance, RL can be employed to craft intelligent fault detection and diagnosis methods that optimize the delicate balance between detection accuracy and timeliness [5]. Neural networks are valuable tools for approximating nonlinear functions and estimating unknown states and parameters [6]. Deep learning, on the other hand, proves instrumental in extracting features and patterns from high-dimensional data, thereby improving fault classification [7].

Another prospective direction involves the integration of diverse methods and models for fault detection and state estimation. This integration enhances the robustness and reliability of solutions by leveraging the strengths of different approaches. For instance, an interval analysis can adeptly handle uncertain state perturbations and measurement noise by computing guaranteed bounds on state estimates [8]. Kalman filtering facilitates the fusion of multiple information sources and continuously updates state estimates based on measurement data [9]. Hybrid systems provide a modeling framework for capturing interactions between continuous and discrete dynamics and events. This facilitates control, verification, state estimation, and fault detection in complex systems [10].

Nevertheless, numerous challenges still require attention, including scalability, computational efficiency, online implementation, fault isolation, fault recovery, and fault-tolerant control. Hence, additional research efforts are imperative to propel this critical field forward, both in theory and practice.

2. An Overview of Published Articles

With the current state of science and technology, the modern industrial production scale and the complexity of automation in control systems have greatly improved. Additionally, state estimation and fault detection are particularly important in the production process of these systems before a fault causes any damage; further, testing and maintenance can reduce the risk of accidents, improve the system's security, and reduce the economic loss of production. Therefore, the purpose of this Special Issue is to introduce the latest fault detection algorithms and state estimation methods.

This Special Issue focuses on intelligent control, intelligent modeling, computational intelligence, artificial intelligence, machine learning, and fault detection. This fits within the scope of Applied Sciences, as the practical applications of fault detection and machine learning are incredibly extensive and important. The research areas of this Special Issue include (but are not limited to) the design and application of fault detection algorithms, state estimation methods, machine learning algorithms, intelligent control systems, and analyzing the characteristics of automatic control systems.

An analysis of the papers published in this Special Issue is shown in Table 1. Many studies have been conducted by scholars on fault detection and state estimation in the context of automatic control, covering many related research areas. Studies of contribution 1, 3 and 5 are related to automation engineering research; studies of contribution 2, 8, 10, and 11 are related to aircraft control; studies of contribution 4, and 5 are related to sensor control; studies of contribution 6, 12, and 13 are related to robot control; and other studies are more related to system control research.

Interestingly, studies of contribution 1, and 9 are mostly focused on the prediction of working conditions, where contribution 1 uses a support vector regression algorithm to predict vibrational amplitudes, and contribution 9 uses a long short-term memory network to predict the exhaust temperature of a diesel engine. Studies contribution 3 and contribution 10 both focus on monitoring specific working conditions in real-time as well as analyzing and optimizing the system stability. The authors of contribution 3 analyze the stability of a system under different working conditions based on several factors. The authors of contribution 10 establish a continuous trajectory planning model combined with the ant colony optimization algorithm to monitor the optimal trajectory of an unmanned aerial vehicle.

Furthermore, in contribution 2, abnormal or faulty behavior is detected by efficiently encoding information about a target pose to recognize various human actions more accurately. The authors of contribution 7 focus on fault identification under specific working conditions and use Fisher's discriminant analysis to diagnose the faults of dissolved oxygen sensors in wastewater treatment plants while evaluating both environmental and economic factors. Meanwhile, the authors of contribution 12 deal with obstacle detection in automatic control applications to achieve real-time obstacle avoidance during a multi-target search by swarm robots.

Table 1. Analysis of the published contributions in the Special Issue.

No.	DOI	Research Area	Focus	Type of Research	Industry	Country
S1	10.3390/ app132111878	Automation Engineering	riserless mud recovery technology, mud pump, ANSYS software, SVR algorithm	Mathematical Modeling	Geological Exploration	China
S2	10.3390/ app13169384	Aircraft Control	deep neural network, convolutional LSTM, action recognition, body pose keypoints; aerial surveillance	Mathematical Modeling	Automation	Pakistan
S3	10.3390/ app13169320	Automation Engineering	riserless mud-recovery technology, ABAQUS software, SVR-DSWA algorithm	Mathematical Modeling	Geological Exploration	China
S4	10.3390/ app13158778	Sensor Control	multi-sensor systems, robust fusion estimation, event-triggered, random packet drops; d-step state delay, deterministic control inputs	Simulation	Manufacturing	China
S5	10.3390/ app13127212	Automation Engineering	crack propagation, machine learning, dynamic load, random forest regressor, support vector regression, gradient boosting regressor, ridge, lasso, k-nearest neighbors	Comparisons	Materials	UK
S6	10.3390/ app13042675	Robot Control	fault-tolerant control, FTC, robot manipulators, artificial intelligence	Literature Review	Manufacturing	Poland
S7	10.3390/ app13042554	Sensor Control	fault identification, fisher discriminant analysis, dissolved oxygen sensor, energy costs assessment, GHG emissions assessment	Simulation	Chemical Engineer- ing	Romania
S8	10.3390/ app13021214	Aircraft Control	dissipativity, large-scale system, linear matrix inequality, networked system, sparseness	Numerical Simulations	Aerospace	China
S9	10.3390/ app13021099	Power System Control	diesel engine heat load, intelligent detection, long short-term memory network, prediction model, evaluation index	Mathematical Modeling	Transportation	China
S10	10.3390/ app122312111	Aircraft Control	persistent monitoring, privacy protection, path planning, monitoring frequency, overdue time	Mathematical Modeling + Simulation	Aerospace	China

Table 1. *Cont.*

No.	DOI	Research Area	Focus	Type of Research	Industry	Country
S11	10.3390/ app122312069	Aircraft Control	powered parafoil, altitude control, sliding mode backstepping, fractional calculus, LESO	Mathematical Modeling + Simulation	Manufacturing	China
S12	10.3390/ app13031969	Robot Control	swarm robot, unknown complex environment, multitarget cooperative search, simplified virtual force model, particle swarm optimization	Mathematical Modeling + Simulation	Geological Exploration	China
S13	10.3390/ app12178511	Robot Control	flexible-joint manipulators, reduced-order extended state observer, backstepping, command filter, error compensation	Numerical Simulations	Manufacturing	China

Overall, these studies cover a wide range of industries, including manufacturing, transportation, aerospace, materials, chemicals and geological exploration. The authors are mainly from China, but there are also scholars from Pakistan, the UK, Poland and Romania who have contributed to our Special Issue.

3. Conclusions

This editorial letter describes the roles of fault detection and state estimation in automatic control systems and highlights the applications of advanced methods and cutting-edge technologies in recent years. In particular, with the development of artificial intelligence, the convergence of deep learning and integration methods has attracted significant attention. These powerful AI-driven methods are designed for fine-grained action recognition, predictive models for fault identification, and fault-tolerant control in complex multi-sensor systems. Other research efforts in the Special Issue advance fault detection and state estimation in both theory and practice. These studies show that fault detection and state estimation have become particularly important in industrial production processes as the level of technology and the scale, complexity, and automation of modern industrial production increase.

Conflicts of Interest: The authors declare no conflict of interest.

List of Contributions:

1. He, G.; Xu, B.; Chen, H.; Qin, R.; Li, C.; Yin, G. Study of the Relationships among the Reverse Torque, Vibration, and Input Parameters of Mud Pumps in Riserless Mud Recovery Drilling *Appl. Sci.* **2023**, *13*, 11878. https://doi.org/10.3390/app132111878.
2. Saeed, S.M.; Akbar, H.; Nawaz, T.; Elahi, H.; Khan, U.S. Body-Pose-Guided Action Recognition with Convolutional Long Short-Term Memory (LSTM) in Aerial Videos. *Appl. Sci.* **2023**, *13*, 9384. https://doi.org/10.3390/app13169384.
3. Qin, R.; Lu, Q.; He, G.; Xu, B.; Chen, H.; Li, C.; Yin, G.; Wang, J.; Wang, L. Quantitative Analysis of the Stability of a Mud-Return Circulation System in a Riserless Mud-Recovery Drilling System. *Appl. Sci.* **2023**, *13*, 9320. https://doi.org/10.3390/app13169320.
4. Du, X.; Liu, H.; Yu, H. Event-Triggered Robust Fusion Estimation for Multi-Sensor Time-Delay Systems with Packet Drops. *Appl. Sci.* **2023**, *13*, 8778. https://doi.org/10.3390/app13158778.
5. Omar, I.; Khan, M.; Starr, A. Comparative Analysis of Machine Learning Models for Predicting Crack Propagation under Coupled Load and Temperature. *Appl. Sci.* **2023**, *13*, 7212. https://doi.org/10.3390/app13127212.
6. Milecki, A.; Nowak, P. Review of Fault-Tolerant Control Systems Used in Robotic Manipulators *Appl. Sci.* **2023**, *13*, 2675. https://doi.org/10.3390/app13042675.

7. Luca, A.-V.; Simon-Várhelyi, M.; Mihály, N.-B.; Cristea, V.-M. Fault Type Diagnosis of the WWTP Dissolved Oxygen Sensor Based on Fisher Discriminant Analysis and Assessment of Associated Environmental and Economic Impact. *Appl. Sci.* **2023**, *13*, 2554. https://doi.org/10.3390/app13042554.
8. Sun, Y.; Wang, J.; Liu, H. Dissipativity Analysis of Large-Scale Networked Systems. *Appl. Sci.* **2023**, *13*, 1214. https://doi.org/10.3390/app13021214.
9. Zhou, R.; Cao, J.; Zhang, G.; Yang, X.; Wang, X. Heat Load Forecasting of Marine Diesel Engine Based on Long Short-Term Memory Network. *Appl. Sci.* **2023**, *13*, 1099. https://doi.org/10.3390/app13021099.
10. Chen, Y.; Shu, Y.; Hu, M.; Zhao, X. Multi-UAV Cooperative Path Planning with Monitoring Privacy Preservation. *Appl. Sci.* **2022**, *12*, 12111. https://doi.org/10.3390/app122312111.
11. Zhu, E.; Du, Y.; Song, W.; Gao, H. Altitude Control of Powered Parafoil Using Fractional Sliding-Mode Backstepping Control Combined with Extended State Observer. *Appl. Sci.* **2022**, *12*, 12069. https://doi.org/10.3390/app122312069.
12. Zhou, Y.; Zhou, S.; Wang, M.; Chen, A. Multitarget Search Algorithm Using Swarm Robots in an Unknown 3D Mountain Environment. *Appl. Sci.* **2023**, *13*, 1969. https://doi.org/10.3390/app13031969.
13. Pan, C.; Fei, X.; Xiao, J.; Xiong, P.; Li, Z.; Huang, H. Model-Assisted Reduced-Order ESO Based Command Filtered Tracking Control of Flexible-Joint Manipulators with Matched and Mismatched Disturbances. *Appl. Sci.* **2022**, *12*, 8511. https://doi.org/10.3390/app12178511.

References

1. Cheliotis, M.; Lazakis, I.; Theotokatos, G. Machine learning and data-driven fault detection for ship systems operations. *Ocean Eng.* **2020**, *216*, 107968. [CrossRef]
2. Zamzam, A.S.; Fu, X.; Sidiropoulos, N.D. Data-driven learning-based optimization for distribution system state estimation. *IEEE Trans. Power Syst.* **2019**, *34*, 4796–4805. [CrossRef]
3. Sutton, R.S.; Barto, A.G. *Reinforcement Learning: An Introduction*, 2nd ed.; MIT Press: Cambridge, MA, USA, 2018.
4. Jiang, W.; Yang, W.; Zhou, J.; Ding, W.; Luo, Y.; Liu, Y. Reinforcement learning-based detection for state estimation under false data injection. *IEEE Access* **2021**, *9*, 66498–66508. [CrossRef]
5. Zhang, W.; Zhu, J. A reinforcement learning system for fault detection and diagnosis in mechatronic systems. *Comput. Model. Eng. Sci.* **2020**, *124*, 1119–1130. [CrossRef]
6. Talebi, H.A.; Abdollahi, F.; Patel, R.V.; Khorasani, K. *Neural Network-Based State Estimation of Nonlinear Systems: Application to Fault Detection and Isolation*; Springer: Berlin/Heidelberg, Germany, 2009.
7. Yu, J.; Zhang, Y. Challenges and opportunities of deep learning-based process fault detection and diagnosis: A review. *Neural Comput. Appl.* **2023**, *35*, 211–252. [CrossRef]
8. Raka, S.A.; Combastel, C. Fault detection based on robust adaptive thresholds: A dynamic interval approach. *Annu. Rev. Control* **2013**, *37*, 119–128. [CrossRef]
9. Ondel, O.; Boutleux, E.; Blanco, E.; Clerc, G. Coupling pattern recognition with state estimation using Kalman filter for fault diagnosis. *IEEE Trans. Ind. Electron.* **2011**, *59*, 4293–4300. [CrossRef]
10. Wang, W.; Li, L.; Zhou, D.; Liu, K. Robust state estimation and fault diagnosis for uncertain hybrid nonlinear systems. *Nonlinear Anal. Hybrid Syst.* **2007**, *1*, 2–15. [CrossRef]

applied sciences

MDPI

Review

Review of Fault-Tolerant Control Systems Used in Robotic Manipulators

Andrzej Milecki * and Patryk Nowak

Division of Mechatronic Devices, Poznan University of Technology, 60-965 Poznan, Poland
* Correspondence: andrzej.milecki@put.poznan.pl

Abstract: Control systems that ensure robot operation during failures are necessary, particularly when manipulators are operating in hazardous or hard-to-reach environments. In such applications, fault-tolerant robot controllers should detect failures and, using fault-tolerant control methods, be able to continue operation without human intervention. Fault-tolerant control (FTC) is becoming increasingly important in all industries, including production lines in which modern robotic manipulators are used. The use of fault-tolerant systems in robotics can prevent the production line from being immobilized due to minor faults. In this paper, an overview of the current state-of-the-art methods of fault-tolerant control in robotic manipulators is provided. This review covers publications from 2003 to 2022. The article pays special attention to the use of artificial intelligence (AI), i.e., fuzzy logic and artificial neural networks, as well as sliding mode and other control methods, in the FTC of robotic manipulators. The cited and described publications were mostly found using Google Scholar.

Keywords: fault-tolerant control; FTC; robot manipulators; artificial intelligence

Citation: Milecki, A.; Nowak, P. Review of Fault-Tolerant Control Systems Used in Robotic Manipulators. *Appl. Sci.* **2023**, *13*, 2675. https://doi.org/10.3390/app13042675

Academic Editors: Sheng Du, Xiongbo Wan, Wei Wang and Hao Fu

Received: 29 December 2022
Revised: 14 February 2023
Accepted: 16 February 2023
Published: 19 February 2023

1. Introduction

The basic concepts and techniques of fault tolerance in computer systems were described in [1], which presented the different error classifications and the application of redundancy techniques to ensure the reliable operation of computers. The modeling and prediction of faults were also described, along with examples of fault-tolerant computers. In the early 1990s, Stengel [2], Veillette [3], and Patton [4] published the results of different applications of FTC systems. In [5], state-of-the-art methods for fault-tolerant control applications up until 1997 were presented.

In a fault-tolerant control system, it is necessary to distinguish concepts such as fault definition and fault tolerance. A fault is defined as the deviation of a parameter from an acceptable value. Fault tolerance is the ability of the system to continue operation regardless of faults. The origins of FTC systems are related to the need to use special control in critical safety devices and equipment. Examples of such devices are aircraft or nuclear power plants. In [6], which was published in 1990, one of the first applications of fault-tolerant control in robot manipulators was described and the measure of joint failure influences on the remaining dexterity of a kinematically redundant manipulator was quantified. This measure was used as a criterion and technique for calculating optimal configurations of fault-tolerant redundant manipulators. An example of a three-dimensional robot was shown that used a fourth actuator, which guaranteed half of the original dexterity, after the joint failure.

In the same period, i.e., in the 1990s, extensive research was undertaken to develop FTC systems for applications in manipulators intended for space research [7–9]. The foundations for the design of fault-tolerant manipulators were presented in [10]. In this article, the modeling tools and reliable evaluation methods of robots were described. The design aspects, together with a reliability evaluation, were presented. Based on the performed analyses, a methodology for designing fault-tolerant manipulators was proposed. Particular attention was paid to the development of recommendations for the design of the arm with 10 degrees of freedom. It was proposed to use two actuators on each joint.

Techniques for a robust residual generation have been proposed by several authors In order to ensure the successful operation of autonomous robots in remote or hazardous environments, fault tolerance is essential. In [11], different fault tolerance algorithms were proposed, where the possible failures of the robot components and the interdependence of these failures must be determined first. To achieve this, fault tree analysis can be a very useful tool. The tree structures developed using this type of analysis make it possible to both determine a block diagram of possible occurrences of robot faults and define the cause-and-effect relationships between failures.

In [12], a control method was proposed in which a model reference algorithm and a proportional integral derivative controller were applied in the fault-tolerant operation of robotic manipulators. This method was tested in a simulation that showed the effectiveness of the recovery algorithm, which enabled it to continue the movement on the assumed trajectory and reach the given end effect.

In 1987, Selkäinaho and Halme [13] proposed the use of artificial intelligence, i.e. a real-time expert system, in a fault control solution, which was used in an algorithm to supervise fault detection and localization. In the case of sensor damage, the system automatically replaced the faulty measurement with an updated predictor model output signal. The investigated system was successfully tested in a pilot process using artificial intelligence to overcome physical faults.

The authors of [14] proposed the use of intelligent fault-tolerant control in a flexible assembly cell, where the fault-tolerant controller would immediately recognize errors and react in real time to such situations. In the described solution, the combination of an advanced autonomous supervision system was combined with a sensor-guided action generator. Experimental tests were conducted using a mobile two-arm robot system.

Groom et al. [15] described the real-time fault-tolerant control developed for a kinematically redundant manipulator. The authors used a fault-tolerance measure to enable the end effector of the manipulator to continuously follow a given trajectory. Finally, an algorithm was developed and used in real time for the control of a seven-degree-of-freedom commercially available manipulator.

The purpose of using FTC systems is to assure reliability, which should be maintained despite the occurrence of faults of an acceptable nature. Intensive work is currently underway to develop this type of control in equipment used during manufacturing processes. One of the advantages of using this type of control in the production process is not only the safety of critical equipment but also cost reduction. The reduction in production costs is influenced by the continued operation of the devices used. At the time of an acceptable FTC failure, there is no need to stop production for maintenance purposes. This results in a reduction in the costs associated with service calls and also saves time by preventing a halt in the production process.

Shin and Lee [16] presented a robust fault-tolerant control system, which can be used to overcome the failure of the robot manipulator actuators. The controller uses the algorithms for normal control (non-failed), fault detection, and control in case of faults to achieve the assumed task completion.

Fault-tolerant control systems can be divided into three main categories: passive FTCS (PFTCS), active FTCS (AFTCS), and hybrid FTCS (HFTCS). PFTCS are characterized by the fact that they operate offline and are only able to adapt to faults that are defined at the design stage. Passive fault-tolerant steer systems do not require fault detection and isolation and are less computationally complex than active systems. AFTCS consist of a reconfiguration mechanism and a fault defection identification (FDI) mechanism. The FDI subsystem provides information to the controller about the fault and the controller responds by reconfiguring itself or the controlled device. The FDI subsystem is a key component of an active FTCS as it detects the fault and immediately sends information to the controller The immediate reconfiguration of the controller upon receiving information from the FDI allows this method to be called active. There are also hybrid HFTCS controllers, which are a combination of PFTCS and AFTCS.

In this paper, all the above-mentioned methods used in applications for robotic arm steering are reviewed. The methodology for selecting articles for this paper was limited to the time frame of the proposed solutions, that is, mainly the last 20 years. In addition, it was decided to place the main emphasis only on solutions that use artificial intelligence and sliding-mode controllers. Furthermore, all the selected papers deal with the use of FTC methods in robot control. In addition, the papers that focus on the control of manipulators with up to 7 degrees of freedom were chosen for our review. Articles about other robots such as hexapods or mobile robots were excluded from this review. This review is presented in six chapters, covering the available solutions of various FTC systems used to control robotic arms. In Section 1, the history and background of FTC systems are described. Section 2 is devoted to presenting the state-of-the-art FTCS that use artificial intelligence methods. Section 3 presents the variants of FTCS that use sliding-mode control. Section 4 deals with other FTCS used in robots that have been developed in recent years. Section 5 presents the summary and Section 6 presents the conclusions.

2. Artificial Intelligence Methods in FTCS

Artificial intelligence methods are highly developed techniques that have applications in fault-tolerant strategies for controlling robotic arms. Among the artificial intelligence methods used for fault-tolerant control systems are fuzzy logic controllers, which are able to effectively deal with non-linear systems using membership functions, which assess analog input signals using logical variables that take values between 0 and 1. The authors of [17] described the so-called adaptive fuzzy type-2 backstepping control (AFT2BC) method. The backstepping method was also described in [18]. AFT2BC was tested on a model of a PUMA560 robotic arm in a MATLAB environment in which the kinematic configuration of the robot could be changed simultaneously to simulate axis failure and changes in the load on the arm. The proposed control algorithm did not require a priori knowledge of the dynamic robot model so the controller could operate when the faulty condition was due to both model uncertainty and external disturbances. Figure 1 shows a block scheme of the proposed adaptive control method. On the description of the graph is a set of Lyapunov Equations (1)–(6) [17]:

$$\dot{V}_1 = e_1\dot{e}_1 = e_1(\dot{q}_d - x_2), \tag{1}$$

where $\dot{V}_1$ is the derivative of the Lyapunov function [19], e_1 is the tracking error, $\dot{q}_d$ is the derivative of the desired signal, x_2 is the vector of the current state $x_2 = \dot{x}_1$, $x_1 = [q_1 q_2 q_3]^T$,

$$u_a = W^T(e_1\dot{e}_1)\Theta, \tag{2}$$

where u_a is the fuzzy type-2 adaptive control expression that is designed to estimate the ideal backstepping control law, W^T represents the average basis functions obtained by a fuzzy type-2 system, where each basis function is given by the average of the corresponding left and right basis functions, Θ denotes the adapted vector parameters,

$$\dot{\Theta} = \gamma e_2 W(e_1\dot{e}_1) - \sigma_1\Theta, \tag{3}$$

where $\gamma > 0$, $\sigma_1 > 0$, $e_2 = \dot{q}_d + c_1e_1 - x_2$, c_1 is a positive constant vector, $\Theta(0) = 0$,

$$u_r = \hat{\varepsilon}\tanh\left(\frac{e_2}{\chi}\right), \tag{4}$$

where u_r is the robust control expression that is introduced to reduce the effects of fuzzy type-2 estimation errors,

$$\dot{\hat{\varepsilon}} = \eta\hat{e}_2\tanh\left(\frac{e_2}{\chi}\right) - \sigma_2\hat{\varepsilon}, \tag{5}$$

where $\eta > 0, \sigma_2 > 0, \chi > 0, \hat{\varepsilon}(0) = 0$ are parameters,

$$u_p = c_2 e_2, \tag{6}$$

where $c_2 > 0$.

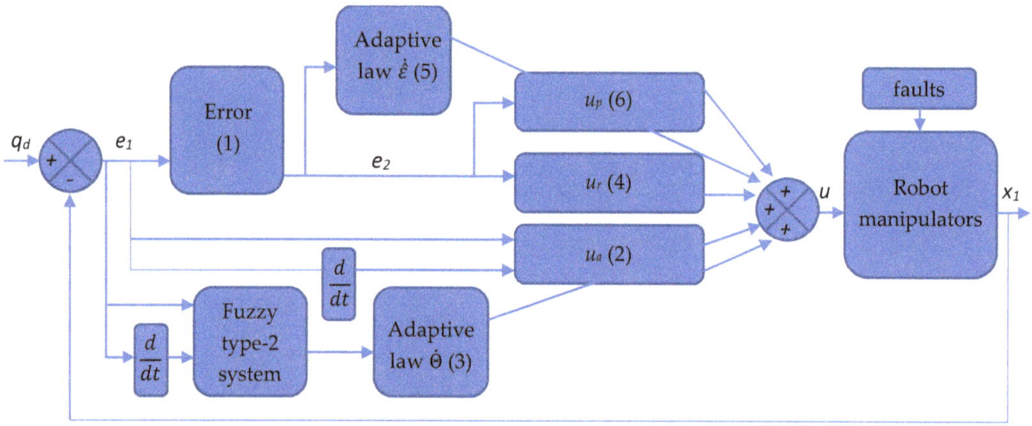

Figure 1. AFT2BC control scheme [17], where u are the desired robot axes' positions.

The authors of [20] presented a fault-tolerant control scheme in the form of non-singular terminal synergetic control (NTSC) combined with interval type-2 fuzzy satin bowerbird optimization (IT2FSBO). The paper also described an adaptive augmented extended Kalman filter (A-AEKF) that detected, identified, and isolated actuator faults, even in the presence of noise [21].

The authors of [22] presented a neuro-fuzzy (NF) robot fault-detection algorithm that allowed for the control of a robot with an SRI [23] controller despite sensor or axis actuator failures. The detection and fault-tolerance architecture of the proposed solution was built with a multilayer perceptron trained with the backpropagation of errors and a fuzzy logic block. The input layer of the artificial neural network consisted of 15 neurons organized into 3 groups, each consisting of 5 neurons, 1 for each axis. The three groups were used in turn for the preliminary assessment of the position, velocity, and acceleration. The network had two groups of output neurons, each with five neurons, one for each axis. These groups generated the positions and velocities for each axis in turn. The results of the multilayer perceptron (MLP) in the form of the product of the position and velocity, i.e., $q\dot{q}$, was summed with the output parameters of the robot. This sum was sent to the inference block, along with the results of the fuzzy logic block. The inference block receiving the data provided information about the axis that was damaged. The algorithm was successfully tested on a simulation of the ER5u robot in a MATLAB environment.

The authors of [24] presented a system in which an artificial neural network in the form of a multilayer perceptron (MLP) was responsible for monitoring and fault detection (FD). This system consisted of a block in which the MLP provided information about the system failure while the fuzzy logic rule base made decisions about the type of fault and its location. The authors demonstrated the performance of fault detection based on a neuro-fuzzy (NF) application for a robotic arm with 5 degrees of freedom when one of the robot's axes fails. The architecture of the system was similar to the solution described in [23] but in this solution, the inference by the fuzzy logic system was based on the output of the neural network.

The authors of [25] proposed a new solution in the form of self-tuning fuzzy proportional–integral–derivative non-singular fast terminal sliding-mode control (SFT-PID-NFTSM) with time delay estimation (TDE). The proposed method involved the self-tuning of the proportional–integral–derivative (PID) block of the controller using fuzzy logic algorithms. Compared to other well-known methods described in the literature, such as the PID controller or the controller with a fuzzy logic block responsible for the self-tuning of the PID block, i.e., (PID-NFTSM), the results of [25] showed that the controller proposed in this work was more stable, had less overshoot, and improved transient responses. In addition, the integration of TDE reduced the required computational power of the controller and also helped to eliminate the requirement to know the exact dynamics of the system. The architecture of this system is shown in Figure 2. In order to fully understand the operation of the presented architecture, it is necessary to explain Formulas (7)–(15), which underpin the system [25]:

$$s = e + k_1 e^{[\lambda]} + k_2 \dot{e}^{[\frac{p}{q}]}, \tag{7}$$

where s is the NFTSM sliding surface, e is the error, $k_1 = diag(k_{11}, k_{12}, \ldots, k_{1n}) \in R^{nxn}$ and $k_2 = diag(k_{21}, k_{22}, \ldots, k_{2n}) \in R^{nxn}$ are two positive definite matrices, p and q are two positive odd numbers selected to satisfy the conditions $1 < \frac{p}{q} < 2 \; and \; \lambda > p/q$,

$$S_{PID}(t) = K_p s(t) + K_i \int_0^t s(t)dt + K_d \frac{ds(t)}{dt}, \tag{8}$$

where S_{PID} is the sliding surface based on PID-NFTSM, K_p, K_i, K_d are the proportional, integral, and derivative gains,

$$|\Delta(x_1, x_2, u, \tau_d)| \leq \Delta_0, \tag{9}$$

where $x_1 = q, x_2 = \dot{q}, u = \tau$ are the state variables, τ_d is the disturbance torque, Δ_0 is a constant, $\Delta(x_1, x_2, u, \tau_d)$ covers all the effects of the uncertainties, disturbances, and faults,

$$\left|\dot{\Delta}(x_1, x_2, u, \tau_d)\right| \leq \Delta_1, \tag{10}$$

where Δ_0 is a constant,

$$\wedge(\dot{e}, \Delta)_{(t)} \cong \wedge(\dot{e}, \Delta)_{(t-L)}, \tag{11}$$

where $\wedge(\dot{e}, \Delta)_{(t)}$ is the unknown function, $(t - L)$ is the time delay,

$$u_{TDE_t} \cong \widehat{\wedge}(\dot{e}, \Delta)_{(t)}, \tag{12}$$

where $u_{TDE_t}, \widehat{\wedge}(\dot{e}, \Delta)_{(t)}$ is the estimation of the unknown function,

$$\dot{u}_{ar} = \left(\widehat{K} + a\right) sign(S_{PID}), \tag{13}$$

where $\dot{u}_{ar}$ is the adaptive law, K is an unknown constant, $\widehat{K}$ is the estimation of the bounded value K,

$$u_{PID} = -\Omega^+(\dot{e}x_1)(u_{eq} + u_{TDE} + u_{ar}), \tag{14}$$

where u_{PID} is the controller output signal, $-\Omega^+(\dot{e}x_1)$ is a lumped unknown function,

$$V = \frac{1}{2} S_{PID}^T S_{PID} + \frac{1}{2} C \check{K}^T \check{K}, \tag{15}$$

where V is the Lyapunov function, $\check{K} = K - \widehat{K}$ is the adaption error, and C is the adaption gain.

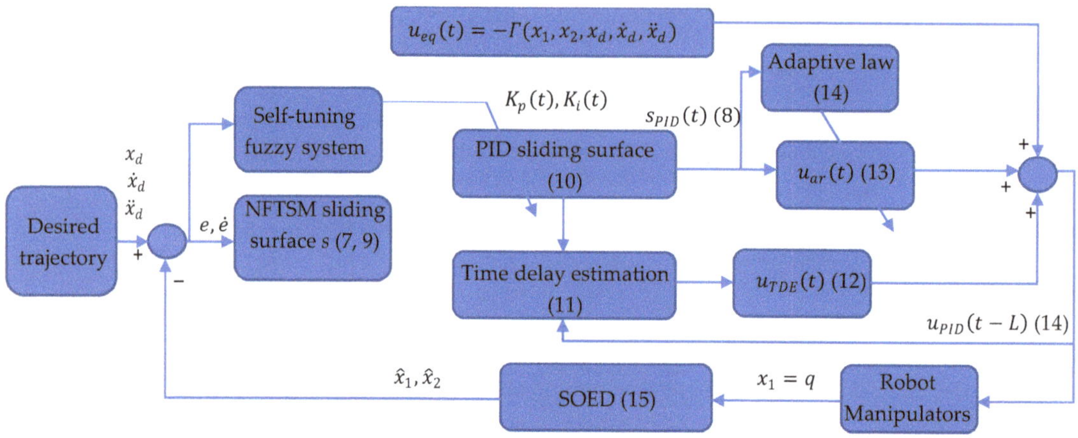

Figure 2. (SFT-PID-NFTSM) with (TDE) [25].

The authors of [26] presented a new method of fault diagnosis and fault-tolerant control for a manipulator using a combination of a support vector machine (SVM)-based neural adaptive high-order variable structure observer (ANHWSO) and an adaptive modern fuzzy backstepping variable structure controller (AMFBVSC). The architecture of this control system is shown in Figure 3. An SVM machine learning technique was used in this method for error detection and identification and the control process was based on AMFBVSC. The proposed control method, i.e., ANHVSO, improved the performance of fault identification significantly compared to the neural high-order variable, structure observer (NHVSO), and variable structure observer VSO. The work described in [26] is another example of the application of artificial intelligence methods in fault identification. In this case, neural networks and machine learning were used in control, where fuzzy logic was used to tune the proposed method.

In [27], the authors proposed the use of a dual neural network (DNN) [28] for fault diagnosis and fault tolerance. This was the first study found in the literature to present the possibility of controlling a redundant manipulator when more than two axes fail. The paper presented the realization of the task of chalking a circle when as many as three of the available seven axes fail at different time periods. Prior to this, the control task at the time of axis failure was to exclude a single axis or a single sensor.

The authors of [29] proposed and compared two approaches to FD error detection in a welding robot based on the prediction of accelerations of individual axes. The first acceleration prediction approach to fault detection was the use of a self-organizing map neural network (SOMNN). This network is shown in Figure 4a. The second approach used a radial basis function neural network (RBFNN), which is shown in Figure 4b. The paper demonstrated the feasibility of using RBFNN and SOMNN to predict the acceleration of individual axes for either detecting robot faults or predicting the remaining life of the device.

The authors of [30] presented decentralized control strategies based on a radial basis function neural network (RBFNN). In this control method, each joint was treated as a separate subsystem. A fault in the joint was detected using a velocity observer. The presented law of adaptive control was based on information from individual joints. It can be applied to any robotic arm configuration.

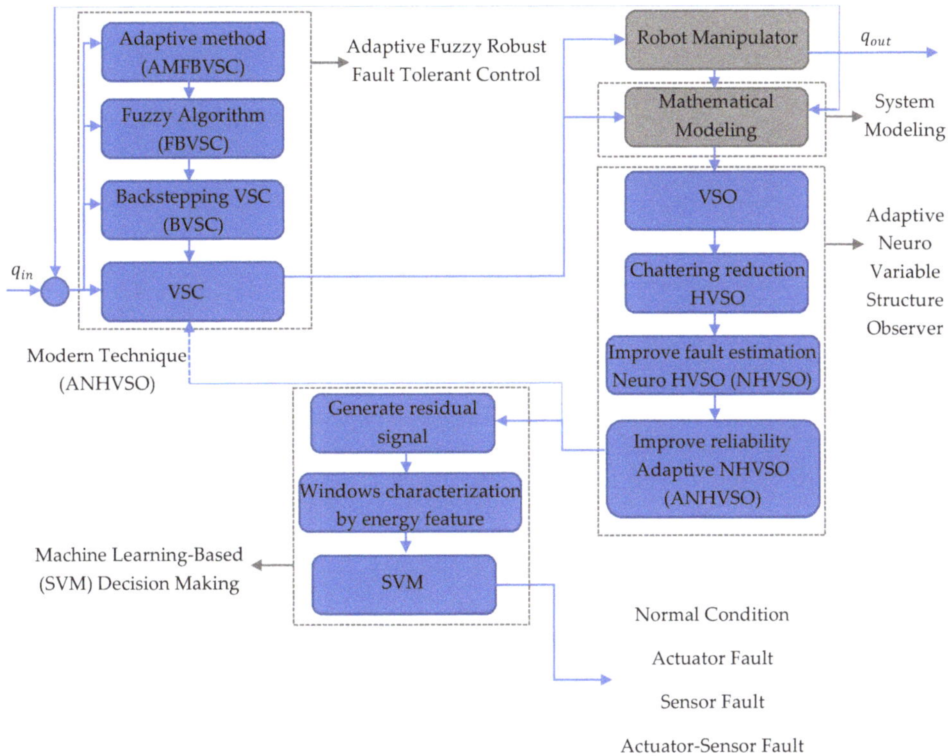

Figure 3. Control architecture with observer (ANHVSO) and controller (AMFBVSC) [26].

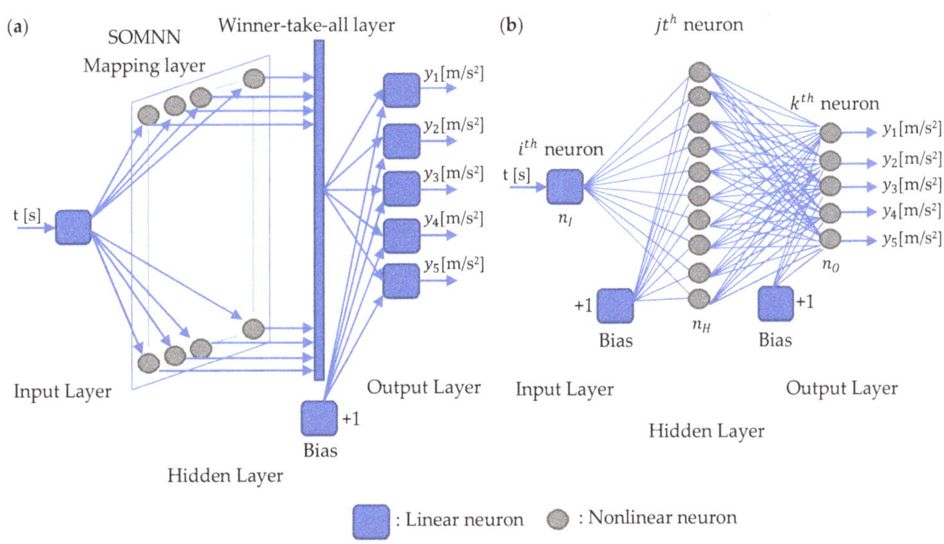

Figure 4. (**a**) Self-organizing map neural network (SOMNN) [29]. (**b**) Radial basis function neural network (RBFNN) [29].

The authors of [31] proposed a radial basis function neural network (RBFNN) for estimating system failures in order to avoid a diagnostic error. The task of the neural network was to compensate for the external disturbances and actuator failures, which were recognized in the proposed method through an adaptive disturbance observer.

In [32], an adaptive fixed-time fault-tolerant constraint control (AFTFTCC) for trajectory tracking was proposed. The system was tested by the authors on two manipulators working together. The described method differs from other methods in that it does not need components such as a state estimator or a fault observer. The method used (RBFNN) to tease out the upper limit of uncertainty.

The authors of [33] presented a control system that is robust to four types of faults: free-swinging joint faults, locked joint faults, incorrectly measured joint positions, and incorrectly measured joint velocities. Artificial neural networks in the form of MLPs and RBFNs were used to detect free-swinging defects and locked robot axes. The operation of neural networks in this solution was based on the detection of failures by the MLP, where each axis was represented by a separate MLP. The information in the form of a mapped error vector was then passed to the RBFNN, which used the vector to classify the error information. Due to the fact that the proposed control system detects the fault and then isolates it and reconfigures the controller, it was able to control the arm in case of multiple faults. However, this was not confirmed by tests in this work. The described control system is shown in Figure 5.

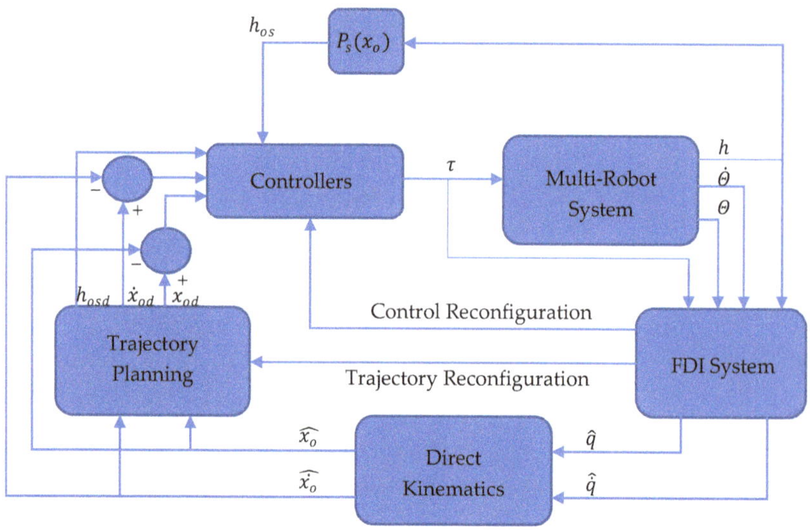

Figure 5. Robust control system architecture [33].

3. Sliding-Mode Control in FTCS

One of the most popular methods used to control robotic arms is sliding-mode control (SMC). Sliding-mode control is based on variable structure systems consisting of independent structures with different properties and switching logic between them. In SMC, a discontinuous control signal, for example, $+k$ and $-k$, are used, which forces the system to "slide" along the cross-section of the normal characteristics of the system. When a system moves along a curved or sliding surface, the system is said to be sliding. An up-to-date overview of all the available methods of control using the sliding mode was presented in [34]. The simplest sliding-mode control method can be described by the formula:

$$u = k \cdot \text{sign}(e), \tag{16}$$

where u is the calculated signal, k is a constant that is the magnitude of change, e is the control signal error, and sign() is the sign function.

Equation (16) describes the simple sliding-mode control method. The advantage of this type of control is its simplicity and insensitivity to system uncertainty. Due to the fact that the method is discontinuous and most control systems are discrete, the main disadvantage of the presented method (16) is the system output signal oscillations called chattering. Nevertheless, the method has become attractive to researchers due to its simplicity; therefore, various sliding-mode control models are currently being designed and implemented for different objects. Due to its robustness to uncertainties, the SMC method is also often used in fault-tolerant control systems. It is applicable to manipulators for which fault tolerance is required.

One of the SMC fault-tolerant methods is the use of a sliding-mode observer (SMO) [35] for fault detection. The use of a non-linear observer makes it possible to estimate the immeasurable state and model the uncertainty, which allows the construction of an error estimation algorithm.

In [36], a fault-tolerant control system in the form of a super-twisting third-order sliding-mode (STW-TOSM) observer and a super-twisting second-order sliding mode (STW-SOSM) controller was presented. This method makes it possible to estimate errors and uncertainties without measuring the speed of the robot's members. For the correct operation of the system, only position measurements are needed. In the paper, the authors presented a comparison of the following systems: the traditional computed torque controller (CTC) [37], the active CTC-FTC controller [38], the passive SM-FTC [39] controller, and the proposed STW-SOSM-FTC controller in both passive and active forms. The test of the proposed controller showed fewer errors, better stability, and no vibrations on the obtained output waveforms. A block diagram of the arrangement of each block in the control process is presented in Figure 6.

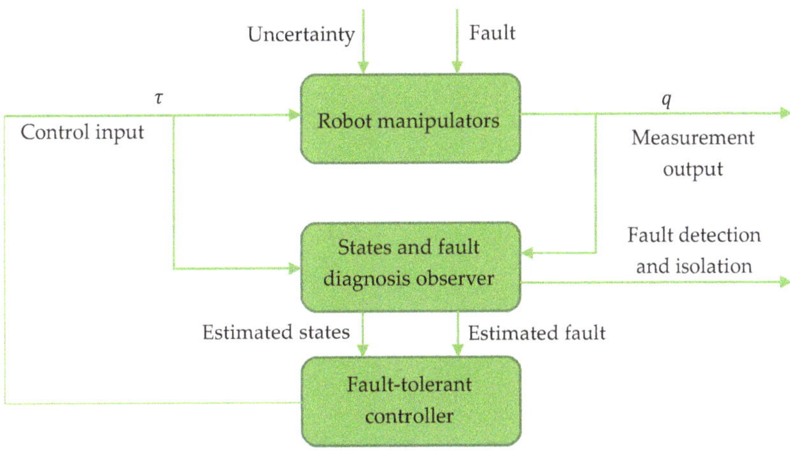

Figure 6. Block diagram of the method presented by Mien Van, Pasquale Franciosa, and Dariusz Ceglarek [36].

The authors of [40] presented a system in the form of a fixed-time second-order sliding-mode observer (FxTSOSMO) and a fixed-time sliding-mode controller (FxTSMC). The work compared the proposed observer (FxTSOSMO) with the fixed-time sliding-mode observer (FxSMO) designed in an earlier work. The work compared the computed torque controller, non-singular fast terminal sliding-mode control (NFTSMC) controllers [41], and the proposed method (FxTSMC). The latter system achieved higher estimation precision and the error of the proposed solution converged within a fixed time frame. The solution proposed in [40] is described by the following Formulas (17)–(20):

$$u_{stw} = \mu_1 |e|^{\frac{1}{2}} \text{sign}(e) + \xi, \tag{17}$$

where u_{stw} is an approximate unknown component, $\mu_1 > 0$, $e = x_2 - v$ is a velocity error, $\xi = k(t)\text{sign}(e)$,

$$\dot{v} = \wedge u + f(x_1, x_2) + u_{stw} + k_1 [e]^{\gamma_1} + k_2 [e]^{\gamma_2}, \tag{18}$$

where $\dot{v}$ is the observer that was designed based on a super-twisting high-order sliding-mode algorithm, $\wedge = M^{-1}(q)$, M is the inertia matrix, $k_1 > 0$, $k_2 > 0$, $[e]^{\gamma_1} = [e]^{\gamma_1}\text{sign}(e)$, v is the estimate of the velocity x_2, $\gamma_1 = \frac{1}{2}$, $\gamma_2 > 2$,

$$s = e + \frac{1}{k_2^{\gamma}} \left[\dot{e} + k_1 [e]^{\alpha} \right]^{\frac{1}{\gamma}}, \tag{19}$$

where s is the proposed sliding surface, $\alpha > 1$, $\frac{1}{2} < \gamma < 1$,

$$u = u_0 + u_c + u_s, \tag{20}$$

where u is the proposed controller output, $u_0 = \wedge^{+}(-f(x_1, x_2) - u_{stw} + \ddot{x}_d - k_1 |e|^{\alpha - 1} \dot{e})$, $\wedge^{+}$ is the pseudo-inverse of the $\wedge$, $u_c = - \wedge^{+} k_2^{\gamma} \left| T(e, \dot{e}) \right|^{1 - \frac{1}{\gamma}} \dot{e}$, $u_s = \wedge^{+}(-(\Gamma + \alpha) sign(s) - \lambda_1 [s]^{m_1} - \lambda_2 [s]^{m_2})$, Γ is a positive constant, $0 < m_1 < 1$, $m_2 > 1$.

The proposed algorithm was designed to follow a preset trajectory. The block for generating the preset trajectory is thus the first block of control. The signal from this block, together with the signal feedback from the FxTSOSMO block described by Equations (17) and (18), goes to the summing node. The feedback signal is the current positions and velocities The position and velocity errors go from the summation node to the sliding surface block described by Equation (19). The signal s from the sliding surface block and u_{stw} from the FxTSOSMO block go to the FxTSMC block described by the equation. The output of the FxTSMC block u described by Equation (20) goes to both the FxTSOSMO block and the robot. Finally, the FxTSOSMO block retrieves information about the robot in the form of positions and velocities.

In [42,43], the authors presented a fault-tolerant control system for a single-link flexible joint manipulator (SFJM). This work proved that it is possible to control the flexible coupling in a robot joint using SMC with SMTO. The presented results show that SMC is a method that can be applied to objects characterized by non-linearity and uncertainty. However, the presented solution was validated only for a single-axis manipulator, but it is possible to develop a similar system for a manipulator with more degrees of freedom.

The authors of [44] proposed a novel methodology for manipulator adaptive backstepping non-singular fast terminal sliding-mode control (ABNFTSMC). The proposed approach combined NFTSMC with a backstepping design mechanism. The combination resulted in few tracking errors and low "chattering" on the output and provided fast response transients. The described system was compared to CTC, PID, PID-SMC, and NFTSMC controllers. The comparison of these systems showed improvements in the above-mentioned parameters.

In [45], a system consisting of combined non-singular fast terminal sliding-mode control (NFTSMC) and a high-order sliding-mode (HOSM) controller was proposed. The presented control system also used an algorithm based on time delay estimation (TDE) for fault estimation. The results described in the work showed that fast terminal sliding-mode control (FTSMC) and NFTSMC systems have faster convergence compared to the non-singular terminal sliding-mode control method. FTSMC systems may encounter the problem of singularity during operation. However, the work shows the advantage of the NFTSMC system over the above-mentioned ones. Other methods using both active fault-tolerant control systems (AFTCS) and passive fault-tolerant control systems (PFTCS) were also rated in the paper.

In [46], a combination of a controller in the form of non-singular fast terminal sliding-mode control (NFTSMC) with an observer in the form of third-order sliding mode (TOSM) was presented. The research results showed that the TOSM observer can estimate the speed of the system so that the system does not need to measure this. This method was compared with the SMC, NTSMC, and NFTSMC methods. The proposed NFTSMC method had the fewest trajectory-tracking errors. The TOSM observer, on the other hand, proved more accurate and had less "chattering" than the SOSM observer.

The authors of [47] presented a combination of a fast terminal sliding-mode surface (FTSMS), super-twisted reaching control law (STRCL), and disturbance observer (DO). The proposed method was compared with the SMC and NFTSMC methods. The method described in the paper provided convergence in finite time and effectively combatted the chattering phenomenon. By using a disturbance observer, the complexity of the calculations was reduced. A DO was also used to estimate uncertainties in the form of dynamics, external disturbances, and failures. A diagram of the proposed method is shown in Figure 7. This work presents the new fast terminal sliding surface described by the formula in [47]:

$$\sigma_i = \dot{e}_i + \frac{2\gamma_1}{1+E^{\mu_1(|e_i|-\Phi)}}e_i + \frac{2\gamma_2}{1+E^{\mu_2(|e_i|-\Phi)}}|e_i|^{\alpha}\text{sign}(e_i), \tag{21}$$

where $\sigma \in R$ is the FTSMS, E is an exponential function, e_i represents the positional control error, $\dot{e}_i$ represents the velocity control error, $\gamma_1, \gamma_2, \mu_1, \mu_2$ are the positive constants, $0 < \alpha < 1, \Phi = \left(\frac{\gamma_1}{\gamma_2}\right)^{1/(1-\alpha)}$,

$$u_r = Y_1|\sigma|^{0.5}\text{sign}(\sigma) + \eta, \tag{22}$$

where u_r is STRCL, $Y_1 = \text{diag}(Y_{11}, \ldots, Y_{1n})$.

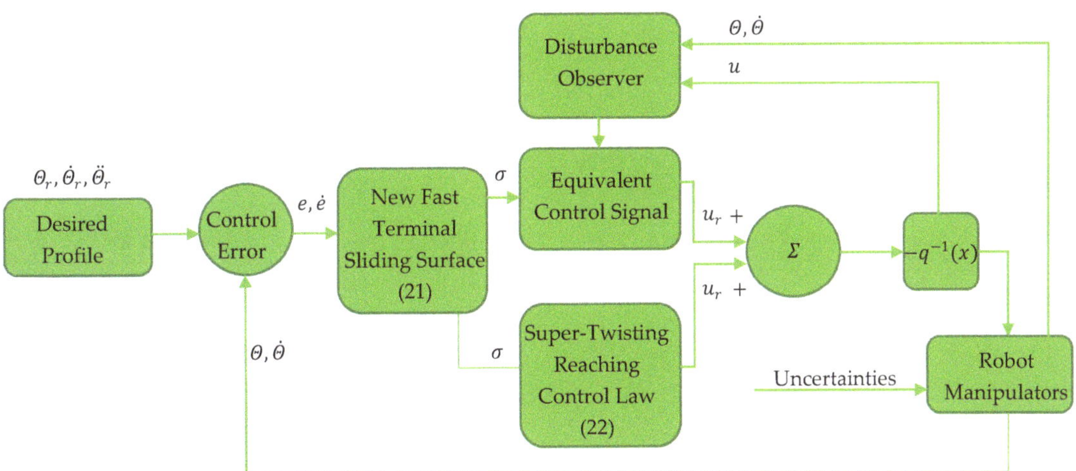

Figure 7. Diagram FTSMS-STRCL-DO [47].

The authors of [48] proposed a new sliding-mode control technique, namely active fault-tolerant control with synchronous sliding-mode control AFTC-SSMC. This control relied on the fact that with a traditional SMC controller, only the position error converges to zero. In synchronous control, on the other hand, both the position error and kinematic relationship between the errors converge to zero. The method was compared with standard active and passive sliding-mode control, namely AFTC-SMC and PFTC-SMC.

It was shown that the pro-rated method had better fault tolerance and provided better trajectory-tracking performance.

The authors of [49] presented a fault-tolerant manipulator control strategy in the form of a combination of NFTSMC and the proposed high-speed third-order sliding-mode observer TOSMO. This combination was able to handle unknown input data, further reducing the vibration phenomenon and improving the accuracy of trajectory tracking. The proposed high-speed TOSMO could estimate both the velocity signal and input data faster than a standard TOSMO. The structure of such a system is similar to that of the FTSMS-STRCL-DO [47] but it has a different observer model. The structure of the proposed method is shown in Figure 8. The solution proposed by the authors is described by the following formulas [49]:

$$\hat{s} = \hat{\dot{e}} + \int \left[\beta_1 |e|^{\gamma_1} sign(e) + \beta_2 \hat{\dot{e}}^{\gamma_2} sign\left(\hat{\dot{e}}\right) + \beta_3 e + \beta_4 e^3\right] dt, \tag{23}$$

where $\hat{s}$ is the NFTS surface, $\beta_1, \beta_2, \beta_3, \beta_4$ are positive constants, $0 < \gamma_1 < 1$, $\gamma_2 = 2\gamma_1/(1+\gamma_1)$ are constants,

$$u_{eq} = \psi(x) + k_2 |\tilde{x}_1|^{\frac{1}{3}} sign\left(\tilde{x}_1\right) + \Gamma\left(\hat{x}_1 - x_2\right) + \int k_3 sign\left(\tilde{x}_1\right) + A - \ddot{x}_d, \tag{24}$$

where u_{eq} is the equivalent control law, $A = \beta_1 |e|^{\gamma_1} sign(e) + \beta_2 \hat{\dot{e}}^{\gamma_2} sign\left(\hat{\dot{e}}\right) + \beta_3 e + \beta_4 e^3$, $\psi(x) = M(q)^{-1}[-C(q,\dot{q}) - G(q)]$ is the nominal model of the robot manipulator, Γ is a positive constant, k_2, k_3 denotes the sliding gains, $\ddot{x}_d$ is an expected acceleration

$$u_{sw} = (\Delta_\delta + \mu) sign(\hat{s}), \tag{25}$$

where u_{sw} is the switching control law, μ is a small positive constant, and Δ_δ is the estimation error as a positive value.

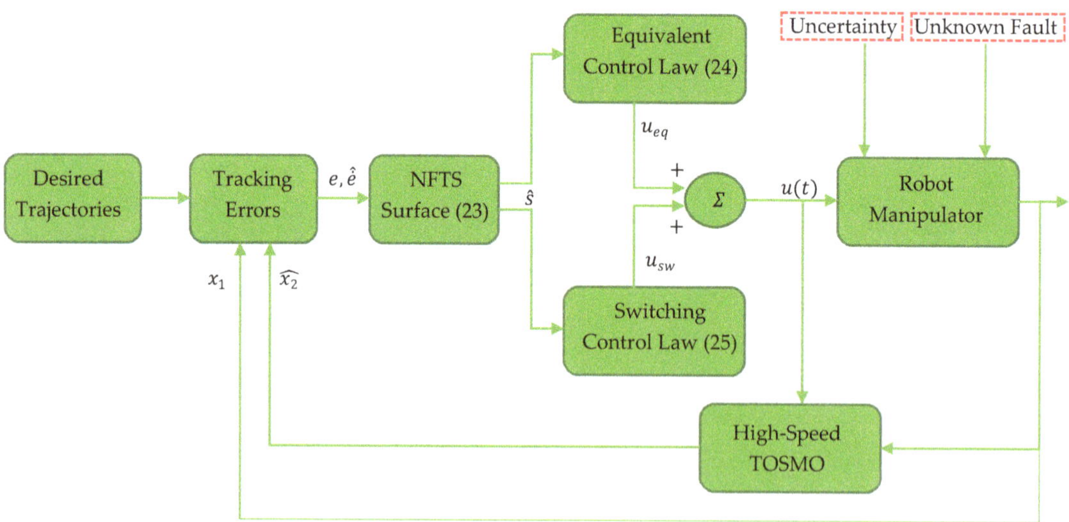

Figure 8. Structure of the NFTSMC-TOSMO control method [49].

4. Other Robot-Dedicated FTCS

In addition to the leading fault-tolerant manipulator control methods such as SMC and artificial intelligence, other dedicated manipulator control methods can also be found in the literature. Some of them are presented in this section of the article.

The authors of [50] presented the use of the well-known Newton–Raphson Method (NRM) [51] to control a manipulator when a robot axis fails. The authors of this paper attached a QR code to the robot's end arm. This code was read by a camera located at the base of the robot and then the position of the robot's end was fed into the control software. This allowed the algorithm to determine the moment at which the failure occurred by not matching the position tracked by the camera with the calculated position. Then, the system, which used the Newton–Raphson method in the inverse kinematics calculation to find a good approximation for the root of a real-valued function, determined which joint had failed. To determine the faulty axis, the position measured by the camera was compared with the position calculated by the NRM. If a difference was detected, the system checked which axis was damaged. If a faulty joint was detected, the failed joint was treated as rigid and the robot's inverse kinematics were recalculated using the Newton–Raphson method. Using this solution, the authors of the paper proved that if the robot was able to reach the set position despite the loss of an axis, the algorithm was able to help the robot achieve its goal.

In [52], a new method of hybrid fault-tolerant predictive control (HFTPC) was proposed. This method was tested on a hybrid manipulator that was a combination of hydraulic and mechanical elements. The paper compared the proposed HFTPC method, one of the components of which was a predictive control (MPC) model [53], with control in which only the MPC part was running. The presented system was able to control a hybrid manipulator when a failure occurred.

The authors of [54] presented robust fault-tolerant tracking control (RFTTC) schemes for uncertain non-linear feedback systems (NFS) using operator theory-based robust right coprime factorization (ORRCF). It was shown that both RFTTC based on an internal model and RFTTC based on operator compensation were effective in dealing with interference and erroneous signals. The proposed system was tested on a trajectory-tracking task by a two-axis robot.

The authors of [55] presented an adaptive fault-tolerant control system for unknown actuator faults. Detailed fault information was not required for the algorithm to work. The proposed method omitted complex logarithmic transformations in the pre-scribed performance control (PPC) structure [56], making the controller efficient and simple to implement.

The authors of [57] proposed a robust fixed-time fault-tolerant control system (RFTFTC) by introducing an additional auxiliary vector that was described by Formula (26) [57]:

$$\chi = \dot{e} + \lambda e, \tag{26}$$

where $\lambda \in R^{+}$, $e = q - q_d$ is the position error and $\dot{e} = \dot{q} - \dot{q}_d$ is the velocity error.

A schematic of the proposed system is shown in Figure 9.

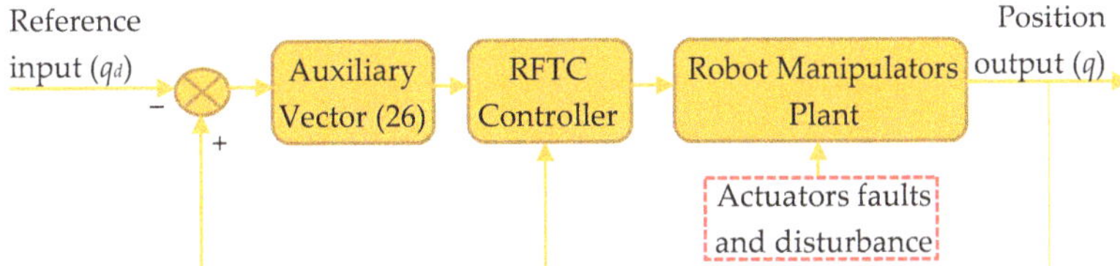

Figure 9. RFTC with auxiliary vector [57].

The authors of [58,59] presented an FTCS for manipulators based on an active inference controller (AIC). AIC makes use of the sensory prediction error in the free energy for the generation of residuals and thresholds for FDI. It does not require additional controllers for fault recovery. In [59], the AIC was improved, and unbiased AIC (u-AIC) was introduced to reduce the probability of false positives and allow for the easy definition of probabilistically robust thresholds for fault detection.

The authors of [43] presented fault-tolerant control using a proportional–integral-derivative (PID) controller. In addition to the PID controller, the system consisted of a Luenberger observer [60], which was able to estimate the errors of the FDI analysis. From the information provided by the FDI, the magnitude of the error was evaluated using singular value decomposition (SVD) [61]. The system in this solution was tested on a single-link flexible manipulator.

In earlier works, i.e., before 2019, the algorithms described in the literature for the control of robots in the case of communication faults dealt with two control types, i.e., state observers or estimators. The works in [62,63] presented a completely different approach to fault-tolerant control. The authors of these works dealt with a problem involving interference in the communication path through which information is sent between the robot's actuators and the controller. The paper proposed to re-construct data corrupted during transmission using a splicing code. The transmission was carried out using the Controller Area Network (CAN) protocol [64].

The authors of [65] presented as many as four techniques for dealing with controlled oscillation and axis damage. These techniques were sinh-cosh, neural compensation, PID gain scheduling, and sinh-cosh gain scheduling. All of them were tested on a manipulator in which the failure was detected by measuring the currents through the robot axis drives. At the moment of axis failure, the current dropped to zero and then the axis was treated as passive, i.e., not participating in control. In the results of the paper, all four methods were compared. The comparison of the methods was conducted using a presentation of the trajectory-tracking error for each method. The best results in terms of the position errors of the individual axes were obtained for neural compensation, whereas the worst results were obtained by the sinh-cosh gain-scheduling method.

5. Summary

This paper presented an overview of the investigated FTC-type solutions used in manipulators. The works that were considered and described were mainly from the last five years. The division of the literature items presented in this article, as proposed by the authors, is shown in Figure 10. The leading solutions are AI (artificial intelligence) and SMC (sliding mode control). A fault-tolerant control system that is a combination of AI and SMC has also been developed, namely SFT-PID-NFTSM [25]. Other presented methods are based on PID controllers [53], NRM [50], MPC [52], and AIC [58,59], or are proposals that deal with interference occurring in the CAN communication protocol [62,63].

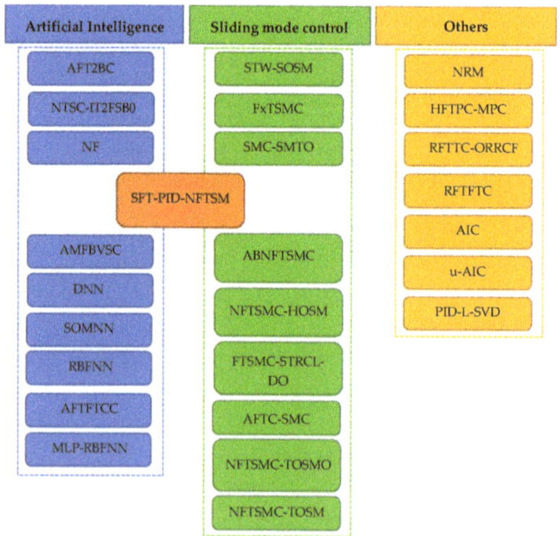

Figure 10. Proposed breakdown of fault-tolerant control system methods used in manipulators. All abbreviations of the control methods and their references to the literature are in the text above and in Table 1 below.

Table 1. Comparison of FTC methods used for robot control described in the cited articles.

	Accuracy	Robustness	Dependence on Model or Training Data	Implementation Complexity	Computational Cost
AFT2BC [17]	2	4	2	3	2
NTSC-IT2FSB0 [20]	5	4	3	3	3
NF [22,24]	2	3	5	4	5
AMFBVSC [26]	5	3	5	5	5
DNN [27]	5	5	4	4	4
SOMNN [29]	3	3	5	4	4
RBFNN [29–31]	5	4	3	4	4
AFTFTCC [32]	4	4	3	4	4
MLP-RBFNN [33]	3	4	5	4	5
SFT-PID-NFTSM [25]	2	4	2	3	3
STW-SOSM [36]	3	4	1	2	2
FXTSMC [40]	4	5	1	2	2
SMC-SMTO [42]	-	-	1	2	2
ABNFTSMC [44]	3	5	1	3	2
NFTSMC-HOSM [45]	3	4	1	2	2
FTSMC-STRCL-DO [47]	4	5	1	2	1
AFTC-SMC [48]	5	5	1	2	2
NFTSMC-TOSMO [49]	4	5	1	2	2
NFTSMC-TOSM [46]	4	5	1	2	2
NRM [50]	2	3	2	3	4
HFTPC-MPC [52]	2	3	3	3	3
RFTTC-ORRCF [54]	5	5	4	4	3
RFTFTC [57]	4	4	2	2	2
AIC [58]	4	4	3	3	3
U-AIC [59]	4	5	3	3	3
PID-L-SVD [43]	-	-	-	2	2

Accuracy: Rating scale (1–5), where 1 indicates low accuracy and 5 indicates high accuracy. **Robustness:** Rating scale (1–5), where 1 indicates low robustness and 5 indicates high robustness. **Dependence on model or training data:** Rating scale (1–5), where 1 indicates no dependence on training data and 5 indicates high dependence on training data. **Implementation complexity:** Rating scale (1–5), where 1 indicates that the implementation was simple and 5 indicates that the implementation was very complex. **Computational cost:** Rating scale (1–5), where 1 indicates that the cost was small and 5 indicates that the cost was high.

The purpose of this review was to present the issue of FTC in the context of manipulators, particularly for the control of a robot in the case of a one- or two-axis failure. We decided to collect works from the last 20 years and then group them. In this way, the three groups created related to the sections of this paper, i.e., AI-based methods, SMC-based methods, and other methods, which, numerically, constituted a similar number to the first two groups. The goal of our review was to systematically collect and summarize the existing research on the topic of FTC in manipulators. After reading this paper, it is expected that readers will have gained a general understanding of FTC in manipulators. We believe that the topic of FTC is very subjective and depends strictly on the application, e.g., if a robot is connected to very long wires, it may be that using FTC algorithms for the communication line is a much better solution than using an AI or SMC-based controller. For manipulators with a large number of degrees of freedom, a computationally simple SMC-based controller is probably better but this may also depend on the required motion speeds or positioning accuracy. The purpose of this article is to direct the reader to the solutions and the valuable works of the cited authors. Therefore, the methods described in the cited articles are compared in Table 1. Each of the proposed methods is valid for a different type of application and their comparison could only be performed for detailed boundary conditions.

6. Conclusions

The study presented in this paper deals with FTCS in relation to manipulators. The authors of the papers discussed in this review deal with failures in both single and multiple axes. In addition, they address aspects of tolerating erroneous position sensor readings and propose algorithms to deal with interference occurring on the communication line between the controller and robot actuators. A review of the literature shows that the topic of FTCS as applied to manipulators is currently being strongly developed. The use of FTCS systems allows the optimization of production by avoiding downtime caused by individual component failures in robotic arms. Looking toward the future, several opportunities for the application of FTC in manipulators have been identified by the authors. The first possibility is for the robot to react to obstacles in its working environment. However, no studies have yet been presented in which the FTC algorithm was implemented and the robot was able to avoid obstacles. Another possible future use of FTC algorithms is the cooperation between a robot and a human. However, there are no studies that indicate how the FTC algorithm would behave in such a scenario. Additionally, there are currently no studies that demonstrate how FTC algorithms could facilitate cooperation between robots such as delegating work to a more efficient robot when another robot is defective or robots communicating with each other about the detection of defects.

Author Contributions: Conceptualization, P.N. and A.M.; methodology, P.N. and A.M.; resources, P.N.; data curation, P.N.; writing—original draft preparation, A.M and P.N.; writing—review and editing, A.M. and P.N.; visualization, P.N.; supervision, P.N. and A.M.; joint first authorship, A.M.; project administration, A.M.; funding acquisition, A.M. All authors have read and agreed to the published version of the manuscript.

Funding: This work was supported by the Polish Ministry of Science and Education grant no. 0614/SBAD/1565.

Institutional Review Board Statement: Not applicable.

Informed Consent Statement: Not applicable.

Data Availability Statement: Not applicable.

Conflicts of Interest: The authors declare no conflict of interest.

References

1. Aviziens, A. Fault-tolerant systems. *IEEE Trans. Comput.* **1976**, *100*, 1304–1312. [CrossRef]
2. Stengel, R.F. Intelligent failure-tolerant control. *IEEE Control. Syst. Mag.* **1991**, *11*, 14–23.
3. Veillette, R.J.; Medanic, J.V.; Perkins, W.R. Design of reliable control systems. In Proceedings of the 29th IEEE Conference on Decision and Control, Honolulu, HI, USA, 5–7 December 1990; pp. 1131–1136.
4. Patton, R. Robustness issues in fault-tolerant control. In Proceedings of the IEE Colloquium on Fault Diagnosis and Control System Reconfiguration, London, UK, 26 May 1993; p. 1.
5. Patton, R.J. Fault-tolerant control: The 1997 situation. *IFAC Proc. Vol.* **1997**, *30*, 1029–1051. [CrossRef]
6. Maciejewski, A.A. Fault tolerant properties of kinematically redundant manipulators. In Proceedings of the IEEE International Conference on Robotics and Automation, Cincinnati, OH, USA, 13–18 May 1990; pp. 638–642.
7. Chladek, J.T. Fault tolerance for space based manipulator mechanisms and control system. In Proceedings of the First International Symposium On Measurement and Control in Robotics, Houston, TX, USA, 20–22 June 1990.
8. Tesar, D.; Sreevijayan, D.; Price, C. Four-level fault tolerance in manipulator design for space operations. In Proceedings of the First International Symposium on Measurement and Control in Robotics, Houston, TX, USA, 20–22 June 1990.
9. Wu, E.; Diftler, M.; Hwang, J.; Chladek, J. A fault tolerant joint drive system for the space shuttle remote manipulator system. In Proceedings of the 1991 IEEE International Conference on Robotics and Automation, Sacramento, CA, USA, 9–11 April 1991; pp. 2504–2509.
10. Tesar, D. On the design of fault-tolerant robotic manipulator systems. In Proceedings of the 6th Annual Workshop on Space Operations Applications and Research (SOAR 1992), Victorville, CA, USA, 4–6 August 1992; Volume 2, p. 223.
11. Visinsky, M.L.; Cavallaro, J.R.; Walker, I.D. *Robotic Fault Tolerance: Algorithms and Architectures*; Prentice Hall: Englewood Cliffs, NJ, USA, 1993.
12. Ting, Y.; Tosunoglu, S.; Tesar, D. A control structure for fault-tolerant operation of robotic manipulators. In Proceedings of the IEEE International Conference on Robotics and Automation, Atlanta, GA, USA, 2–6 May 1993; pp. 684–690.
13. Selkäinaho, J.; Halme, A. An intelligent fault tolerant control system. *IFAC Proc. Vol.* **1987**, *20*, 69–73. [CrossRef]
14. Hörmann, A.; Hugel, T.; Meier, W. A concept for an intelligent and fault-tolerant robot system. *J. Intell. Robot. Syst.* **1988**, *1*, 259–286. [CrossRef]
15. Groom, K.N.; Maciejewski, A.A.; Balakrishnan, V. Real-time failure-tolerant control of kinematically redundant manipulators. *IEEE Trans. Robot. Autom.* **1999**, *15*, 1109–1115. [CrossRef]
16. Shin, J.H.; Lee, J.J. Fault detection and robust fault recovery control for robot manipulators with actuator failures. In Proceedings of the 1999 IEEE International Conference on Robotics and Automation (Cat. No. 99CH36288C), Detroit, MI, USA, 10–15 May 1999; Volume 2, pp. 861–866.
17. Rahali, H.; Zeghlache, S.; Benyettou, L. Fault Tolerant Control of Robot Manipulators Based on Adaptive Fuzzy Type-2 Backstepping in Attendance of Payload Variation. *Int. J. Intell. Eng. Syst.* **2021**, *14*, 312–325. [CrossRef]
18. Ali, K.; Mehmood, A.; Iqbal, J. Fault-tolerant scheme for robotic manipulator—Nonlinear robust back-stepping control with friction compensation. *PLoS ONE* **2021**, *16*, e0256491. [CrossRef]

19. Tanaka, K.; Hori, T.; Wang, H.O. A multiple Lyapunov function approach to stabilization of fuzzy control systems. *IEEE Trans. Fuzzy Syst.* **2003**, *11*, 582–589. [CrossRef]
20. Hagh, Y.S.; Asl, R.M.; Fekih, A.; Wu, H.; Handroos, H. Active fault-tolerant control design for actuator fault mitigation in robotic manipulators. *IEEE Access* **2021**, *9*, 47912–47929. [CrossRef]
21. Asl, R.M.; Hagh, Y.S.; Handroos, H. Adaptive extended Kalman filter designing based on non-singular fast terminal sliding mode control for robotic manipulators. In Proceedings of the 2017 IEEE International Conference on Mechatronics and Automation (ICMA), Takamatsu, Japan, 6–9 August 2017; pp. 1670–1675.
22. Dev Anand, M.; Selvaraj, T.; Kumanan, S.; Ajitha, T. A neuro-fuzzy-based fault detection and fault tolerance methods for industrial robotic manipulators. *Int. J. Adapt. Innov. Syst.* **2010**, *1*, 334–371. [CrossRef]
23. Deo, A.S. Application of Optimal Damped Least-Squares Method to Inverse Kinematics of Robotic Manipulators. Master's Thesis, Rice University, Houston, TX, USA, 1991.
24. Dev Anand, M.; Selvaraj, T.; Kumanan, S. Fault Detection and Fault Tolerance Methods for Industrial Robot Manipulators Based on Hybrid Intelligent Approach. *Adv. Prod. Eng. Manag.* **2012**, *7*, 225–236. [CrossRef]
25. Van, M.; Do, X.P.; Mavrovouniotis, M. Self-tuning fuzzy PID-nonsingular fast terminal sliding mode control for robust fault tolerant control of robot manipulators. *ISA Trans.* **2020**, *96*, 60–68. [CrossRef] [PubMed]
26. Piltan, F.; Prosvirin, A.E.; Sohaib, M.; Saldivar, B.; Kim, J.M. An SVM-based neural adaptive variable structure observer for fault diagnosis and fault-tolerant control of a robot manipulator. *Appl. Sci.* **2020**, *10*, 1344. [CrossRef]
27. Li, Z.; Li, C.; Li, S.; Cao, X. A fault-tolerant method for motion planning of industrial redundant manipulator. *IEEE Trans. Ind. Inform.* **2019**, *16*, 7469–7478. [CrossRef]
28. Zhang, Y.; Wang, J.; Xia, Y. A dual neural network for redundancy resolution of kinematically redundant manipulators subject to joint limits and joint velocity limits. *IEEE Trans. Neural Netw.* **2003**, *14*, 658–667. [CrossRef] [PubMed]
29. Eski, I.; Erkaya, S.; Savas, S.; Yildirim, S. Fault detection on robot manipulators using artificial neural networks. *Robot. Comput.-Integr. Manuf.* **2011**, *27*, 115–123. [CrossRef]
30. Zhao, B.; Li, Y. Local joint information based active fault tolerant control for reconfigurable manipulator. *Nonlinear Dyn.* **2014**, *77*, 859–876. [CrossRef]
31. Zhang, S.; Yang, P.; Kong, L.; Chen, W.; Fu, Q.; Peng, K. Neural networks-based fault tolerant control of a robot via fast terminal sliding mode. *IEEE Trans. Syst. Man Cybern. Syst.* **2019**, *51*, 4091–4101. [CrossRef]
32. Hwang, C.L.; Yu, W.S. Tracking and cooperative designs of robot manipulators using adaptive fixed-time fault-tolerant constraint control. *IEEE Access* **2020**, *8*, 56415–56428. [CrossRef]
33. Tinós, R.; Terra, M.H.; Bergerman, M. A fault tolerance framework for cooperative robotic manipulators. *Control. Eng. Pract.* **2007**, *15*, 615–625. [CrossRef]
34. Yu, X.; Feng, Y.; Man, Z. Terminal sliding mode control—An overview. *IEEE Open J. Ind. Electron. Soc.* **2020**, *2*, 36–52. [CrossRef]
35. Khireddine, M.S.; Boutarfa, A. Fault tolerant control on robotic manipulator using sliding mode observers. In Proceedings of the 2013 International Conference on Computer Applications Technology (ICCAT), Sousse, Tunisia, 20–22 January 2013; pp. 1–6.
36. Van, M.; Franciosa, P.; Ceglarek, D. Fault diagnosis and fault-tolerant control of uncertain robot manipulators using high-order sliding mode. *Math. Probl. Eng.* **2016**, *2016*, 7926280. [CrossRef]
37. Piltan, F.; Mirzaei, M.; Shahriari, F.; Nazari, I.; Emamzadeh, S. Design baseline computed torque controller. *Int. J. Eng.* **2012**, *6*, 129–141.
38. Tosunoglu, S. Fault-tolerant control of mechanical systems. In Proceedings of the IECON'95-21st Annual Conference on IEEE Industrial Electronics, Orlando, FL, USA, 6–10 November 1995; Volume 1, pp. 127–132.
39. Merheb, A.R.; Noura, H.; Bateman, F. Design of passive fault-tolerant controllers of a quadrotor based on sliding mode theory. *Int. J. Appl. Math. Comput. Sci.* **2015**, *25*, 561–576. [CrossRef]
40. Van, M.; Ceglarek, D. Robust fault tolerant control of robot manipulators with global fixed-time convergence. *J. Frankl. Inst.* **2021**, *358*, 699–722. [CrossRef]
41. Xu, B.; Zhang, L.; Ji, W. Improved non-singular fast terminal sliding mode control with disturbance observer for PMSM drives. *IEEE Trans. Transp. Electrif.* **2021**, *7*, 2753–2762. [CrossRef]
42. Chen, Y.; Guo, B. Sliding mode fault tolerant tracking control for a single-link flexible joint manipulator system. *IEEE Access* **2019**, *7*, 83046–83057. [CrossRef]
43. Abd Latip, S.F.; Husain, A.R.; Ahmad, M.N.; Mohamed, Z. Fault tolerant control for sensor fault of a single-link flexible manipulator system. *J. Teknol.* **2016**, *78*. [CrossRef]
44. Van, M.; Mavrovouniotis, M.; Ge, S.S. An adaptive backstepping nonsingular fast terminal sliding mode control for robust fault tolerant control of robot manipulators. *IEEE Trans. Syst. Man Cybern. Syst.* **2018**, *49*, 1448–1458. [CrossRef]
45. Van, M.; Ge, S.S.; Ren, H. Finite time fault tolerant control for robot manipulators using time delay estimation and continuous nonsingular fast terminal sliding mode control. *IEEE Trans. Cybern.* **2016**, *47*, 1681–1693. [CrossRef] [PubMed]
46. Nguyen, V.C.; Vo, A.T.; Kang, H.J. A finite-time fault-tolerant control using non-singular fast terminal sliding mode control and third-order sliding mode observer for robotic manipulators. *IEEE Access* **2021**, *9*, 31225–31235. [CrossRef]
47. Vo, A.T.; Kang, H.J. A novel fault-tolerant control method for robot manipulators based on non-singular fast terminal sliding mode control and disturbance observer. *IEEE Access* **2020**, *8*, 109388–109400. [CrossRef]

48. Le, Q.D.; Kang, H.J. Implementation of fault-tolerant control for a robot manipulator based on synchronous sliding mode control *Appl. Sci.* **2020**, *10*, 2534. [CrossRef]
49. Nguyen, V.C.; Tran, X.T.; Kang, H.J. A Novel High-Speed Third-Order Sliding Mode Observer for Fault-Tolerant Control Problem of Robot Manipulators. *Actuators* **2022**, *11*, 259. [CrossRef]
50. Zhang, Y.; Zhu, W.; Rosendo, A. QR code-based self-calibration for a fault-tolerant industrial robot arm. *IEEE Access* **2019**, *7*, 73349–73356. [CrossRef]
51. Ge, D. Kinematics modeling of redundant manipulator based on screw theory and Newton-Raphson method. *J. Phys. Conf. Ser.* **2022**, *2246*, 012068. [CrossRef]
52. Zahaf, A.; Bououden, S.; Chadli, M.; Chemachema, M. Robust fault tolerant optimal predictive control of hybrid actuators with time-varying delay for industrial robot arm. *Asian J. Control.* **2022**, *24*, 1–15. [CrossRef]
53. Altan, A.; Hacıoğlu, R. Model predictive control of three-axis gimbal system mounted on UAV for real-time target tracking under external disturbances. *Mech. Syst. Signal Process.* **2020**, *138*, 106548. [CrossRef]
54. Bu, N.; Pang, J.; Deng, M. Robust fault tolerant tracking control for the multi-joint manipulator based on operator theory. *J. Frankl. Inst.* **2020**, *357*, 2696–2714. [CrossRef]
55. Zhang, G.; Cheng, D. Adaptive fault-tolerant guaranteed performance control for euler-Lagrange systems with its application to a 2-link robotic manipulator. *IEEE Access* **2020**, *8*, 184160–184171. [CrossRef]
56. Ma, H.; Zhou, Q.; Li, H.; Lu, R. Adaptive prescribed performance control of a flexible-joint robotic manipulator with dynamic uncertainties. *IEEE Trans. Cybern.* **2021**, *52*, 12905–12915. [CrossRef] [PubMed]
57. Liu, L.; Zhang, L.; Wang, Y.; Hou, Y. A novel robust fixed-time fault-tolerant tracking control of uncertain robot manipulators *IET Control. Theory Appl.* **2021**, *15*, 195–208. [CrossRef]
58. Pezzato, C.; Baioumy, M.; Corbato, C.H.; Hawes, N.; Wisse, M.; Ferrari, R. Active inference for fault tolerant control of robot manipulators with sensory faults. In Proceedings of the International Workshop on Active Inference, Ghent, Belgium, 14 September 2020; Springer: Cham, Switzerland, 2020; pp. 20–27.
59. Baioumy, M.; Pezzato, C.; Ferrari, R.; Corbato, C.H.; Hawes, N. Fault-tolerant control of robot manipulators with sensory faults using unbiased active inference. In Proceedings of the 2021 European Control Conference (ECC), Delft, The Netherlands, 29 June–2 July 2021; pp. 1119–1125.
60. Brivadis, L.; Andrieu, V.; Serres, U. Luenberger observers for discrete-time nonlinear systems. In Proceedings of the 2019 IEEE 58th Conference on Decision and Control (CDC), Nice, France, 11–13 December 2019; pp. 3435–3440.
61. Epps, B.P.; Krivitzky, E.M. Singular value decomposition of noisy data: Noise filtering. *Exp. Fluids* **2019**, *60*, 126. [CrossRef]
62. Urrea, C.; Kern, J.; López-Escobar, R. Design and implementation of a fault-tolerant system for industrial robots under hostile operating conditions. *Comput. Electr. Eng.* **2021**, *90*, 106951. [CrossRef]
63. Urrea, C.; Kern, J.; López, R. Fault-tolerant communication system based on convolutional code for the control of manipulator robots. *Control. Eng. Pract.* **2020**, *101*, 104508. [CrossRef]
64. Lokman, S.F.; Othman, A.T.; Abu-Bakar, M.H. Intrusion detection system for automotive Controller Area Network (CAN) bus system: A review. *EURASIP J. Wirel. Commun. Netw.* **2019**, *2019*, 184. [CrossRef]
65. Urrea, C.; Kern, J.; Álvarez, E. Design and implementation of fault-tolerant control strategies for a real underactuated manipulator robot. *Complex Intell. Syst.* **2022**, *8*, 5101–5123. [CrossRef]

applied sciences

MDPI

Article

Study of the Relationships among the Reverse Torque, Vibration, and Input Parameters of Mud Pumps in Riserless Mud Recovery Drilling

Guolei He [1,2,†], Benchong Xu [1,2,†], Haowen Chen [1,2], Rulei Qin [1,2,*], Changping Li [3] and Guoyue Yin [1,2]

1 Institute of Exploration Techniques, Chinese Academy of Geological Sciences, Langfang 065000, China; hguolei@mail.cgs.gov.cn (G.H.); xbenchong@mail.cgs.gov.cn (B.X.); chaowen@mail.cgs.gov.cn (H.C.); yguole@mail.cgs.gov.cn (G.Y.)
2 Technology Innovation Center for Directional Drilling Engineering, Ministry of Natural Resources, Langfang 065000, China
3 School of Mechanical Engineering and Electronic Information, China University of Geosciences, Wuhan 430074, China; lichangpingcug@126.com
* Correspondence: qrulei@mail.cgs.gov.cn
† These authors contributed equally to this work.

Abstract: Compared with traditional deepwater drilling, riserless mud recovery (RMR) drilling technology has the advantages of improving drilling efficiency, reducing risks, and minimizing environmental effects. Therefore, RMR drilling technology has been widely applied in recent years. This study primarily investigates the relationships among reverse torque, vibration, and input parameters of mud pumps in riserless mud recovery drilling. Firstly, the operating principle and the structure of the mud pump module are analyzed, and an analytical model for the reverse torque and the vibration of the mud pump is established. Secondly, relevant data are derived from theoretical calculations and experiments, and the relationships among the reverse torque, vibration, and input parameters of the mud pump are analyzed using ANSYS (Version 2020 R1) software. Furthermore, the SVR (support vector regression) algorithm is employed to predict and analyze the amplitude of the mud pump's vibration. Finally, the conclusions are drawn based on the findings of the relationships among the reverse torque, vibration, and input parameters of the mud pump. The findings show that the reverse torque of the mud pump increases approximately linearly with an increase in rotational speed, and the vibration of the mud pump increases and then decreases with an increase in rotational speed. The predicted values obtained through the prediction algorithm closely match the actual values. The findings provide a valuable reference for the application of RMR technology.

Keywords: riserless mud recovery technology; mud pump; ANSYS software; SVR algorithm

Citation: He, G.; Xu, B.; Chen, H.; Qin, R.; Li, C.; Yin, G. Study of the Relationships among the Reverse Torque, Vibration, and Input Parameters of Mud Pumps in Riserless Mud Recovery Drilling. *Appl. Sci.* **2023**, *13*, 11878. https://doi.org/10.3390/app132111878

Received: 10 August 2023
Revised: 19 October 2023
Accepted: 25 October 2023
Published: 30 October 2023

1. Introduction

Although marine oil and gas resources are abundant, deepwater drilling is facing many technical challenges. RMR technology can improve the efficiency of drilling processes, reduce environmental effects, and increase operational flexibility [1–3]. Dual-gradient drilling technology is an unconventional drilling technology, which changes the annular pressure gradient of the riser through the pump or changes the fluid density. This technology can address some issues related to formation pressure in deepwater drilling [4–8]. Riserless mud recovery drilling (RMR) [9] is an implementation solution for dual-gradient drilling. It forms a closed mud circulation system, allowing mud to be recovered and significantly reducing the amount of mud used. Additionally, it avoids discharging mud into the seawater, thereby protecting the marine environment. RMR is widely used in deep-sea offshore drilling.

Li et al. [10] analyzed the effects of seawater depth, equivalent static density (ESD) of drilling fluid, drilling chip concentration, and discharge on the annular pressure and

ECD of an RMR system based on drilling data from a vertical well in the South China Sea. Reynolds et al. [11] presented a pioneering case study for the Luiperd Mine and one of the measures for reducing environmental risk was to adopt a new articulation tool to minimize the bending stress applied to the subsea wellhead running tool (WHRT) and landing rope in running through the inflow pipe and surface casing. Claudey et al. [12] employed a controlled mud level technique in the Barents Sea, which controlled the liquid position within the riser, optimized the downhole pressure, prevented mud leakage, and maintained pressure balance. Claudey et al. [13] successfully deployed a riserless mud recovery (RMR) system at a water depth of 854 m, the deepest operation of its kind in the North Sea to date. Mud-lifting circulation was achieved without wellbore stability or shallow hazard issues, and no mud contamination occurred. Rosenberg et al. [14] introduced a design method for oil wells and used casing drilling to eliminate the need for multiple deep-water riserless series connections, enabling the high-pressure wellhead casing and its conductors to be set deeper. In the land and shallow sea environment, the practice of casing drilling has become very mature to improve drilling efficiency and reduce drilling hazards. RMR technology was also applied to achieve the successful cementing of the surface casing in the Malikai deepwater field in Sabah, Malaysia [15]. A model was devised to measure the bottomhole pressure of the RMR system during gas intrusion conditions using the gas–liquid two-phase flow theory [16]. In the Gulf of Mexico, RMR was initially implemented, leading to an enhanced improved operational efficiency and better wellbore structure [17]. The successful commercial application of dual-gradient drilling was first achieved in the ultra-deepwater Gulf of Mexico [18]. In the UK Continental Shelf (UKCS) oil-field development project, a RMR top-hole operation was employed for the first time [19]. Riserless dual-gradient technology was used for drilling highly deviated wells in the Santos Basin offshore Brazil, addressing issues related to formation erosion and wellbore collapse [20]. RMR has been utilized offshore multiple times, successfully resolving problems of loose sandstone soil and deepening surface casing, completing over 40 wells with a water depth of 450 m [21–23]. When drilling in overpressured zones [24–27], RMR can also compensate for the absence of a riser and blowout preventer (BOP). It has been utilized successfully in various offshore areas with loose formations and losses.

Mud pumps for drilling have also been studied by some researchers, in addition to RMR technology. In order to develop a system to monitor their technical conditions, Bejger et al. [28] investigated these pumps under operational conditions. Deng et al. [29] presented a practical diagnosis method that could effectively improve the fault diagnosis level for the fluid end of high-pressure reciprocating mud pumps. Khademi-Zahediet al. [30] utilized analytical studies and finite element methods to examine how different loading conditions impact the behaviors of mud pumps. Piasecki et al. [31] investigated mud pumps to develop an effective diagnostic tool that could help pump operators with their daily maintenance routines. ANSYS is a large-scale general-purpose finite element analysis software that can simulate and optimize a wide range of complex engineering problems. In this paper, pumps are analyzed using the ANSYS software. Additionally, the ANSYS software has numerous applications in other fields. ANSYS software was used by Bhatti et al. [32] to numerically model and evaluate the performance of a standing-wave thermoacoustic refrigerator with multiple stacks. The ANSYS Fluent software package was used by Hamdamov et al. [33] to conduct mathematical modeling of vertical axis wind turbines. ANSYS Fluent software was used by Ezzat et al. [34] to numerically simulate the phase transition process. Sohn et al. [35] used ANSYS and ANSYS pre-processor/post-processor composite material to create the layered composite material structure described in the article. Tjitra et al. [36] used ANSYS software to perform numerical simulation analysis of three-dimensional reinforced concrete beams under various collapse mechanisms.

However, few scholars have conducted analytical studies on the stability of mud pumps and their manifold connection employed in RMR technology, which is the core condition for the stable operation of RMR in the deep sea. During the drilling process, the deepwater RMR pump is mounted using a mid-hanging configuration, which results in reverse torque and vibration during normal operation. Hence, it is crucial to conduct a pertinent examination and investigation of the mud pump's reverse torque and vibration in the mud return circulation system. This study examines the relationships among reverse torque, vibration, and input parameters of the mud pump within the mud return circulation system. The study's contributions can be summarized as follows:

(I) An analysis is conducted of the working principle and the structure of the mud pump module, and an analytical model for the reverse torque and vibration of the mud pump is developed.
(II) Theoretical calculations and experiments are used to generate relevant data. The relationships among the reverse torque, vibration, and input parameters of the mud pump are analyzed using the ANSYS (Version 2020 R1) software.
(III) Predicting and analyzing the amplitude of the mud pump vibration is achieved by employing the SVR algorithm.
(IV) Relevant conclusions are drawn from the research results on the relationships among the reverse torque, vibration, and input parameters of the mud pump.

The findings show that the reverse torque of the mud pump increases approximately linearly with an increase in the rotational speed, and the vibration of the mud pump increases and then decreases with the increase in the rotational speed. The predicted values obtained through the prediction algorithm closely match the actual values.

2. Analysis of the Reverse Torque and Vibration in the Mud Pump

Due to the mid-hanging configuration of the deepwater RMR pump, it generates reverse torque and vibration in the mud return pipeline connected at both the upper end and the lower end during startup and operation. The return pipeline for the mud may be damaged, which will affect the stability and safety of both the mud pump and the return pipeline. In this section, the reverse torque and the vibration of the mud pump were analyzed using the ANSYS (Version 2020 R1) software. Since the mud pump used in the RMR technique is a ten-stage mud pump, a single-stage pump was first used for the analysis, and the error between the theoretical calculations and the simulation analysis was compared to ensure the error was within a reasonable range, thus proving the correctness of the analysis. Then, the mud pump (ten-stage pump) was analyzed based on the setup adopted for analyzing the single-stage pump.

Figure 1 shows the flowchart of the main analysis process. Firstly, the reverse torque of the single-stage mud pump was theoretically calculated and compared with the numerical simulation results as well as the error analysis. Secondly, the fluid model was established; its mesh division, materials, and boundary conditions were set; and the mud pump's reverse torque and rotational speed were derived from the changes in the relationships. Finally, the rotor model of mud pump was established; its mesh division, materials, and boundary conditions were set; and the relationships between the mud pump's vibration and the motor's input parameters were obtained.

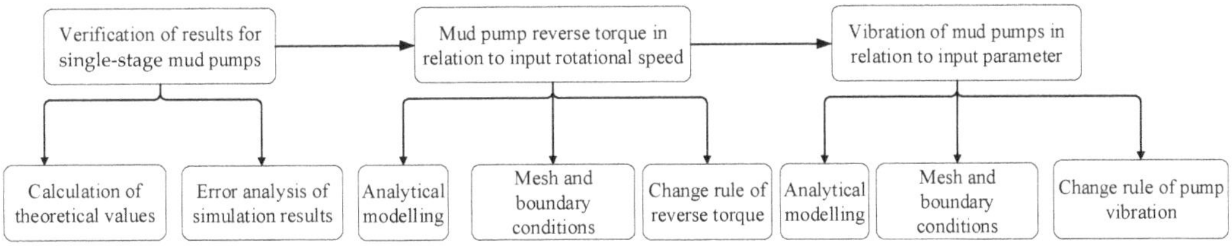

Figure 1. Analysis flowchart.

2.1. Working Principle and Structure of the Mud Pump Module

The riserless mud recovery (RMR) closed-loop drilling system generally consists of a drilling unit, a mud return system unit, a mud treatment unit, and a power monitoring unit.

The operational mode of the RMR closed-loop system is shown in Figure 2. The mud pump module is located within the mud return system unit and is installed using a mid-hanging configuration on the mud-lifting pipeline. The main function of the mud pump module is to transport the cuttings generated by the suction module through the mud return pipeline to the surface mud treatment equipment. With the suction action of the subsea mud pump, drilling fluid is drawn from the lower outlet of the suction module, enters the subsea mud pump through the subsea drilling fluid delivery pipeline, and is then lifted to the drilling vessel for processing and subsequent reuse.

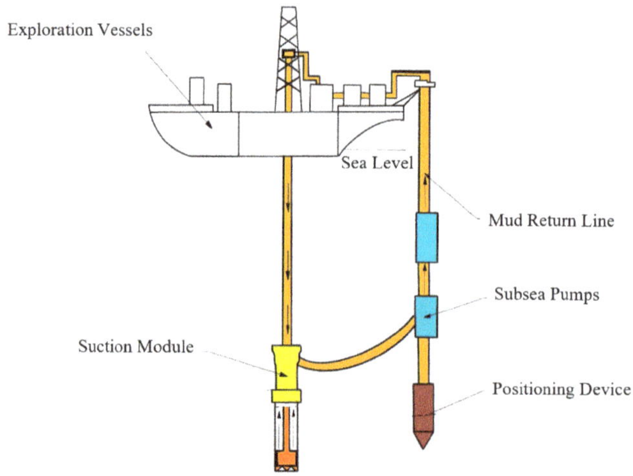

Figure 2. Diagram of the riserless mud recovery closed-loop operating mode [37].

The mud pump module is used to circulate a mixture of drilling fluid, cuttings, and even gas. The external view of the mud pump module is shown in Figure 3. The mud pump is powered by the platform and all equipment is only subjected to the hydrostatic pressure of seawater. The mud pump module is a critical component of the riserless mud recovery closed-loop system. It controls the pressure balance inside the suction module using the subsea pump and facilitates the return of drilling mud. The subsea pump module is fixed on the mud return pipeline and not only needs to ensure the reliability of the entire pipeline structure but also to maintain the smooth flow of the mud return pipeline. To prevent a collision between the lower subsea pump module and the suction module during operation, a subsea anchor is used to secure the lower end of the mud return pipeline, and a tension force is applied to restrict the horizontal movement of the mud return pipeline.

The lifting module of the mud pump unit mainly consists of drill pipe joints, corrective blocks, ball valves, centrifugal pumps, a four-way connector, a skid assembly, a ball valve control device, a skid, a trolley adapter, a supporting mechanism, and other components. The skid is the foundation of the mud pump module and serves mainly for connection and support. During drilling operation, the skid is installed on the mud lifting pipeline using a center-hanging arrangement, providing support for the internal components of the mud pump module. The skid remains stationary during the mud-lifting process and does not undergo significant vibration. The centrifugal pump used in the module is a multistage centrifugal pump, which increases the power of the pump module, enabling it to meet the requirements of deepwater operation. The mud return riser provides a conduit for drilling mud to flow from the seabed wellhead to the drilling rig. It also serves to fix and support two lifting pump module units, bearing the weight of the seabed anchor and resisting the force applied by sea currents and the movement of the drilling rig.

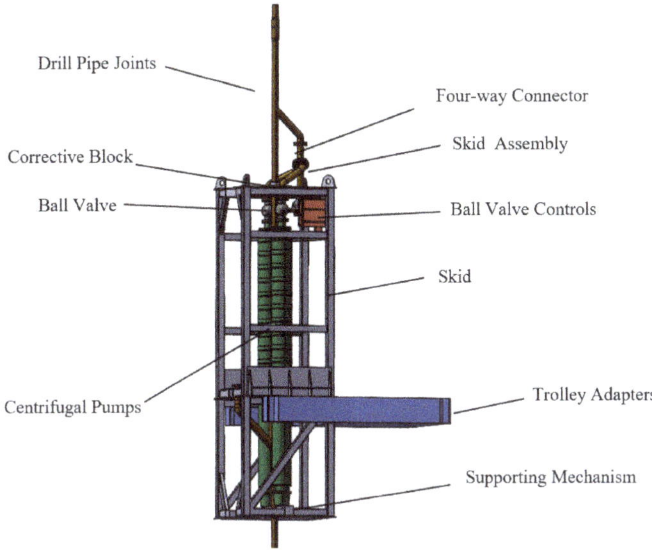

Figure 3. Exterior diagram of the mud pump module.

2.2. *Validation of the Theoretical Calculations and the Numerical Simulation Results for the Reverse Torque of the Single-Stage Mud Pump*

2.2.1. Theoretical Calculations of the Reverse Torque for a Single-Stage Mud Pump

According to the principle of energy conservation, the torque work conducted by the single-stage mud pump on the mud can be divided into two parts: one portion is converted into an increase in the energy of the mud and another portion is lost due to friction in the pipeline. The portion that contributes to the increase in mud energy is considered useful, while the portion lost due to friction in the pipeline is considered useless. The energy of the mud includes pressure energy, potential energy, and kinetic energy. Since the single-stage mud pump selected has the same inlet and outlet pipe diameters, the kinetic energy of the mud at the inlet and outlet remains constant. The calculation formula for the increase in the pressure energy of the mud within a time unit is shown as follows [38], where ΔE_P is the pressure energy from the increase in mud within a time unit, ΔP is the pressure difference between the inlet and outlet, and Q is the mud volume flow.

$$\Delta E_P = \Delta P Q \quad (1)$$

The formula for calculating the increase in potential energy of mud per unit time is shown as follows [38], where ΔE_h is the potential energy from the increase in mud within a time unit, Δh is the height difference between the inlet and outlet, ρ is the density of the mud, and g is the gravitational acceleration.

$$\Delta E_h = Q\rho g \Delta h \quad (2)$$

Based on Equations (1) and (2), the formula can be derived for calculating the useful work of the single-stage mud pump per unit time, as shown in the following equation, where W_e is the useful work of the single-stage mud pump within a time unit.

$$W_e = \Delta E_P + \Delta E_h \quad (3)$$

In the three formulas above, the mud flow rate Q is 2000 L/min, the mud density ρ is 1200 kg/m^3, the acceleration of gravity g is 9.81 m/s^2, the height difference Δh between the inlet and outlet is 0.32 m, and the speed is 1500 rpm. In addition, in order to find the useful work value, it is necessary to know the pressure difference between the inlet and outlet of the single-stage pump. For this reason, a three-dimensional model of the single-stage pump flow channel was established by using the three-dimensional modeling software Solidworks (Version 2018), and its cutaway view is shown in Figure 4a. The model was meshed by using the Mesh tool in ANSYS Workbench (version 2020 R1), and automatic meshing was used for the division. A total of 94,412 grids were divided by setting the mesh size to 0.005 m, as shown in Figure 4b. It is worth mentioning that when it is more finely meshed, the analysis results are consistent with those obtained using the current mesh; hence, the mesh independence is verified. Finally, the fluid model was analyzed by using the Fluent tool in ANSYS Workbench (Version 2020 R1). The pressure inlet–mass flow outlet boundary was used for this analysis. Based on the mud parameters shown in Table 1 and Equation (4), the outlet mass flow Q_m can be obtained, which is equal to 40 kg/s. In order to calculate the inlet and outlet pressure difference, the inlet pressure was 0 MPa, and the pressure difference between the inlet and outlet was the outlet pressure. Since the impeller was set as the rotating domain and the guide shell was set as the stationary domain, the interface between the rotating domain and stationary domains needed to establish the interaction surface so as to achieve the exchange of data between the rotating and stationary domains. Domain surfaces are all wall surfaces.

$$Q_m = Q\,\rho \tag{4}$$

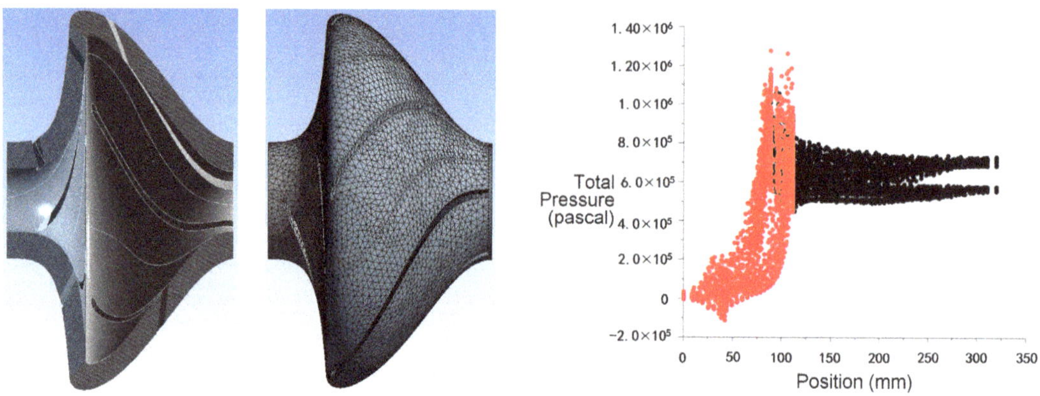

(a) (b) (c)

Figure 4. (**a**) The cutaway view of the three-dimensional model of the single-stage pump flow channel, (**b**) meshing of the three-dimensional model of the single-stage pump flow channel, and (**c**) pressure distribution diagram of the single-stage mud pump flow path at 1500 rpm.

Table 1. Mud parameters.

Parameter	Value
Density ρ (kg/m^3)	1200
Volume flow Q (L/min)	2000
Flow rates v (m/s)	1.88628
Viscosity μ (Pa·s)	0.12681

Based on the fluid analysis, the pressure difference Δp between the inlet and outlet of the single-stage mud pump at 1500 rpm is 0.756 MPa, as shown in Figure 4c, which represents the pressure distribution along the flow path of the pump. In this figure, the red dots represent the impeller and the black dots represent the guide shell. Similarly, the inlet–outlet pressure difference Δp at different speeds can be obtained, as shown in Table 2. Therefore, the relationship between the useful work per unit time and the single-stage mud pump rotational speed, n, can be determined, as shown in Columns 1 and 2 in Table 3. Based on Equation (5) [39] of the relationship between the torque exerted by mud on the single-stage mud pump and the useful work per unit time, and according to the data of the useful work and rotational speed shown in Columns 1 and 2 in Table 3, the relationship between the calculated torque T and the rotational speed n can be obtained, as shown in Columns 1 and 3 in Table 3.

$$W_e = Tn\eta \tag{5}$$

Table 2. Relationship between inlet and outlet pressure difference and rotational speed.

Rotational Speed (rpm)	1400	1410	1420	1430	1440	1450	1460	1470	1480	1490	1500
Pressure difference between inlet and outlet (MPa)	0.658	0.662	0.672	0.685	0.695	0.705	0.715	0.725	0.734	0.746	0.756

Table 3. Relationship between useful work and calculated reverse torque with rotational speed.

Rotational Speed (rpm)	Useful Work (kW)	Calculated Reverse Torque (N·m)
1400	22.06	273.17
1410	22.19	272.83
1420	22.53	275.06
1430	22.96	278.35
1440	23.29	280.39
1450	23.63	282.53
1460	23.96	284.51
1470	24.29	286.46
1480	24.59	288.04
1490	24.99	290.76
1500	25.33	292.76

2.2.2. Error Analysis of Torque Simulation Results for a Single-Stage Mud Pump

Fluent is a computational fluid dynamics software that utilizes numerical methods to solve fluid mechanics problems. The numerical methods employed by Fluent are based on finite volume and finite element methods, enabling a numerical simulation and analysis of various fluid flow problems. In this analysis, Fluent was used to analyze the torque of a single-stage mud pump. Based on the analysis in Section 2.2.1, it was straightforward to obtain the single-stage pump's reverse torque at 1500 rpm. However, for other rotational speeds, it was sufficient to set different rotational speeds for the analysis. Based on the fluid simulation results, the relationship between the simulated reverse torque and rotational speed can be obtained. Figure 5 illustrates the variation curve of the calculated reverse torque and simulated reverse torque concerning the rotational speed, while Figure 6 shows the variation curve of the relative error concerning the rotational speed. It can be observed that the average relative error of the simulation is 3.41%, which demonstrates the correctness of the simulation results.

2.3. Relationship between Mud Pump Reverse Torque and Rotational Speed

To establish the relationship between the mud pump's reverse torque and rotational speed, a three-dimensional model of the mud pump flow channel was established by using the three-dimensional modeling software Solidworks (version 2018), and its cutaway view is shown in Figure 7. The model was meshed by using the Mesh tool in ANSYS Workbench (Version 2020 R1) and automatic meshing was used for the division. A total of 239,462 grids were divided, as shown in Figure 8. Except for the difference in the analysis model, the rest of the settings are identical to those used when analyzing single-stage pumps and will not be repeated here.

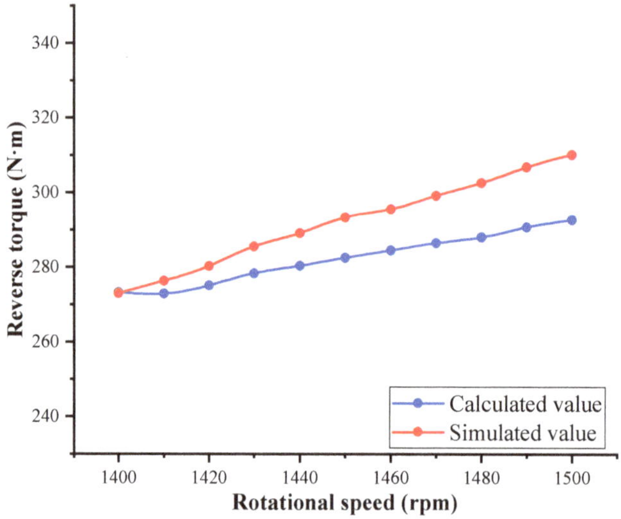

Figure 5. Relationship between the calculated and simulated values as they vary with the rotational speed.

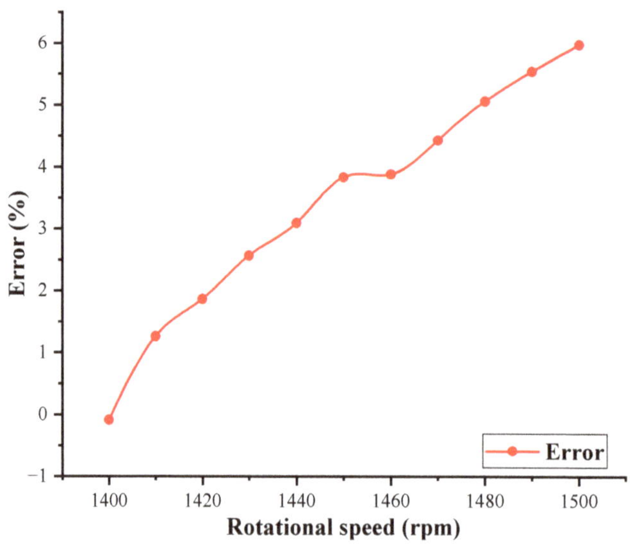

Figure 6. Relationship between the error and rotational speed as they vary.

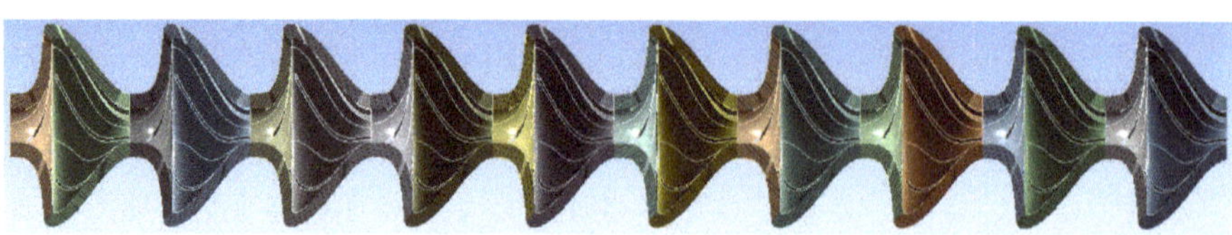

Figure 7. The cutaway view of the mud pump flow channel.

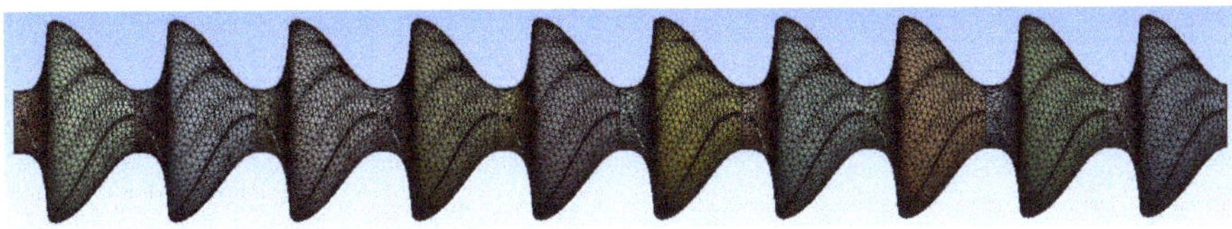

Figure 8. Meshing of the three-dimensional model of the mud pump flow channel.

To establish the relationship between the reverse torque and rotational speed of the mud pump within the speed range from startup to normal operation, 24 sets of analytical experiments were conducted at different pump speeds. The reverse torque values at 24 different speeds were obtained, and a torque–speed relationship curve was plotted based on the analytical experimental data, as shown in Figure 9. The analysis results indicate that the reverse torque of the mud pump increases approximately linearly with an increase in the rotational speed. According to the actual operating conditions, as the pump speed increases, the fluid in the pump moves rapidly accordingly. At a high speed, the excitation force of the fluid becomes higher, resulting in an increase in the reverse torque. This effect is especially significant in the areas with a high fluid velocity. The analysis results are consistent with the actual operating conditions. The minimum reverse torque is −320.4 N·m at 0 rpm, while the maximum reverse torque is 3620.28 N·m at 1500 rpm. The value of the maximum reverse torque provides a reference for the selection of motors.

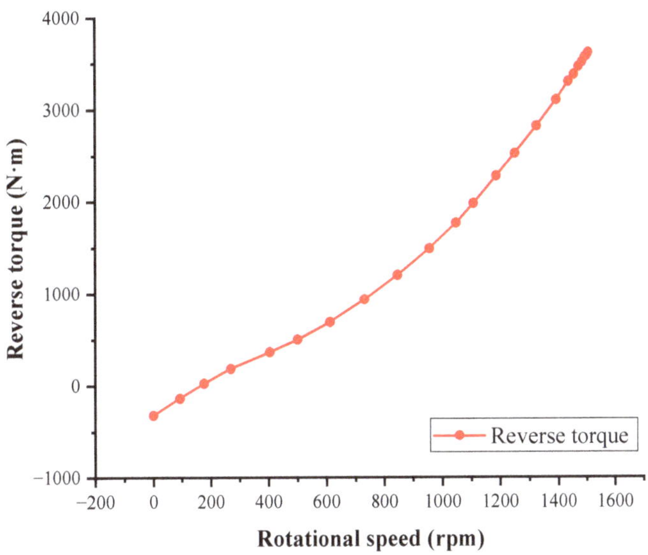

Figure 9. Relationship between reverse torque and rotational speed.

2.4. Relationship between Mud Pump Vibration and Input Parameters

2.4.1. Analysis Modeling and Pre-Processing Setting

Due to the unstable force exerted by the mud on the mud pump, the operation of the mud pump inevitably generates certain vibrations. To analyze the effect of vibration on the safety and stability of the mud pump and the return pipeline, a mud pump rotor model, as shown in Figure 10, was established to investigate the effect of mud-induced vibration on the mud pump rotor using a one-way fluid-structure coupling analysis.

Figure 10. Three-dimensional model of the mud pump rotor.

In the one-way fluid-structure coupling analysis, the calculations of the fluid and solid were performed independently. Firstly, in the fluid calculation, the fluid parameters, such as velocity, pressure, and temperature, were computed to obtain the force exerted by the fluid on the solid. Then, in the solid calculation, the force was applied as the external load on the solid to compute parameters, such as stress, strain, and deformation. In the ANSYS software, the workflow for the one-way fluid-structure coupling analysis mainly involved using ANSYS Fluent for the fluid calculation and transferring the results to ANSYS Mechanical for the solid calculation. Specifically, in ANSYS, the input for the fluid calculation included the initial state of the fluid, boundary conditions, mesh, and so on. The output included fluid velocity, pressure, temperature, and other related parameters. For the solid calculation, the input included material properties, geometric shapes, boundary conditions, and so on, while the output included stress, strain, deformation, and other relevant parameters.

The mud pump rotor was divided using a hexahedral meshing method, with a total of 247,947 cells. The material used for the mud pump rotor was stainless-steel 022Cr22Ni5Mo3N, and its material properties are shown in Table 4.

Table 4. Performance parameters of stainless-steel 022Cr22Ni5Mo3N.

Property	Value
Density (kg/m^3)	7800
Tensile strength (MPa)	620
Yield strength (MPa)	420
Thermal conductivity (W/(m·K))	10
Elastic modulus (GPa)	210
Poisson's ratio	0.3

2.4.2. Relationship between Different Input Parameters and Mud Pump Vibration

The mud pump exhibited different mechanical characteristics when operated at different voltages. Figure 11 shows the mechanical characteristic curves of the mud pump at rated voltage values of 60%, 80%, 90%, 100%, and 110%.

When the voltage is Un, the relationship between the rotor amplitude and rotational speed is analyzed based on the data of the speed and torque. When the rotational speed is 0 rpm, the displacement of the rotor over time is shown in Figure 12. From the figure, it can be observed that the rotor amplitude is 2.2411 mm. By conducting the same analysis, the amplitude corresponding to each rotational speed on the mechanical characteristic curve can be obtained. The analysis results are shown in Figure 13, indicating that the rotor amplitude increases with an increase in the rotational speed initially, followed by a decrease. For a motor speed between 0 and 1440 rpm, the amplitude increases slowly with an increase in the speed. When the motor speed is between 1440 and 1500 rpm, the amplitude decreases rapidly with an increase in the motor speed. The maximum amplitude of 2.8179 mm was observed at a rotational speed of 1440 rpm. According to the relationship between mud pump torque and rotational speed on the mechanical characteristic curve, the mud pump torque also exhibited an initial increase followed by a decrease with an increase in the rotational speed. Furthermore, both the maximum torque and vibration occurred at a rotational speed of 1440 rpm.

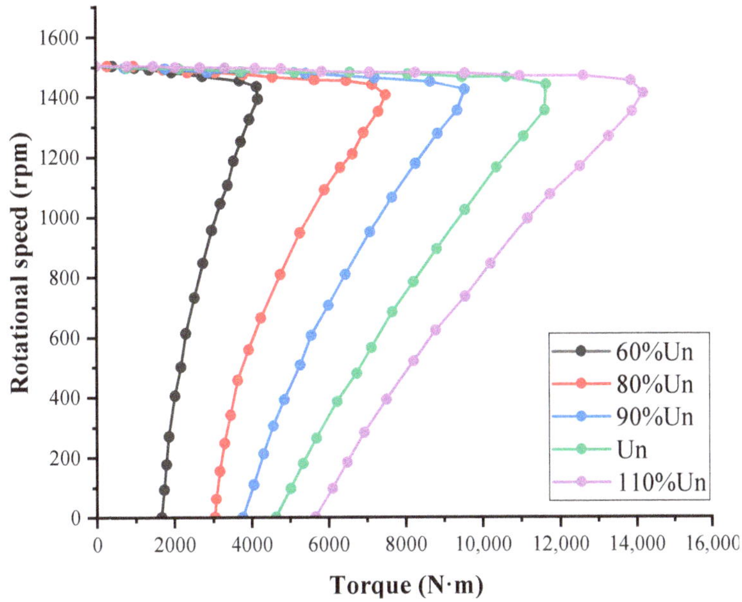

Figure 11. Mechanical characteristic curves of the mud pump at different voltages.

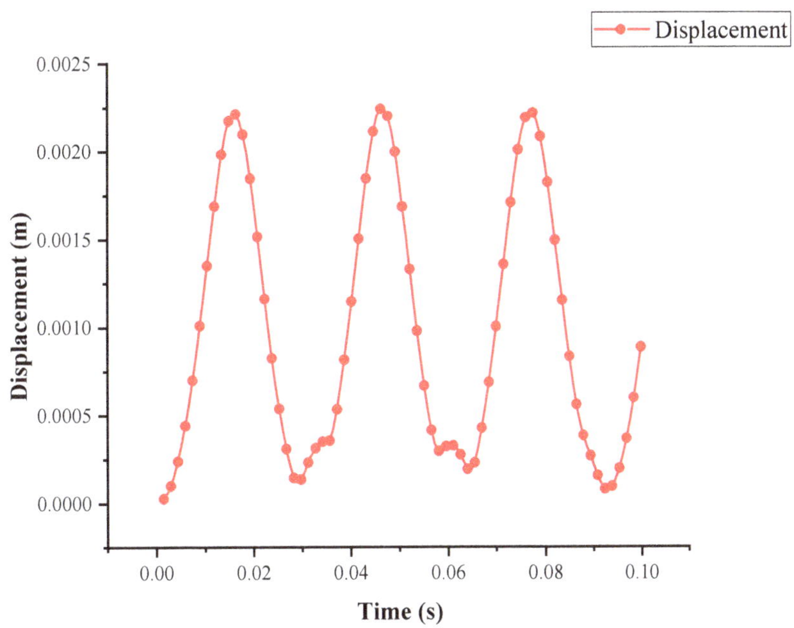

Figure 12. Relationship between rotor displacement and time when the rotational speed is 0 rpm.

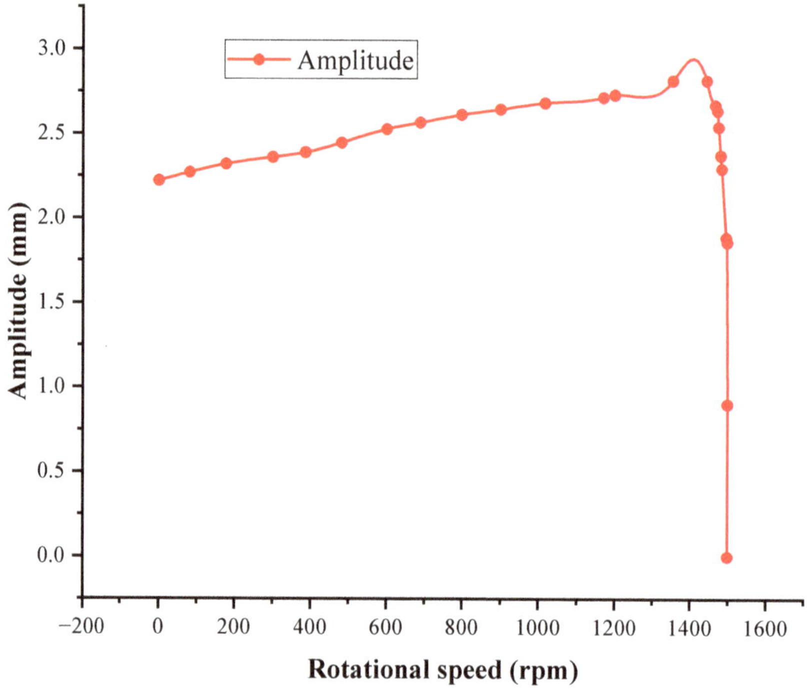

Figure 13. Variation of rotor amplitude with rotational speed under rated voltage.

3. Prediction of Pump Amplitude Based on the Support Vector Regression Algorithm

To accurately determine the relationship between the mud pump's amplitude and input parameters, it was necessary to conduct a predictive analysis of the pump's amplitude. In this section, the support vector regression (SVR) algorithm is used to predict the amplitude of the mud pump, enabling the prediction of amplitude conditions using a small amount of data on speed, torque, and amplitude.

When using support vector machines for the regression analysis, nonlinear mapping $\Phi(x)$ was employed to map the input vectors to a high-dimensional feature space. Subsequently, linear regression was performed in the high-dimensional feature space to find the optimal hyperplane that minimized the errors of all samples from the optimal hyperplane.

Assuming the training sample set is $\{(x_i, y_i), i = 1, 2, \cdots, N\}$, where N is the number of samples, x_i represents the input values, and y_i represents the desired output values. The regression model of the SVM can be expressed as the following equation [40]:

$$f(x) = w \times \Phi(x) + b \tag{6}$$

where w and b are the weight vector and bias, respectively. Introducing the penalty factor c and slack variables $\xi_i (1 = 1, 2, \cdots, N)$ under the insensitive loss function ε, the solution of the SVR is an optimization problem.

$$\begin{cases} \min\limits_{w,b} \frac{1}{2}\|w\|^2 + c \sum\limits_{i=1}^{N} (\xi_i + \xi_i^*) \\ s.t \begin{cases} y_i - w \cdot \Phi(x) - b \leq \varepsilon + \xi_i \\ y_i - w \cdot \Phi(x) - b \geq -\varepsilon - \xi_i{}^* \end{cases} \\ \xi_i, \xi_i^* \geq 0, (i = 1, 2, \cdots, N) \end{cases} \tag{7}$$

By introducing the Lagrange multiplier $\{\alpha_i, \alpha_i^*, \beta_i, \beta_i^*, (i = 1, 2, \cdots, N)\}$, Equation (7) can be rewritten as Equation (8):

$$\begin{aligned} &\min_{w,b} L(w, b, \xi, \xi^*) = \tfrac{1}{2}\|w\|^2 + c\sum_{i=1}^{N}(\xi_i + \xi_i^*) \\ &-\sum_{i=1}^{N}\alpha_i[\varepsilon + \xi_i - y_i + w \cdot \Phi(x_i) + b] \\ &-\sum_{i=1}^{N}\alpha_i^*\left[\varepsilon + \xi_i^* + y_i - w \cdot \Phi(x_i) - b\right] \\ &-\sum_{i=1}^{N}\left(\beta_i\xi_i + \beta_i^*\xi_i^*\right) \end{aligned} \tag{8}$$

By taking the partial derivatives of W, b, ξ_i, and ξ_i^* in Equation (8), the following can be obtained:

$$\begin{cases} \frac{\partial L}{\partial w} = w - \sum_{i=1}^{N}(\alpha_i - \alpha_i^*) \cdot \Phi(x) = 0 \\ \frac{\partial L}{\partial b} = -\sum_{i=1}^{N}(\alpha_i - \alpha_i^*) = 0 \\ \frac{\partial L}{\partial \xi_i} = c - \alpha_i - \beta_i = 0 \\ \frac{\partial L}{\partial \xi_i^*} = c - \alpha_i^* - \beta_i^* = 0 \end{cases} \tag{9}$$

By substituting Equation (9) into Equation (8), the dual form of Equation (7) can be obtained:

$$\begin{aligned} &\max_{\alpha,\alpha^*}\sum_{i=1}^{N}(\alpha_i - \alpha_i^*)y_i - \sum_{i=1}^{N}(\alpha_i + \alpha_i^*)\varepsilon \\ &-\tfrac{1}{2}\sum_{i=1}^{N}\sum_{j=1}^{N}(\alpha_i - \alpha_i^*)(\alpha_j - \alpha_j^*)\Phi(x_i)\Phi(x_j) \\ &s.t.\sum_{i=1}^{N}(\alpha_i - \alpha_i^*) = 0 \\ &\alpha_i \geq 0, \alpha_i^* \leq c, (i = 1, 2, \cdots, N) \end{aligned} \tag{10}$$

By introducing the kernel function $K(x_i, x_j) = \Phi(x_i)\Phi(x_j)$, the data are mapped from a low-dimensional space to a high-dimensional space, thereby transforming the linearly inseparable problem into a linearly separable problem. This allows for the calculation of inner products in the high-dimensional space to be performed as function calculations in the low-dimensional space. By employing quadratic programming methods, the optimal solution (α_i, α_i^*) for Equation (11) can be obtained, resulting in the decision function of the SVR:

$$f(x) = \sum_{i=1}^{N}(\alpha_i - \alpha_i^*)K(x_i, x_j) + b \tag{11}$$

Due to the advantages of high accuracy and low computational complexity, the radial basis function (RBF) is employed for training and prediction. It is defined as:

$$K(x_i, x_j) = \exp(-\gamma\|x_i - x_j\|^2) \tag{12}$$

where γ is the undetermined kernel parameter. The prediction results are evaluated using the mean relative error (MRE), mean absolute error (MAE), mean squared error (MSE), root mean squared error (RMSE), and coefficient of determination (R^2).

The amplitude prediction process is illustrated in Figure 14, and the specific steps are as follows:

Step 1: Construct the training and testing sets. From the given n data samples, select N data samples as the training set. The remaining (n − N) test samples are used as the testing set.

Step 2: Normalize the data samples. The 'mapminmax' function is used to normalize the data samples to the range of [−1, 1] to improve the convergence speed and accuracy of the SVR.

Step 3: Parameter optimization. The Gaussian radial basis kernel function is used for training and prediction. The grid search method combined with ten-fold cross-validation is employed to find the optimal penalty factor c and kernel parameter γ. The training set is randomly divided into ten subsets, with nine subsets as the training set and one subset as the validation set. The training and validation sets iterate alternately for ten rounds. In each iteration, the 'svmtrain' function is used to train the prediction model, and the RMSE e_i of the prediction model is recorded. The average value $E_i = \frac{1}{10}\sum_{i=1}^{10} e_i$ is calculated. Then, change the parameters, traverse the grid to train the model for m rounds, and obtain the average RMSE E_m of each round of training. Obtain the minimum value of E_m and determine the optimal parameters of the model.

Step 4: Find out the predicted results. Train the predictive model with the optimal parameters and use the model to predict the output values for the (n − N) test samples, obtaining the model's output values.

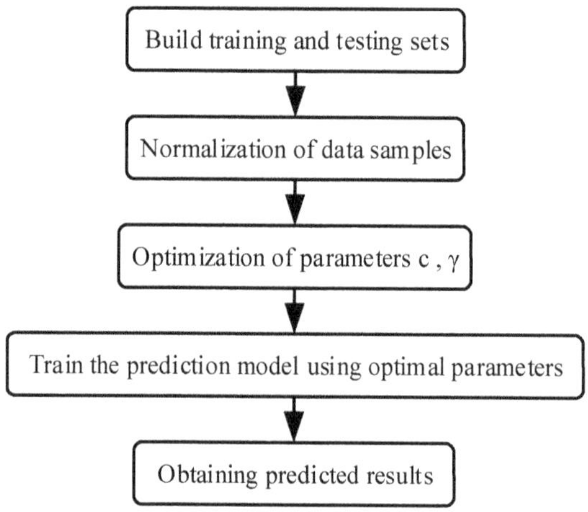

Figure 14. Amplitude prediction process based on the SVR algorithm.

The parameter settings for the SVR algorithm are presented in Table 5. The input data for the amplitude prediction based on the SVR algorithm is shown in Appendix A. The results of the amplitude prediction based on the SVR algorithm are displayed in Figure 15, demonstrating a close match between the actual and predicted values. This indicates a high level of reliability in the prediction results.

Table 5. SVR algorithm parameter settings.

Values of Penalty Factor C	[−10, 0.5, 10]
Values of kernel parameter γ	[−10, 0.5, 10]
Number of cross-validation folds v	5
Training set	50
Testing set	10

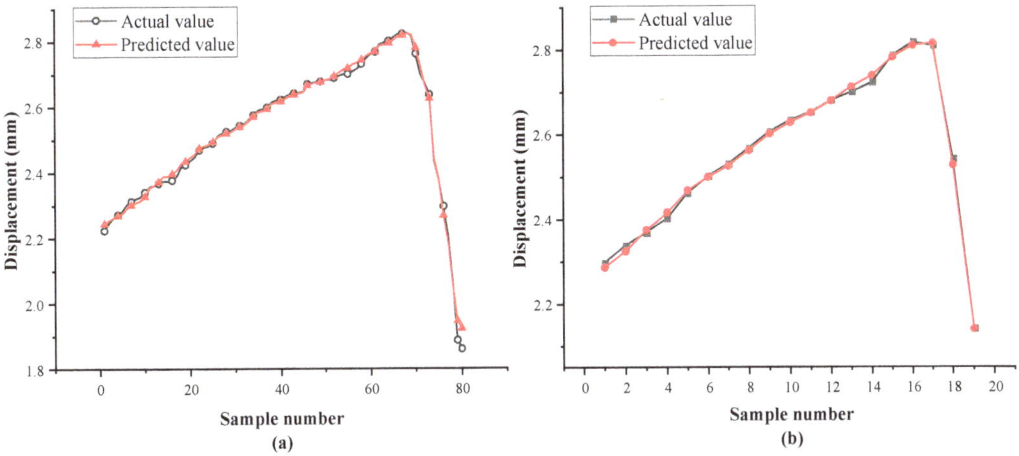

Figure 15. (**a**) Comparison of prediction results for training set: mse = 0.00082288, R^2 = 0.99588, (**b**) comparison of prediction results for testing set: mse = 0.00033816, R^2 = 0.99775.

4. Conclusions and Recommendations

Based on the study of the relationships among the reverse torque, vibration, and input parameters of the mud pumps in the riserless mud recovery drilling, the following conclusions can be drawn:

(1) The error between the theoretical calculation and the actual simulation of the reverse torque of the mud pump is inevitable, and the average relative error obtained from the experiments in this paper is only 3.41%, which is within the permissible range.

(2) When the speed of the mud pump is 0 rpm, the reverse torque of the mud to the pump is the lowest, at −320.4 N·m, and when the speed of the mud pump is 1500 rpm, the reverse torque of the mud to the pump is the highest, at 3620.28 N·m. As the speed of the mud pump increases, the reverse torque of the mud to the pump also increases, which is in line with the actual working condition.

(3) The amplitude of the mud pump exhibits a pattern of initially increasing and then decreasing with an increase in speed. In the range of 0–1440 rpm, the amplitude increases slowly with the speed. When the motor speed is in the range of 1440–1500 rpm, the amplitude decreases rapidly with an increase in speed. At a speed of 1440 rpm, the amplitude reaches its maximum value of 1.5499 mm. The variation of the rotor amplitude with speed is consistent with the variation of torque with speed in the mechanical characteristics curve of the mud pump.

(4) The SVR algorithm is applied to predict the amplitude of the mud pump and the predicted values closely match the actual values, which indicates a high level of reliability in the prediction results. The prediction results provide a reference for the stability of the mud pump.

Author Contributions: Conceptualization, G.H. and B.X.; data curation, H.C. and R.Q.; formal analysis, G.Y. and B.X.; methodology, C.L.; project administration, G.H. and R.Q.; software, G.Y. and B.X.; writing—original draft, C.L. and G.H. G.H. and B.X. contributed equally to this work and should be considered co-first authors. All authors have read and agreed to the published version of the manuscript.

Funding: This research was funded by the National Key Research and Development Program of China (2021YFC2800802; 2021YFC2800803), the Hi-tech Ship Project of Ministry of Industry and Information Technology (CJ05N20), the Marine Economy Development Foundation of Guangdong Province (GDNRC [2022]44), and Technical Support for the Stimulation and Testing of Gas Hydrate Reservoirs.

Institutional Review Board Statement: Not applicable.

Informed Consent Statement: Not applicable.

Data Availability Statement: Not applicable.

Conflicts of Interest: The authors declare no conflict of interest.

Nomenclature

Δp	The pressure difference between the inlet and outlet, MPa
Q	Volume flow, L/min
Q_m	Mass flow, kg/s
ΔE_P	The pressure energy added by the increase in mud within a time unit, kW
ρ	The density of the mud, kg/m^3
g	Gravitational acceleration, m/s^2
Δh	The height difference between the inlet and outlet, m
ΔE_h	The potential energy gained by the increase in mud within a time unit, kW
W_e	The useful work of the mud pump within a time unit, kW
T	The reverse torque exerted by the mud on the mud pump, N·m
n	The rotational speed of the mud pump, rpm
η	The efficiency of the mud pump

Appendix A

Table A1. Input data for pump amplitude prediction based on the SVR algorithm.

Number	Rotational Speed (rpm)	Torque (N·m)	Amplitude (mm)
1	0	4546.54055	2.2229
2	41.99719	4691.81983	2.2454
3	68.69492	4812.88589	2.2625
4	80.83025	4885.52552	2.2721
5	102.67384	4982.37837	2.2842
6	126.9445	5103.44444	2.2982
7	143.93397	5151.87086	2.3034
8	168.20463	5248.72371	2.3127
9	177.16368	5335.02618	2.3215
10	197.98253	5381.18077	2.3257
11	221.11457	5473.48995	2.3385
12	230.36739	5496.56725	2.3404
13	283.5711	5727.3402	2.3574
14	300	5796.57208	2.3618
15	322.89559	5888.88127	2.3672
16	329.8352	5911.95856	2.3685
17	357.59366	6027.34504	2.3744
18	360.01548	6033.51788	2.3747
19	371.54121	6091.01012	2.3773
20	385.94837	6205.99459	2.3901
21	397.4741	6292.23295	2.4031
22	423.40698	6378.4713	2.4158

Table A1. *Cont.*

Number	Rotational Speed (rpm)	Torque (N·m)	Amplitude (mm)
23	437.81415	6435.96354	2.4240
24	466.62847	6522.2019	2.4361
25	481.03563	6608.44025	2.4477
26	504.08708	6723.42473	2.4626
27	509.84995	6780.91697	2.4698
28	521.37568	6838.4092	2.4769
29	535.78284	6867.15532	2.4804
30	544.42713	6953.39368	2.4905
31	564.59716	7010.88592	2.5031
32	573.24146	7068.37815	2.5096
33	587.64862	7154.61651	2.5192
34	590.53005	7212.10875	2.5254
35	600	7240.85487	2.5285
36	607.81864	7269.60099	2.5315
37	627.98867	7327.09322	2.5375
38	645.27726	7384.58546	2.5433
39	653.92156	7442.0777	2.5489
40	676.97301	7585.80829	2.5625
41	688.49874	7643.30053	2.5677
42	705.78733	7729.53889	2.5753
43	728.83879	7902.0156	2.5897
44	740.36452	7959.50784	2.5942
45	754.77168	8017.00007	2.5987
46	769.17884	8131.98455	2.6072
47	797.99316	8218.22291	2.6148
48	803.75603	8304.46126	2.6206
49	815.28175	8333.20738	2.6225
50	832.57035	8448.19186	2.6299
51	838.33321	8505.68409	2.6334
52	855.6218	8591.92245	2.6385
53	875.79183	8649.41469	2.6418
54	890.19899	8678.16081	2.6435
55	900	8735.65304	2.6466
56	919.01331	8850.63752	2.6526
57	953.5905	9138.09871	2.6712
58	970.87909	9224.33706	2.6749
59	988.16768	9281.8293	2.6774
60	991.04911	9310.57542	2.6785
61	996.81198	9368.06766	2.6807
62	1011.21914	9425.55989	2.6829

Table A1. *Cont.*

Number	Rotational Speed (rpm)	Torque (N·m)	Amplitude (mm)
63	1016.982	9483.05213	2.6850
64	1045.79633	9655.52884	2.6899
65	1074.61065	9885.49779	2.6963
66	1100.54354	10000.48227	2.6994
67	1106.3064	10057.97451	2.7008
68	1123.59499	10115.46674	2.7021
69	1158.17218	10345.43569	2.7088
70	1169.69791	10431.67405	2.7171
71	1184.10507	10489.16629	2.7227
72	1200	10590.07644	2.7322
73	1235.97085	10776.62748	2.7492
74	1244.61514	10920.35807	2.7618
75	1256.14087	11006.59643	2.7692
76	1290.71806	11207.81926	2.7859
77	1299.36235	11294.05761	2.7928
78	1305.12522	11351.54985	2.7974
79	1325.29524	11409.04209	2.802
80	1339.7024	11552.77268	2.813
81	1354.10956	11610.26492	2.8173
82	1385.80532	11667.75716	2.8217
83	1391.56818	11696.50328	2.8239
84	1429.0268	11725.24939	2.8262
85	1443.43396	11610.26492	2.8179
86	1454.95969	11495.28044	2.8094
87	1460.72255	10920.35807	2.7632
88	1463.60398	10029.22839	2.7025
89	1466.48542	9080.60647	2.672
90	1472.24828	8505.68409	2.6374
91	1475.12971	7327.09322	2.5427
92	1478.01115	6493.45578	2.438
93	1480.89258	5918.5334	2.3752
94	1483.77401	5027.40372	2.2967
95	1486.65544	4394.98911	2.2046
96	1489.53687	4078.7818	2.1426
97	1492.41831	3417.62107	2.0532
98	1495.29974	2670.22198	1.8874
99	1498.18117	2497.74527	1.8608
100	1500	0	0.0012923

References

1. Wang, Y.; Luan, T.; Gao, D.; Wang, J. Research Progress on Recoil Analysis and Control Technology of Deepwater Drilling Risers. *Energies* **2022**, *15*, 6897. [CrossRef]
2. Wilson, A. RFID Technology for Deepwater Drilling and Completions Challenges. *J. Pet. Technol.* **2017**, *69*, 62–64. [CrossRef]
3. Fisher, A.; Kaura, J.; Kratzer, M.; Oyler, K.; Reutzel, R.; Teoh, M.; Thomassen, S.; Vives, J. Innovative Technologies Deliver Operational Objectives in a Multi-Well, Ultra-Deepwater, Managed Pressure Drilling and Completions Campaign. In Proceedings of the SPE Annual Technical Conference and Exhibition, Dubai, United Arab Emirates, 21–23 September 2021; OnePetro: Richardson, TX, USA, 2021.
4. Guo, X.; Zhang, Z. Study on the pressure wave velocity model of multiphase fluid in the annulus of dual-gradient drilling. *Therm. Sci.* **2023**, 136. [CrossRef]
5. Li, J.; Zhang, H.; Li, X.; Ding, S.; Tao, Q. Dual-Gradient Drilling of Deepwater in South China Sea: What is the Well's Maximum Measured Depth and How to Improve It. In Proceedings of the SPE/IADC Managed Pressure Drilling and Underbalanced Operations Conference and Exhibition, New Orleans, LA, USA, 17–18 April 2018; OnePetro: Richardson, TX, USA, 2018.
6. Yang, H.; Li, J.; Zhang, G.; Zhang, H.; Guo, B.; Chen, W. Wellbore multiphase flow behaviors during gas invasion in deepwater-downhole dual-gradient drilling based on oil-based drilling fluid. *Energy Rep.* **2022**, *8*, 2843–2858. [CrossRef]
7. Cohen, J.H.; Stave, R.; Hauge, E.; Molde, D.O. Field trial of well control solutions with a dual gradient drilling system. In Proceedings of the SPE/IADC Managed Pressure Drilling and Underbalanced Operations Conference & Exhibition, Dubai, United Arab Emirates, 13–14 April 2015; OnePetro: Richardson, TX, USA, 2015.
8. Wilson, A. Taking a Fresh View of Riser Margin for Deepwater Wells Potentially Boosts Safety. *J. Pet. Technol.* **2015**, *65*, 93–96. [CrossRef]
9. Chen, Q.; Deng, Q.; Tan, L.; Qi, G.; Zhang, J. Status and Prospects of Dual-Gradient Drilling Technologies in Deep-Water Wells. *Front. Energy Res.* **2022**, *10*, 675. [CrossRef]
10. Li, X.; Zhang, J.; Zhang, M.; Wang, P.; Sun, X. Study on the Factors Affecting the Annulus Pressure of Riserless Mud Recovery System. *CIREA J. Energy* **2019**, *4*, 62–82.
11. Reynolds, T.R.; Cavé, A.; Horn, T.P. Deepwater Operator Saves Multiple Days of Riserless Drilling Operations Using Novel Articulation Tool. In Proceedings of the Offshore Technology Conference, Houston, TX, USA, 1–4 May 2023; OTC: Hollister, MO, USA, 2023.
12. Claudey, E.; Fossli, B.; Elahifar, B.; Qiang, Z.; Olsen, M.; Mo, J. Experience using managed pressure cementing techniques with riserless mud recovery and controlled mud level in the barents sea. In Proceedings of the SPE Norway One Day Seminar, Bergen, Norway, 18 April 2018; OnePetro: Richardson, TX, USA, 2018.
13. Claudey, E.; Maubach, C.; Ferrari, S. Deepest deployment of riserless dual gradient mud recovery system in drilling operation in the north sea. In Proceedings of the SPE Bergen One Day Seminar, Bergen, Norway, 20 April 2016; OnePetro: Richardson, TX, USA, 2016.
14. Rosenberg, S.; Kotow, K.; Wakefield, J.; Sampietro, S.; Hulett, B. Riserless Casing Drilling for Mitigation of Shallow Hazards-A Paradigm Shift in Deepwater Well Construction. In Proceedings of the IADC/SPE International Drilling Conference and Exhibition, Galveston, TX, USA, 5–7 March 2022; OnePetro: Richardson, TX, USA, 2022.
15. Goenawan, J.; Goncalves, R.; Dooply, M.; Pasteris, M.; Heu, T.; Chan, L.; Bhaskaran, S.; Hinoul, W. Overcoming shallow hazards in deepwatermalikai batch-set top-hole sections with engineered trimodal particle-size distribution cement. In Proceedings of the Offshore Technology Conference Asia, Kuala Lumpur, Malaysia, 22–25 March 2016; OnePetro: Richardson, TX, USA, 2016.
16. Zhang, W.; Zhang, J.; Wang, Z.; Sun, R.; Yu, S.; Yang, Z. Flow regulation and bottom hole pressure control under gas influx condition based on RMR system. *J. Phys. Conf. Ser.* **2022**, *2280*, 012050. [CrossRef]
17. Li, X.; Zhang, J.; Tang, X.; Mao, G.; Wang, P. Study on wellbore temperature of riserless mud recovery system by CFD approach and numerical calculation. *Petroleum* **2020**, *6*, 163–169. [CrossRef]
18. Carpenter, C. Dual-Gradient Drilling in Ultradeepwater Gulf of Mexico. *J. Pet. Technol.* **2014**, *66*, 112–115. [CrossRef]
19. Allan, R.M.; Arain, Z.; Fraser, B.; Short, S.; Davidson, S. First Oil Requirements Drive Simultaneous Drilling and Subsea Construction Operations on UK Central North Sea Development. In Proceedings of the SPE/IADC Drilling Conference, Amsterdam, The Netherlands, 5–7 March 2013; OnePetro: Richardson, TX, USA, 2013.
20. Pessanha, W.; Indio, M.; Miranda, A. Riserless Drilling Technique for Maximizing Bore Hole Stability: Dynamic Kill Drilling Application in Offshore Brazil. In Proceedings of the Offshore Technology Conference Brasil, Rio de Janeiro, Brazil, 27–29 October 2015; OTC Brasil. OnePetro: Richardson, TX, USA, 2015.
21. Bui, H.; Hasanov, A.; Mendoza, L.; Rowe, S.; Ontiveros, P.; Mennie, J. Geohazards in Riserless Drilling for an Exploration Well in Deepwater GoM: Identification and Mitigations. In Proceedings of the Offshore Technology Conference, Houston, TX, USA, 1–4 May 2023; OTC: Hollister, MO, USA, 2023.
22. Daniel, M. Use of Riserless Mud Recovery for Protection of Cold Water Corals While Drilling in Norwegian Sea. In Proceedings of the SPE International Conference and Exhibition on Health, Safety, Security, Environment, and Social Respon-Sibility, Stavanger, Norway, 11–13 April 2016; OnePetro: Richardson, TX, USA, 2016.
23. Cheng, W.; Ning, F.; Sun, J.; Liu, Z.; Jiang, G.; Li, X. A porothermoelastic wellbore stability model for riserless drilling through gas hydrate-bearing sediments in the Shenhu area of the South China Sea. *J. Nat. Gas Sci. Eng.* **2019**, *72*, 103036. [CrossRef]

24. Godhavn, J.; Gaassand, S.; Hansen, K.H.; Morris, R.; Nott, D. Development and First Use of Controlled Mud Level System in US Deepwater GoM. In Proceedings of the SPE/IADC Managed Pressure Drilling and Underbalanced Operations Conference & Exhibition, Dubai, United Arab Emirates, 13–14 April 2015; OnePetro: Richardson, TX, USA, 2015.
25. Ali, W.; Hashmi, A.A.; Al-Turki, F.A.; Abbas, A. Successful Deployment of Managed Pressure Cementing (MPC) System Across a Challenging Narrow Pressure Window Condition for a Critical Liner Cement Job: A Case Study From Saudi Arabia. In Proceedings of the SPE/IADC International Drilling Conference and Exhibition, Stavanger, Norway, 7–9 March 2023; OnePetro: Richardson, TX, USA, 2023.
26. Mao, L.; Cai, M.; Liu, Q.; Wang, G. Dynamical Simulation of High-Pressure Gas Kick in Ultra-Deepwater Riserless Drilling. *J. Energy Resour. Technol.* **2021**, *143*, 063001. [CrossRef]
27. Liu, J.; Zhao, H.; Yang, S.X.; Liu, Q.; Wang, G. Nonlinear dynamic characteristic analysis of a landing string in deepwater riserless drilling. *Shock Vib.* **2018**, *2018*, 8191526. [CrossRef]
28. Bejger, A.; Piasecki, T. Technical problems of mud pumps on ultra deepwater drilling rigs. *Zesz./Akad. W Szczecinie* **2013**, *36*, 13–16.
29. Deng, S.; Pei, J.; Wang, Y.; Liu, B. Research on drilling mud pump fault diagnosis based on fusion of acoustic emission and vibration technology. *Insight—Non-Destr. Test. Cond. Monit.* **2017**, *59*, 415–423.
30. Khademi-Zahedi, R.; Makvandi, M.; Shishesaz, M. Technical analysis of the failures in a typical drilling mud pump during field operation. In Proceedings of the 22nd Annual International Conference on Mechanical Engineering—ISME2014, Ahvaz, Iran, 22–24 April 2014.
31. Piasecki, T.; Bejger, A.; Kozak, M. The assessment of the operational suitability of mud pumps on deep water drillships depends on maintenance strategy. *Zesz. W Szczecinie* **2016**, *46*, 49–53.
32. Bhatti, U.N.; Bashmal, S.; Khan, S.; Ben-Mansour, R. Numerical Modeling and Performance Evaluation of Standing Wave Thermoacoustic Refrigerators with a Multi-Layered Stack. *Energies* **2020**, *13*, 4360. [CrossRef]
33. Hamdamov, M.M.; Ishnazarov, A.I.; Mamadaliev, K.A. Numerical Modeling of Vertical Axis Wind Turbines Using ANSYS Fluent Software. In Proceedings of the Internet of Things, Smart Spaces, and Next Generation Networks and Systems, Tashkent, Uzbekistan, 15–16 December 2022; Koucheryavy, Y., Aziz, A., Eds.; Springer Nature: Cham, Switzerland, 2023; pp. 156–170.
34. Ezzat, Y.; Abdel-Rehim, A.A. Numerical modelling of lauric acid phase change material using iterative and non-iterative time-advancement schemes. *J. Energy Storage* **2022**, *53*, 105173. [CrossRef]
35. Sohn, J.M.; Hirdaris, S.; Romanoff, J.; Kim, S.J. Development of Numerical Modelling Techniques for Composite Cylindrical Structures under External Pressure. *J. Mar. Sci. Eng.* **2022**, *10*, 466. [CrossRef]
36. Tjitradi, D.; Eliatun, E.; Taufik, S. 3D ANSYS Numerical Modeling of Reinforced Concrete Beam Behavior under Different Collapsed Mechanisms. *Int. J. Mech. Appl.* **2017**, *7*, 14–23.
37. Huang, F.; Yang, J.; Li, B.; Gan, L.; Feng, Q. Study on dynamic characteristics of drill string in deep-water riserless drilling. *Adv. Mech. Eng.* **2023**, *15*, 168781322211421. [CrossRef]
38. Kara Omar, A.; Khaldi, A.; Ladouani, A. Prediction of centrifugal pump performance using energy loss analysis. *Aust. J. Mech. Eng.* **2017**, *15*, 210–221. [CrossRef]
39. Prawira, A.; Rini, E. Size and configuration of mud motor drilling affects the optimum power outputs. *Eng. Solid Mech.* **2017**, *5*, 93–102. [CrossRef]
40. Yang, Y.; Zhu, W.; Li, Y.; Zhao, B.; Zhang, L.; Song, J.; Deng, Z.; Shi, Y.; Xie, C. Modeling of PEMFC and analysis of multiple influencing factors on output characteristics. *J. Electrochem. Soc.* **2022**, *169*, 034507. [CrossRef]

applied sciences

MDPI

Article

Body-Pose-Guided Action Recognition with Convolutional Long Short-Term Memory (LSTM) in Aerial Videos

Sohaib Mustafa Saeed [1], Hassan Akbar [1,2], Tahir Nawaz [1,2], Hassan Elahi [1,*] and Umar Shahbaz Khan [1,2]

1 Department of Mechatronics Engineering, National University of Sciences and Technology (NUST), Islamabad 44000, Pakistan; smsaeed.mts20ceme@student.nust.edu.pk (S.M.S.); hassanakbar2013@yahoo.com (H.A.); tahir.nawaz@ceme.nust.edu.pk (T.N.); u.shahbaz@ceme.nust.edu.pk (U.S.K.)

2 National Centre of Robotics and Automation, Islamabad 44000, Pakistan

* Correspondence: hassan.elahi@ceme.nust.edu.pk; Tel.: +92-51-54444402

Abstract: The accurate detection and recognition of human actions play a pivotal role in aerial surveillance, enabling the identification of potential threats and suspicious behavior. Several approaches have been presented to address this problem, but the limitation still remains in devising an accurate and robust solution. To this end, this paper presents an effective action recognition framework for aerial surveillance, employing the YOLOv8-Pose keypoints extraction algorithm and a customized sequential ConvLSTM (Convolutional Long Short-Term Memory) model for classifying the action. We performed a detailed experimental evaluation and comparison on the publicly available Drone Action dataset. The evaluation and comparison of the proposed framework with several existing approaches on the publicly available Drone Action dataset demonstrate its effectiveness, achieving a very encouraging performance. The overall accuracy of the framework on three provided dataset splits is 74%, 80%, and 70%, with a mean accuracy of 74.67%. Indeed, the proposed system effectively captures the spatial and temporal dynamics of human actions, providing a robust solution for aerial action recognition.

Keywords: deep neural network; convolutional LSTM; action recognition; body pose keypoints; aerial surveillance

Citation: Saeed, S.M.; Akbar, H.; Nawaz, T.; Elahi, H.; Khan, U.S. Body-Pose-Guided Action Recognition with Convolutional Long Short-Term Memory (LSTM) in Aerial Videos. *Appl. Sci.* **2023**, *13*, 9384. https://doi.org/10.3390/app13169384

Academic Editors: Sheng Du, Xiongbo Wan, Wei Wang and Hao Fu

Received: 5 July 2023
Revised: 12 August 2023
Accepted: 16 August 2023
Published: 18 August 2023

1. Introduction

Action recognition involves automatically identifying and categorizing human actions in video sequences, which is highly beneficial and needed for surveillance applications [1–3]. Action recognition is, indeed, a challenging task due to the presence of various challenges, particularly background clutter occlusions and camera viewpoint [4–6]. Conventional action recognition methods involved hand-crafted feature extraction [7,8], based on manually representing actions, such as motion, shape, or appearance descriptors. The limitations of this approach lie in the fact that hand-crafted features may not be able to effectively capture complex temporal relationships or variations in different action scenarios. Indeed, designing effective features could be challenging; plus, they may not generalize well to different datasets or action classes.

The 3D CNNs extend the concept of traditional 2D CNNs [9,10] to incorporate temporal information by processing video frames as 3D volumes. They, however, require a large amount of training data and computational resources. Additionally, they may struggle with long-term temporal dependencies or capturing fine-grained motion details. Moreover, the training of 3D CNNs from scratch can be challenging due to the limited availability of annotated video datasets.

Recurrent neural networks (RNNs), gated recurrent unit (GRU), or LSTM [11,12] model temporal dependencies by maintaining internal memory states. However, RNNs may struggle with modeling long-term dependencies or capturing complex spatial dynamics.

They could be sensitive to the order and timing of actions within sequences. RNNs are computationally intensive, especially for longer sequences.

Two-stream networks [13,14] consist of the spatial stream (CNN for appearance) as well as the temporal stream (CNN or RNN for motion). They require synchronized and aligned RGB and optical flow inputs, which could be challenging to obtain in practice. Combining the information from two streams can introduce additional complexity and potential performance degradation.

Graph convolutional networks (GCNs) [15,16] represent actions as graphs and exploit graph convolution operations to capture spatial and temporal relationships between body joints or keypoints. However, GCNs rely heavily on accurate and reliable detection and the tracking of skeletal keypoints and also have limitations when dealing with occlusions or missing keypoints in complex action scenarios. Designing appropriate graph structures and defining graph convolution operations are, inevitably, challenging.

The recent introduction of vision transformers has proved to be more efficient in accuracy. There are approaches that utilize transformers for action recognition [13,17,18] in complex scenarios; however, they are generally computationally more resource-consuming.

Aerial videos provide a comprehensive view [5] of the scene, enabling surveillance operators to monitor larger areas and detect events that may otherwise be overlooked. Action recognition from aerial scenarios, however, requires reliable detection of the target in complex backgrounds, with varying camera angle altitudes for an accurate classification of the action [19–22].

Malik et al. [23] proposed a method that relied on extracting 2D skeletal data using OpenPose that are then fed into LSTM for training and testing. Their framework was, however, validated in an indoor multi-view scenario and may not be directly deployable for aerial videos.

Another limitation in human action recognition is that the trained models generally misclassify when provided with unannotated data from new users [24], even after being trained on a large amount of data. This challenge arises as it is impractical to collect data for every new user. Yang et al. [25] aimed to address this problem by presenting a semi-supervised learning action recognition method for training on labeled as well as unlabeled data but not primarily for the aerial camera settings that are under consideration in this paper.

Dai et al. [26] introduced a dual-stream attention-based LSTM containing a visual attention mechanism that enables selectively focusing on key elements in the image frames by applying varying levels of attention to each individual deep feature map's output. The deep feature correlation layer embedded in their framework is, indeed, relevant to our work, and it contributes towards enhancing the robustness of the action recognition. The validation in [26] was, however, in experimental scenarios, different from that considered in this work.

Unlike the existing related methods reviewed above, the proposed research combines the robust pose detection ability of YOLOv8 with temporal sequencing ability of the ConvLSTM to propose an effective and efficient approach aimed specifically at aerial action recognition. In fact, the proposed framework offers a reliable recognition of human actions from an aerial perspective by utilizing the convolutional LSTM's capacity to parse temporal sequences. Specifically, the proposed method extracts the body pose keypoints from the frames and classifies actions at the frame level utilizing the customized convolutional LSTM network model. The reason behind relying on the extraction of the target body pose or keypoints is the lower computational cost as compared to the spatial features. Moreover, we use the LSTM network due to its demonstrated effectiveness for sequential data classification [23,24,27,28]; plus, it is not well explored in the literature for the problem under consideration. We showed the effectiveness of the proposed method in terms of encouraging performance accuracy and computational cost when compared on a public dataset (containing a wide range of action types) with several existing related approaches.

The organization of the paper is as follows. The proposed method for action recognition is described in Section 2. Section 3 provides details of the experimental results and analysis, which is followed by the conclusions in Section 4.

2. Proposed Action Recognition Method

We employed the YOLOv8 pose detection model for the extraction of 17 body keypoints. The extracted keypoints are then passed to the second stage, which is ConvLSTM, to extract spatiotemporal features across the sequence. The sequence length of 30 frames, chosen empirically, is set for the extraction of temporal information. The intuition behind incorporating the body pose with ConvLSTM is a selection of suitable features that are keypoints and performing the memory-based sequence classification using LSTM. Figure 1 illustrates the proposed human action recognition system.

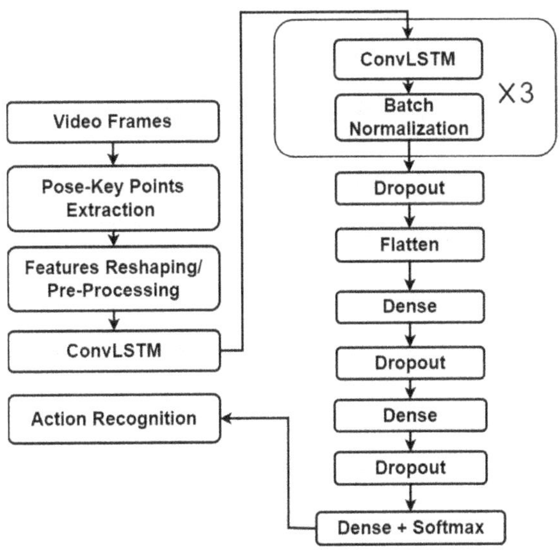

Figure 1. Block diagram illustrating different steps involved in the proposed action system.

The architecture in Figure 1 is designed to process raw keypoints for the analysis of both spatial and temporal aspects. The ConvLSTM architecture shown in Figure 1 is made up of multiple hidden layers that work together to collect spatial and temporal features from frames. For an accurate classification of actions, this extracted feature set is essential. These characteristics ultimately influence how the recognized action is predicted, enabling the system to efficiently analyze actions occurring in successive frames.

Convolutional layers are used in the context indicated above to extract significant features from the body pose keypoints. Convolutional layers apply filters to the keypoints in order to capture significant spatial characteristics, such as the placement of body parts and their interactions. These filters help in finding patterns and correlations among the keypoints.

The network can automatically learn hierarchical representations of the body positions using convolutional layers. The network's capacity to recognize and accurately classify various activities within the video sequences is greatly aided by the extracted characteristics.

To accurately capture the temporal dynamics of activities throughout a series of frames, the use of LSTM is crucial. LSTMs effectively capture patterns and changes that emerge over time by processing the retrieved features or representations from each frame. LSTMs give the network the ability to comprehend how actions develop and classify by keeping track of past frames and taking into account how they affected the current frame.

2.1. Pose Extraction

The YOLOv8 pose extractor is a popular deep learning-based algorithm for keypoint detection. There are several other approaches that can be utilized for this purpose, but the latest YOLOv8 is known to be more efficient in accuracy as well as computationally [29] Figure 2 shows the output of the pose extractor.

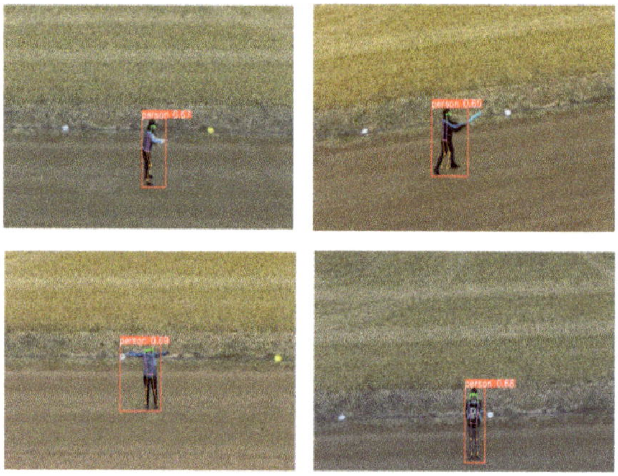

Figure 2. Results of Yolov8 pose estimator on Drone Action dataset [22]: stabbing (**top left**), hitting stick (**top right**), waving hands (**bottom left**), and clapping (**bottom right**).

The keypoint coordinates for a given video can be represented as a (F, Kp), where Kp represents the keypoints of the image and F represents the number of sequential frames or sequence length, which, in our case, is set to 30. The extracted keypoints are made to be aligned with the input of the next stage.

To extract spatiotemporal features from the video sequence, we stack the keypoint tensors for a given person over time. Let Kp_t be the keypoint tensor for the person at time t and let $Kp_1, Kp_2, \ldots, Kp_t$ be the keypoint tensors for the person over T frames of the video sequence. We stack these tensors along the time dimension to obtain a tensor P with dimensions (F,Kp_t):

$$P = [Kp_1, Kp_2, Kp_3, \ldots, Kp_T] \tag{1}$$

The YOLOv8 algorithm uses a fully convolutional neural network (FCN) to predict a heat map for each keypoint, which can be used to estimate the pose of the person in the video. The resulting output yields 17 keypoint coordinates for each detected person at the frame level across the video sequence.

2.2. Custimized Convolutional LSTM Model

We used the LSTM model for action classification in aerial videos. The LSTM model is a type of RNN that can effectively encapsulate the dependencies of the sequential data. In the proposed approach, we first extract the temporal features from the aerial videos using the YOLOv8 pose extractor and then use the LSTM model to classify the actions based on these features. The tensor P in Equation (1) represents the spatiotemporal features of the person over time.

The LSTM model contains a memory cell and three gates, including an input gate, output gate, and forget gate [30–32], defined as follows:

Input gate:

$$i_t = \sigma(W_i x_t + U_i h_{t-1} + b_i) \tag{2}$$

Forget gate:

$$f_t = \sigma\left(W_f x_t + U_f h_{t-1} + b_f\right) \tag{3}$$

Output gate:

$$o_t = \sigma(W_o x_t + U_o h_{t-1} + b_o) \tag{4}$$

Memory cell:

$$C_t = f_t . c_{t-1} + i_t \cdot \tanh\ (W_c x_t + U_c h_{t-1} + b_c) \tag{5}$$

Output:

$$h_t = o_t \cdot\ tanh(c_t) \tag{6}$$

where x_t h_t , and c_t denote the input, the output, and the cell state t, respectively. i_t, f_t, and o_t are the input, the forget, and the output gates, respectively. W_i, W_f, W_o, and W_c refer to the collection of weight matrices used to transform the input data at each time step, whereas U_i, U_f, U_o, and U_c are the weight matrices to transform the hidden state from the previous time step. b_i, b_f, b_o, and b_f represent the bias terms.

The output H is a sequence of hidden states that captures the temporal dependencies in the spatiotemporal features. We can then use the final hidden state of the LSTM as input to a fully connected layer with softmax activation to obtain the probability distribution across the different action classes:

$$P = Softmax(W_h H + b) \tag{7}$$

We designed the custom sequential LSTM model by stacking three ConvLSTM 1D layers, such that each layer is followed by a batch normalization layer, with decreasing filter sizes of 128, 32, and 16, respectively. We added a dropout layer after the third ConvLSTM1D layer to prevent overfitting. Next, we flattened the output and added two fully connected layers with ReLU activation and a dropout layer after each. Finally, we added a dense output layer with the softmax activation function. The LSTM model applied in this research is convolutional LSTM (Figure 3), which combines convolutional layers with LSTM to model spatiotemporal data.

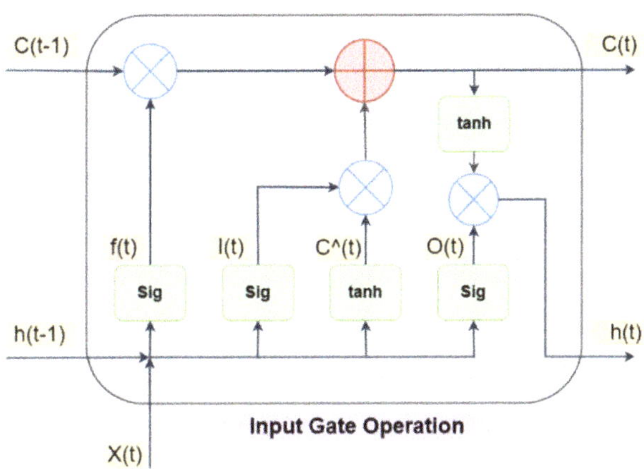

Figure 3. Structure of the convolutional LSTM.

3. Experimental Results and Analysis

This section first describes the dataset in Section 3.1, which is followed by an evaluation of the results in Section 3.2 and performance comparisons with existing related approaches in Section 3.3.

3.1. Dataset

We used the publicly available Drone Action dataset for evaluation [22]. This dataset comprises 240 videos that run for a total duration of approximately 44.6 min, embodying

66,919 frames and containing 13 distinct human action classes. The videos were captured from a low-altitude and slow-moving drone to ensure the details of body pose were reliably extracted. The complexity of this dataset is augmented by the diversity in body size, camera motion, varying target speed, and background clutter, making it a suitable benchmark for human action recognition studies. Figure 4 shows representative frames from the dataset for each action class [22].

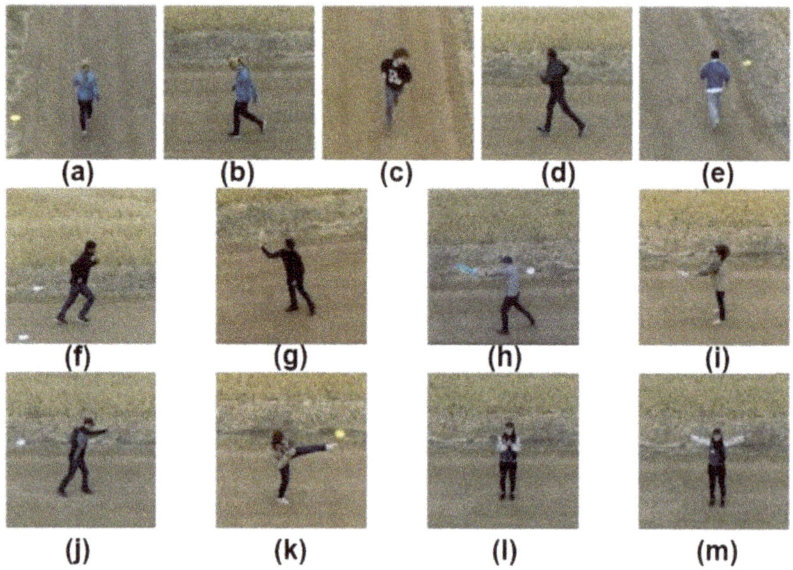

Figure 4. Representative frames from each class of the Drone Action dataset [22]: (**a**) walking front_back, (**b**) walking side, (**c**) jogging front_back, (**d**) jogging side, (**e**) running front_back, (**f**) running side, (**g**) hitting with bottle, (**h**) hitting with stick, (**i**) stabbing, (**j**) punching, (**k**) kicking, (**l**) clapping, (**m**) waving hands.

3.2. *Evaluation of Results*

The proposed action recognition framework for aerial videos demonstrates an improved accuracy and robustness. Indeed, the combination of the YOLOv8-Pose algorithm and customized sequential convolutional LSTM model effectively captures the spatial and temporal information of actions, leading to an encouraging action recognition performance. The proposed model is trained and tested separately on the three dataset splits, as provided by the original paper [22]. In each split, 70% data are used for training and 30% for testing. The training was caried out for 200 epochs (chosen empirically), and network parameters were kept the same for training and testing for each split of the data. Table 1 lists the corresponding values for the validation loss and accuracy on all three splits. A representative graphical representation of the validation loss and validation accuracy is shown in Figure 5 for Split 1.

Table 1. Training details for the 3 splits.

Dataset	Epochs	Validation Loss	Validation Accuracy
Split 1	200	2.75–0.25	0.05–0.88
Split 2	200	2.55–1.00	0.13–0.83
Split 3	200	2.58–1.00	0.12–0.82

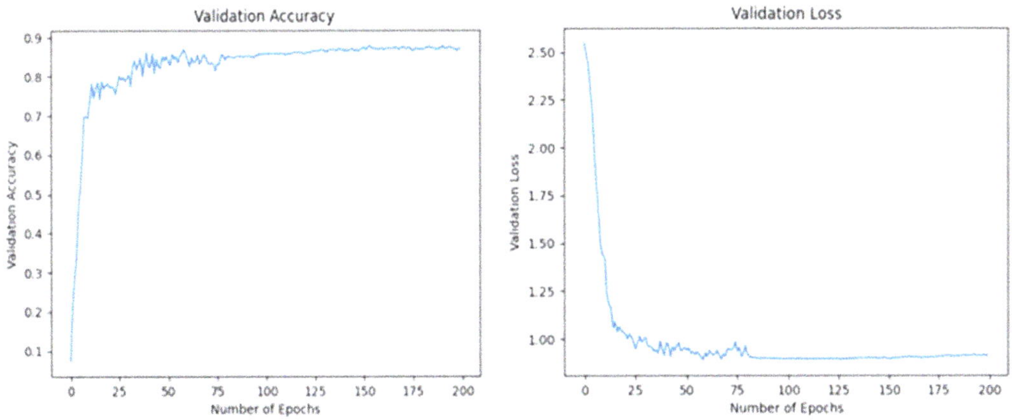

Figure 5. Plots for the validation accuracy (**left**) and validation loss (**right**) during the training for Split 1.

The overall accuracies achieved on Split 1, Split 2, and Split 3 are 74%, 80%, and 70%, respectively. The corresponding confusion matrices are provided in Figures 6–8, respectively. The class-wise results on each split are given in Tables 2–4, respectively, based on the standard well-known evaluation measures, which are precision, recall, and F1-score.

Analyzing the results in more detail, we observe that some actions had consistently high precision, recall, and F1-score values across all dataset splits. For instance, the actions "Clap", "Kick", "Walk_fb", "Walk_side", and "Wave_hands" achieved high scores on all three splits. This suggests that the proposed framework is highly effective in recognizing these actions, even when presented with variations in the data. The high accuracy in these classes can be attributed to the combination of YOLO-Pose and the custom-designed ConvLSTM network, which allows for an efficient extraction of spatial and temporal information in video frames.

Table 2. Performance evaluation of the proposed method on all action types based on precision, recall, and F1-score on Split 1.

Action	Precision	Recall	F1-Score
Clap	1.00	1.00	1.00
Hit_botl	0.19	0.14	0.16
Hit_stick	0.65	0.64	0.65
Jogging	0.73	0.88	0.80
Jog_side	0.91	0.89	0.90
Kick	0.99	1.00	0.99
Punch	0.91	0.99	0.95
Run_fb	0.50	0.40	0.44
Run_side	0.86	0.89	0.87
Stab	0.29	0.40	0.34
Walk_fb	1.00	0.90	0.95
Walk_side	1.00	1.00	1.00
Wave_hands	0.98	1.00	0.99
Average	**0.77**	**0.78**	**0.77**

Table 3. Performance evaluation of the proposed method on all action types based on precision, recall, and F1-score on Split 2.

Action	Precision	Recall	F1-Score
Clap	1.00	1.00	1.00
Hit_botl	0.50	0.36	0.42
Hit_stick	0.72	0.78	0.75
Jog_fb	0.83	0.91	0.87
Jog_side	0.86	0.98	0.91

Table 3. *Cont.*

Action	Precision	Recall	F1-Score
Kick	0.99	0.92	1.00
Punch	0.76	0.99	0.83
Run_fb	0.73	0.53	0.62
Run_side	1.00	0.76	0.86
Stab	0.40	0.55	0.46
Walk_fb	1.00	1.00	1.00
Walk_side	0.98	0.98	0.98
Wave_hands	0.97	1.00	0.99
Average	**0.83**	**0.83**	**0.82**

Table 4. Performance evaluation of the proposed method on all action types based on precision, recall, and F1-score on Split 3.

Action	Precision	Recall	F1-Score
Clap	1.00	0.89	0.94
Hit_botl	0.33	0.29	0.31
Hit_stick	0.59	0.68	0.64
Jog_fb	0.67	0.61	0.63
Jog_side	0.85	0.58	0.69
Kick	0.99	0.85	0.92
Punch	0.83	0.95	0.84
Run_fb	0.28	0.33	0.30
Run_side	0.45	0.77	0.57
Stab	0.37	0.39	0.38
Walk_fb	0.91	0.95	0.93
Walk_side	0.96	1.00	0.98
Wave_hands	1.00	1.00	1.00
Average	**0.71**	**0.71**	**0.70**

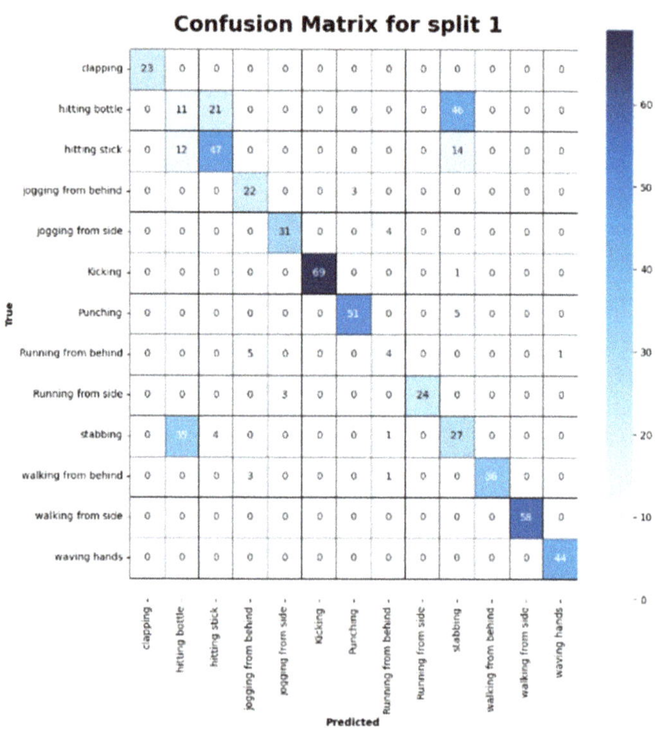

Figure 6. Confusion matrix for Split 1.

On the other hand, some actions demonstrated lower precision, recall, and F1-score values. For example, the "Hit_botl" action achieved lower scores on all the three splits, with the lowest F1-score being 0.16 on Split 1. Similarly, the "Stab" action had an F1-score of 0.34 on Split 1, 0.46 on Split 2, and 0.38 on Split 3.

The lower performance for these actions (Hit_botl, Stab) could be attributed to the higher complexity of the movements and the similarity of these actions with each other and some other classes, making it challenging for the proposed framework to differentiate them from others. Moreover, factors, such as background clutter and variation in viewpoint, could further hinder the recognition of these actions.

It is worth mentioning that there is performance variation for some actions across different splits. For instance, the "Hit_stick" action had an F1-score of 0.65 on Split 1, which increased to 0.75 on Split 2 and then decreased slightly to 0.64 in Split 3. This observation suggests that the performance of the proposed framework is sensitive to the choice of training and testing data.

We also calculated the computational performance of the proposed method. The evaluation was performed in terms of the number of network parameters (in millions) and the number of floating-point operations (FLOPS) (in millions) and the classification time for the proposed customized ConvLSTM network. We practically implemented this model on Intel(R) Core(TM) i5-8250U CPU @ 1.80 GHz with 8.00 GB RAM. The total number of FLOPS was 36.79 million, with 1.03 million trainable parameters. The classification time for the 612 test sequences with 30 frames each on Split 1 was 3.58 s. The per sequence classification time was 5.457 milliseconds. This suggests that the proposed method is lightweight in terms of computational complexity and could be deployable in real-world applications.

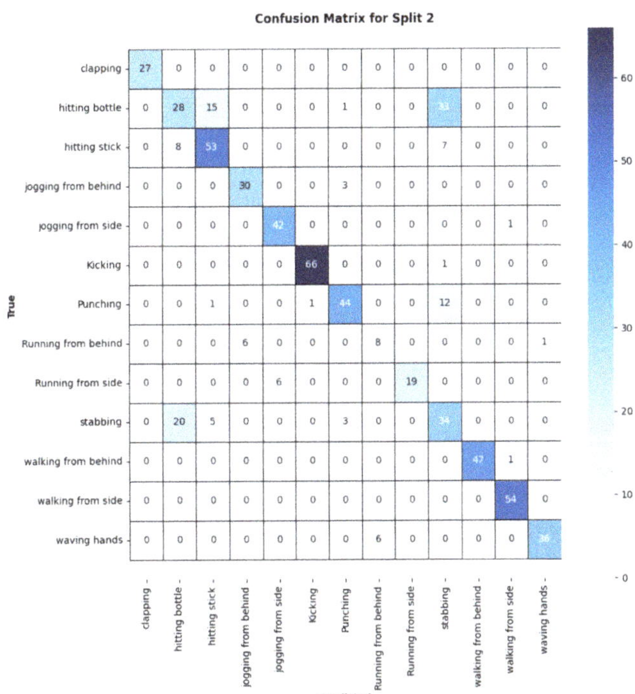

Figure 7. Confusion matrix for Split 2.

3.3. Performance Comparison with Related Approaches

We also compared the performance of the proposed action recognition framework with two existing approaches, as reported in the benchmark paper [22] (see Table 5). The benchmark paper provides an analysis of the classification accuracy of two methods, including the high-level pose features (HLPFs) method and the pose-based convolutional neural networks (P-CNNs) method. The high-level pose features (HLPFs) method uses skeletal information from human poses to represent actions. In P-CNN, at each frame of a video, descriptors are extracted from the body regions. These descriptors encode relevant information, such as motion flow patterns and visual characteristics of the regions, leading to two-streamed information. Over time, these descriptors are aggregated, combining the information from multiple frames, to form a video descriptor. The proposed method shows better or comparable performance as compared to these existing methods (Table 5), owing to its capability to efficiently model temporal information and long-term dependencies in action sequences.

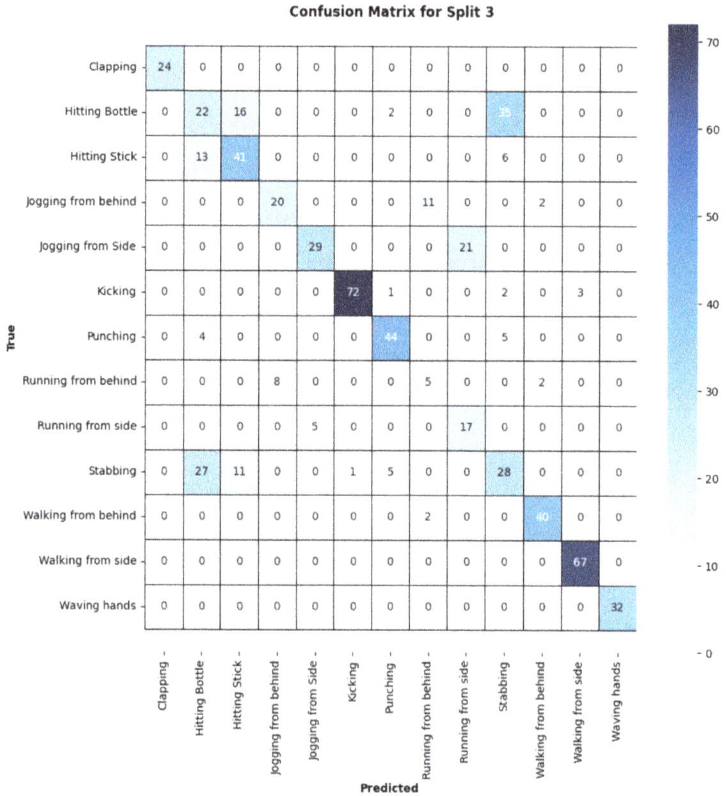

Figure 8. Confusion matrix for Split 3.

Table 5. Comparison of the proposed method with existing approaches on Drone Action dataset.

Method	Accuracy (Split 1)	Accuracy (Split 2)	Accuracy (Split 3)	Mean Accuracy
HLPF	63.89%	68.09%	61.11%	64.36%
P-CNN	72.22%	81.94%	73.61%	75.92%
Pose+ LSTM	74.00%	80.00%	70.00%	74.67%

For a more detailed performance comparison of the proposed approach with other models, we investigated several state-of-the-art deep learning models, such as Action Transformer, ResNet18, ResNet101, 3D ResNet, and ST-GCN. Action Transformer [33] has recently been employed for human action recognition. For evaluation, we set the corresponding parameters as follows, heads: 1, layers: 4, embedding dimensions: 64, MLP: 256, and encoder layers: 5. The reason to keep the parameters at a minimum is to reduce the computational complexity of the model for the application at hand. ResNet is a specific configuration of the architecture that consists of 101 layers in the case of ResNet101 and 18 layers in ResNet18. The network includes residual blocks, which are designed to learn residual mappings that help mitigate the vanishing gradient issue. Each residual block contains multiple convolutional layers and shortcut connections that allow information to flow more effectively through the network. ResNet networks have been widely used for several computer vision tasks [34]. Further, 3D ResNet is an extension of the ResNet architecture designed to tackle video action recognition tasks by considering both spatial and temporal features in videos [28]. It adds a temporal dimension to the standard ResNet architecture, making it well suited for analyzing sequences of frames in videos. Thus, 3D ResNet takes advantage of this temporal aspect by incorporating 3D convolutional layers. These layers consider the spatial relationships within each frame as well as the temporal relationships between consecutive frames, enabling the network to capture motion patterns and changes over time. Finally, the Spatio-Temporal Graph Convolutional Network (ST-GCN) [35] is also a useful architecture used in video action detection applications, especially for addressing the spatial and temporal features present in films. In order to capture both spatial correlations within individual frames and temporal dependencies between successive frames, ST-GCN uses graph convolutional procedures. For evaluation, we replaced the proposed ConvLSTM with each of the above-mentioned models and accordingly trained and tested them on the same lines for all the three splits of the dataset. Figure 9 presents the performance comparison of the proposed approach with these models in terms of the mean accuracy across the three splits. It is clear that the proposed method outperforms all these related approaches, which further validates its effectiveness.

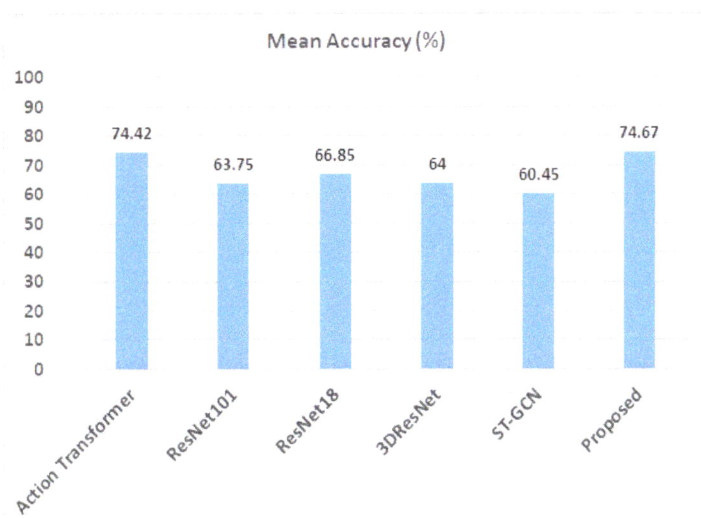

Figure 9. Performance comparison of the proposed method with existing related approaches in terms of mean accuracy across the three splits of the dataset.

4. Conclusions

In this paper, we presented a convolutional LSTM-based model for human action recognition, which was built on the extracted target pose information using YOLOv8 to effectively encode the unique body movements for various action types. The proposed framework aimed to address the challenges associated with aerial action recognition, such as varying viewpoints and background clutter. The study was inspired by the growing interest in drone applications and the need for robust and efficient action recognition methods for various applications, including security and surveillance. The comparisons with numerous existing methods show very encouraging performance through the proposed method. While the proposed framework can effectively classify the single person action in low-altitude aerial video sequences, in future work, the framework could be adapted to classify actions involving multiple objects.

Author Contributions: Conceptualization, S.M.S., H.A. and T.N.; methodology, S.M.S. and T.N. software, S.M.S. and H.A.; validation, S.M.S. and H.A.; formal analysis, S.M.S. and H.A.; investigation S.M.S. and H.A.; resources, T.N. and H.E.; writing—original draft preparation, S.M.S. and H.A. writing—review and editing, T.N., H.E. and U.S.K.; supervision, T.N., H.E. and U.S.K. All authors have read and agreed to the published version of the manuscript.

Funding: This work was partially supported by the Higher Education Commission of Pakistan and the National Centre of Robotics and Automation under Grant DF 1009-0031.

Institutional Review Board Statement: Not applicable.

Informed Consent Statement: Not applicable.

Data Availability Statement: No new data were created or analyzed in this study. Data sharing is not applicable to this article.

Conflicts of Interest: The authors declare no conflict of interest.

References

1. Kumar, R.; Tripathi, R.; Marchang, N.; Srivastava, G.; Gadekallu, T.R.; Xiong, N.N. A secured distributed detection system based on IPFS and blockchain for industrial image and video data security. *J. Parallel Distrib. Comput.* **2021**, *152*, 128–143. [CrossRef]
2. Shorfuzzaman, M.; Hossain, M.S.; Alhamid, M.F. Towards the sustainable development of smart cities through mass video surveillance: A response to the COVID-19 pandemic. *Sustain. Cities Soc.* **2021**, *64*, 102582. [CrossRef] [PubMed]
3. Kashef, M.; Visvizi, A.; Troisi, O. Smart city as a smart service system: Human-computer interaction and smart city surveillance systems. *Comput. Hum. Behav.* **2021**, *124*, 106923. [CrossRef]
4. Özyer, T.; Ak, D.S.; Alhajj, R. Human action recognition approaches with video datasets—A survey. *Knowl.-Based Syst.* **2021**, *222*, 106995. [CrossRef]
5. Sultani, W.; Shah, M. Human Action Recognition in Drone Videos Using a Few Aerial Training Examples. *arXiv* **2021**, arXiv:1910.10027. Available online: http://arxiv.org/abs/1910.10027 (accessed on 15 June 2023).
6. Wang, X.; Xian, R.; Guan, T.; de Melo, C.M.; Nogar, S.M.; Bera, A.; Manocha, D. AZTR: Aerial Video Action Recognition with Auto Zoom and Temporal Reasoning. *arXiv* **2023**, arXiv:2303.01589. Available online: http://arxiv.org/abs/2303.01589 (accessed on 15 June 2023).
7. Hejazi, S.M.; Abhayaratne, C. Handcrafted localized phase features for human action recognition. *Image Vis. Comput.* **2022**, *123*, 104465. [CrossRef]
8. El-Ghaish, H.; Hussein, M.; Shoukry, A.; Onai, R. *Human Action Recognition Based on Integrating Body Pose, Part Shape, and Motion*; IEEE Access: Piscataway, NJ, USA, 2018; pp. 49040–49055. [CrossRef]
9. Arunnehru, J.; Chamundeeswari, G.; Bharathi, S.P. Human Action Recognition using 3D Convolutional Neural Networks with 3D Motion Cuboids in Surveillance Videos. *Procedia Comput. Sci.* **2018**, *133*, 471–477. [CrossRef]
10. Sánchez-Caballero, A.; de López-Diz, S.; Fuentes-Jimenez, D.; Losada-Gutiérrez, C.; Marrón-Romera, M.; Casillas-Pérez, D.; Sarker, M.I. 3DFCNN: Real-time action recognition using 3D deep neural networks with raw depth information. *Multimed Tools Appl.* **2022**, *81*, 24119–24143. [CrossRef]
11. Sánchez-Caballero, A.; Fuentes-Jiménez, D.; Losada-Gutiérrez, C. Real-time human action recognition using raw depth video-based recurrent neural networks. *Multimed Tools Appl.* **2023**, *82*, 16213–16235. [CrossRef]
12. Muhammad, K.; Mustaqeem; Ullah, A.; Imran, A.S.; Sajjad, M.; Kiran, M.S.; Sannino, G.; de Albuquerque, V.H.C. Human action recognition using attention based LSTM network with dilated CNN features. *Future Gener. Comput. Syst.* **2021**, *125*, 820–830. [CrossRef]
13. Xiao, S.; Wang, S.; Huang, Z.; Wang, Y.; Jiang, H. Two-stream transformer network for sensor-based human activity recognition. *Neurocomputing* **2022**, *512*, 253–268. [CrossRef]
14. Zhao, Y.; Man, K.L.; Smith, J.; Siddique, K.; Guan, S.-U. Improved two-stream model for human action recognition. *EURASIP J. Image Video Process.* **2020**, *2020*, 24. [CrossRef]
15. Ahmad, T.; Jin, L.; Zhang, X.; Lai, S.; Tang, G.; Lin, L. Graph Convolutional Neural Network for Human Action Recognition: A Comprehensive Survey. *IEEE Trans. Artif. Intell.* **2021**, *2*, 128–145. [CrossRef]

16. Feng, L.; Zhao, Y.; Zhao, W.; Tang, J. A comparative review of graph convolutional networks for human skeleton-based action recognition. *Artif. Intell. Rev.* **2022**, *55*, 4275–4305. [CrossRef]
17. Yang, J.; Dong, X.; Liu, L.; Zhang, C.; Shen, J.; Yu, D. Recurring the Transformer for Video Action Recognition. In Proceedings of the 2022 IEEE/CVF Conference on Computer Vision and Pattern Recognition (CVPR), New Orleans, LA, USA, 18–24 June 2022; IEEE: New Orleans, LA, USA, 2022; pp. 14043–14053. [CrossRef]
18. Wang, X.; Zhang, S.; Qing, Z.; Shao, Y.; Zuo, Z.; Gao, C.; Sang, N. OadTR: Online Action Detection with Transformers. In Proceedings of the 2021 IEEE/CVF International Conference on Computer Vision (ICCV), Montreal, QC, Canada, 10–17 October 2021; IEEE: Montreal, QC, Canada, 2021; pp. 7545–7555. [CrossRef]
19. Barekatain, M.; Martí, M.; Shih, H.-F.; Murray, S.; Nakayama, K.; Matsuo, Y.; Prendinger, H. Okutama-Action: An Aerial View Video Dataset for Concurrent Human Action Detection. In Proceedings of the 2017 IEEE Conference on Computer Vision and Pattern Recognition Workshops (CVPRW), Honolulu, HI, USA, 21–26 July 2017; pp. 2153–2160. [CrossRef]
20. Liu, C.; Szirányi, T. Real-Time Human Detection and Gesture Recognition for On-Board UAV Rescue. *Sensors* **2021**, *21*, 2180. [CrossRef]
21. Mliki, H.; Bouhlel, F.; Hammami, M. Human activity recognition from UAV-captured video sequences. *Pattern Recognit.* **2020**, *100*, 107140. [CrossRef]
22. Perera, A.G.; Law, Y.W.; Chahl, J. Drone-Action: An Outdoor Recorded Drone Video Dataset for Action Recognition. *Drones* **2019**, *3*, 82. [CrossRef]
23. Malik, N.U.R.; Abu-Bakar, S.A.R.; Sheikh, U.U.; Channa, A.; Popescu, N. Cascading Pose Features with CNN-LSTM for Multiview Human Action Recognition. *Signals* **2023**, *4*, 40–55. [CrossRef]
24. Yang, S.-H.; Baek, D.-G.; Thapa, K. Semi-Supervised Adversarial Learning Using LSTM for Human Activity Recognition. *Sensors* **2022**, *22*, 4755. [CrossRef]
25. Kumar, A.; Rawat, Y.S. End-to-End Semi-Supervised Learning for Video Action Detection. In Proceedings of the 2022 IEEE/CVF Conference on Computer Vision and Pattern Recognition (CVPR), New Orleans, LA, USA, 18–24 June 2022; IEEE: New Orleans, LA, USA, 2022; pp. 14680–14690. [CrossRef]
26. Dai, C.; Liu, X.; Lai, J. Human action recognition using two-stream attention based LSTM networks. *Appl. Soft Comput.* **2020**, *86*, 105820. [CrossRef]
27. Mathew, S.; Subramanian, A.; Pooja, S. Human Activity Recognition Using Deep Learning Approaches: Single Frame CNN and Convolutional LSTM. *arXiv* **2023**, arXiv:2304.14499.
28. Zhang, J.; Bai, F.; Zhao, J.; Song, Z. Multi-views Action Recognition on 3D ResNet-LSTM Framework. In Proceedings of the 2021 IEEE 2nd International Conference on Big Data, Artificial Intelligence and Internet of Things Engineering (ICBAIE), Nanchang, China, 26–28 March 2021; pp. 289–293. [CrossRef]
29. Reis, D.; Kupec, J.; Hong, J.; Daoudi, A. Real-Time Flying Object Detection with YOLOv8. *arXiv* **2023**, arXiv:2305.09972. Available online: http://arxiv.org/abs/2305.09972 (accessed on 16 June 2023).
30. Arif, S.; Wang, J.; Ul Hassan, T.; Fei, Z. 3D-CNN-Based Fused Feature Maps with LSTM Applied to Action Recognition. *Future Internet* **2019**, *11*, 42. [CrossRef]
31. Mateus, B.C.; Mendes, M.; Farinha, J.T.; Cardoso, A.M. Anticipating Future Behavior of an Industrial Press Using LSTM Networks. *Appl. Sci.* **2021**, *11*, 6101. [CrossRef]
32. Khan, L.; Amjad, A.; Afaq, K.M.; Chang, H.-T. Deep Sentiment Analysis Using CNN-LSTM Architecture of English and Roman Urdu Text Shared in Social Media. *Appl. Sci.* **2022**, *12*, 2694. [CrossRef]
33. Mazzia, V.; Angarano, S.; Salvetti, F.; Angelini, F.; Chiaberge, M. Action Transformer: A Self-Attention Model for Short-Time Pose-Based Human Action Recognition. *Pattern Recognit.* **2022**, *124*, 108487. [CrossRef]
34. He, K.; Zhang, X.; Ren, S.; Sun, J. Deep Residual Learning for Image Recognition. *arXiv* **2015**, arXiv:1512.03385. Available online: http://arxiv.org/abs/1512.03385 (accessed on 19 November 2022).
35. Chen, S.; Xu, K.; Jiang, X.; Sun, T. Pyramid Spatial-Temporal Graph Transformer for Skeleton-Based Action Recognition. *Appl. Sci.* **2022**, *12*, 9229. [CrossRef]

applied sciences

MDPI

Article

Quantitative Analysis of the Stability of a Mud-Return Circulation System in a Riserless Mud-Recovery Drilling System

Rulei Qin [1,2], Qiuping Lu [3,*], Guolei He [1,2,*], Benchong Xu [1], Haowen Chen [1], Changping Li [4], Guoyue Yin [1], Jiarui Wang [1] and Linqing Wang [1]

1 Institute of Exploration Techniques, Chinese Academy of Geological Sciences, Langfang 065000, China; qrulei@mail.cgs.gov.cn (R.Q.); xbenchong@mail.cgs.gov.cn (B.X.); chaowen@mail.cgs.gov.cn (H.C.); yguole@mail.cgs.gov.cn (G.Y.); wjiarui@mail.cgs.gov.cn (J.W.); wanglinqing@mail.cgs.gov.cn (L.W.)
2 Technology Innovation Center for Directional Drilling Engineering, Ministry of Natural Resources, Langfang 065000, China
3 Guangzhou Marine Geological Survey, China Geological Survey, Guangzhou 511458, China
4 School of Mechanical Engineering and Electronic Information, China University of Geosciences, Wuhan 430074, China; lichangpingcug@126.com
* Correspondence: luqiupingmail@163.com (Q.L.); hguolei@mail.cgs.gov.cn (G.H.)

Abstract: Riserless mud-recovery (RMR) drilling technology was widely applied in recent years. Compared with traditional deepwater drilling, RMR drilling can improve drilling efficiency, reduce risks, and minimize environmental effects. This paper focuses primarily on the stability of a mud-return circulation system in an RMR system. First, various factors that affect the stability of a mud-return circulation system are analyzed. An analytical model for the skid-and-mud-return line is established. Second, relevant data are derived from theoretical calculations and experiments. ABAQUS software is used to analyze the effects of each factor on the stability of the mud-return circulation system. The influencing patterns of each factor on the stability of the mud-return circulation system are summarized. Furthermore, the stability of the system under different operating conditions is analyzed based on the coupling of multiple factors. The support vector regression with derivative significance weight analysis (SVR-DWSA) algorithm is employed to perform a weight analysis of the effect on the system's stability. Finally, based on the research findings on the stability of the mud-return circulation system, relevant conclusions and recommendations are drawn. The results of this study provide valuable references for the application of RMR technology.

Keywords: riserless mud-recovery technology; ABAQUS software; SVR-DSWA algorithm

Citation: Qin, R.; Lu, Q.; He, G.; Xu, B.; Chen, H.; Li, C.; Yin, G.; Wang, J.; Wang, L. Quantitative Analysis of the Stability of a Mud-Return Circulation System in a Riserless Mud-Recovery Drilling System. *Appl. Sci.* **2023**, *13*, 9320. https://doi.org/10.3390/app13169320

Academic Editors: Sheng Du, Xiongbo Wan, Wei Wang and Hao Fu

Received: 28 June 2023
Revised: 6 August 2023
Accepted: 11 August 2023
Published: 16 August 2023

1. Introduction

In traditional offshore drilling processes, drilling mud needs to be recovered and treated using a riser pipe. However, as oil reservoirs go deeper, the lengths of riser pipes increase, posing a series of challenges and issues [1–4]. The applications of the riser system are mostly limited to oil and gas drilling [5–8] because of its high cost and complex structure and the demand for large-tonnage drill platforms or drillships. In open-circuit drilling conditions, the direct discharge of mud into the sea induces marine environmental pollution and involves a substantial consumption of mud [9]. To address the mud-handling and recovery challenges in a downhole operation, riserless mud recovery (RMR) technology has been widely adopted in deepwater oil production.

RMR technology has undergone continuous improvement and development. It has been successfully applied in different countries and regions. British Petroleum (BP) successfully implemented RMR technology in offshore waters near Egypt, effectively mitigating the effect of formation fractures and improving drilling efficiency [10]. RMR was also deployed near Australia to address the specific challenges associated with soft sandstone in the Browse Basin [11–13]. Its initial application in the Gulf of Mexico enhanced operational efficiency and improved overall wellbore integrity [14,15]. Moreover, RMR was employed in the United Kingdom continental shelf (UKCS) field exploration project, specifically in top-hole operations on the UKCS [16].

Double-gradient technology was used in the Santos Basin in offshore waters near Brazil to drill extended-reach wells. It effectively solved the problems of formation erosion and wellbore collapse in the region [17]. RMR technology was also applied in Sabah Malaysia, successfully cementing the surface casing for the Malikai deepwater oil field [18]. In the Zumba well in Norway, RMR was utilized in a unique manner to prevent drilling fluid and cuttings from being discharged into the seabed. Thereby, it protected a large number of corals near the well [19–21]. RMR was employed in the Dampier sub-basin in offshore waters near Australia, where it successfully stabilized the system's formation, reduced the system's vibration, and prevented leakage or loss [22,23]. In Norway, RMR was utilized in three wells with shallow water flow issues. It led to complete subsea mud recovery and prevented the shallow flow [24].

RMR can also compensate for the lack of intermediate casing and the absence of a blowout preventer (BOP) when drilling in overpressured zones [25–29]. Thus, it has been successful in various offshore areas with loose formations and losses. The Integrated Ocean Drilling Program (IODP), in collaboration with companies such as AGR, conducted research on RMR technology for ultra-deepwater applications [30]. The Elvary Neftegaz Company conducted riserless drilling northeast of the Kola Peninsula shelf using the extensively developed Discflo pump and increasing efficiency by 50% [31]. The DEMO 1800 program, led by AGR Subsea, BP America, Shell, and others, pioneered deepwater mud-circulation systems [32]. Claudey et al. [33] employed controlled mud-level technology in the Barents Sea to prevent mud leakage and maintain pressure balance. Li et al. [34] established a physical model to analyze the heat transfer characteristics of drilling fluid under different RMR system configurations. They also developed a transient simulation model to temporally and spatially simulate the circulating fluid's temperature.

Currently, research on RMR technology focuses mainly on its applications in different regions. During the drilling process, the stability of a mud-return circulation system in an RMR system can be affected by various factors. For example, the actions of sea currents and vessel motion can pose a threat to operational safety. However, there is little research on the stability of mud return-circulation systems; therefore, it is necessary to conduct relevant studies in this area.

Deepwater RMR involves installing a mud lift pump in a mid-suspension configuration on the mud-lift pipeline. Therefore, to analyze the stability of this installation, this paper focuses on the stability of the skid-and-mud-return line in the mud-return circulation system. First, various factors affecting the stability of the mud-return circulation system are analyzed. An analytical model for the skid-and-mud-return line is established. Second, relevant data are derived from theoretical calculations and experiments. ABAQUS software is used to analyze the effect of each factor on the stability of the mud-return circulation system. The influencing patterns of these factors on the system's stability are summarized. Furthermore, the stability of the mud-return circulation system, in different operating conditions with multiple coupled factors, is analyzed. The weight of each factor affecting the stability of the system is analyzed based on the support vector regression with derivative significance weight analysis (SVR-DSWA) algorithm. Finally, based on the research in this paper, relevant conclusions and recommendations are drawn.

2. Analysis of Factors Affecting the Stability of the Skid-and-Mud-Return Line

2.1. Establishment of an Analysis Model for the Skid-and-Mud-Return Line

An RMR drilling system typically consists of several main components: the drilling unit, the mud-return system unit, the mud-processing unit, and the power-monitoring unit. The drilling unit includes the exploration vessels, drill pipes, the derrick, downhole tools, casings, risers, and subsea suction modules. It is used for rock breaking and core extraction. The mud-return system includes the seabed anchoring system, the seabed pump assembly, mud-return pipelines, pipeline docking connectors, and a subsea installation platform. It is used to lift the rock cuttings and mud from the seabed into the exploration vessel to circulate the drilling mud. The skid is typically made of wear-resistant and corrosion-resistant materials, such as high-hardness metal or rubber. Its sealing functions guarantee that there will be no leakage of the pumped mud. The mud-return pipeline transports drilling fluid and other return fluids from the wellhead to the surface equipment for processing and treatment. During the drilling process, the mud-return pipeline plays a role. It brings back the used drilling fluid, rock cuttings, the formation water, and other byproducts from the wellbore to the surface. These returned fluids are processed and separated in surface equipment to recover and treat the drilling fluid.

In mud-pump systems, it is crucial to maintain a seal between the skid-and-mud-return line to prevent mud leakage. Usually, it joins two components together via a threaded connection or a similar rigid connection to form a reliable joint. This connection method can guarantee high sealing performance and pressure resistance. Therefore, it is suitable for working environments involving high-pressure mud pumping. The installation diagram of deepwater RMR equipment is shown in Figure 1. The water depth in the work area is about 2000 m. The lower mud-return pipeline is about 1300 m long, and the upper mud-return pipeline is about 700 m long. This section mainly focuses on the stability of the return pipeline of the mud circulation system.

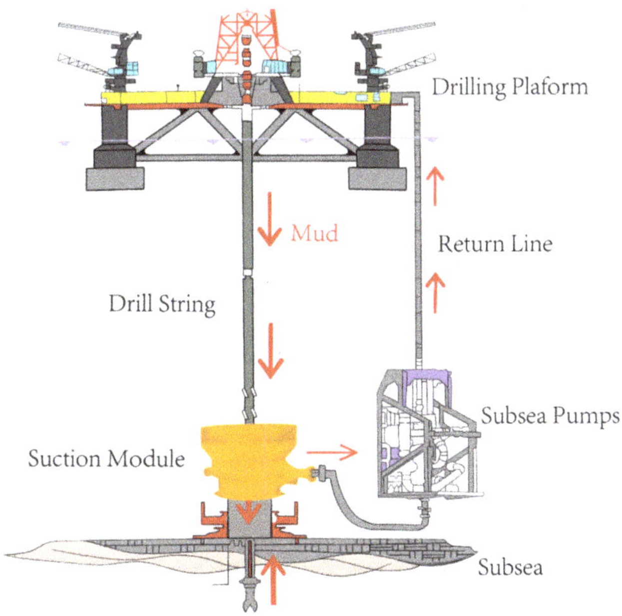

Figure 1. Installation diagram of deepwater RMR equipment.

Due to the rigid connection of the skid-and-mud-return line, the stability of the skid-and-mud-return line is analyzed as a whole. The mechanical model of the skid-and-mud-return line is shown in Figure 2a. The skid-and-mud-return line is fixed to the subsea, with the uppermost part of the mud-return line connected to the drilling vessel. The drilling vessel is subject to wind and sea currents, which can affect the skid-and-mud-return line. The vibration from the operation of the mud pumps in the skid and the reverse torque of the mud against the mud pumps are generated. These can affect the stability of the skid-and-mud-return line. In addition, because the skid-and-mud-return line is in seawater, sea currents can also affect the stability of the skid-and-mud-return line.

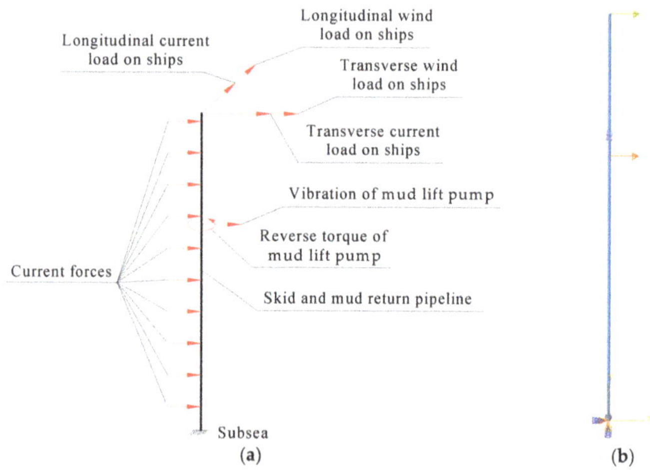

Figure 2. The model of the skid-and-mud return pipeline: (**a**) mechanical model and (**b**) analytical model.

Since the ABAQUS/Aqua module in ABAQUS can be used to calculate drag, buoyancy and fluid inertia load can be calculated from steady and wave-induced flows. ABAQUS is widely used in numerical simulation of offshore petroleum engineering. Based on the mechanical model of the skid-and-mud-return line, an analytical model of the skid-and-mud-return line is established using the ABAQUS 2018 software, as shown in Figure 2b. The bottom of the model is fully fixed. Due to the installation of the skid at 1300 m from the seabed, reverse torque and displacement are applied to the model at this position. Similarly, a load is applied to the top of the model to simulate the effects of the drilling vessel on the skid-and-mud-return line. The sea current force is applied by editing keywords in the ABAQUS/Aqua module. Finally, the model is meshed every one meter for a total of 2000 meshes.

2.2. Analysis of Factors Affecting the Stability of the Skid-and-Mud-Return Line

With the suction action of the subsea mud lift pump, the drilling fluid is drawn from the lower outlet of the suction module. The drilling fluid is returned to the subsea mud lift pump through the subsea mud return line. It is then transported to the drilling vessel through the upper mud return pipeline and recycled after treatment. Therefore, there are three main factors affecting the stability of the skid-and-mud-return line:

(1) The effects of drilling vessels on the skid-and-mud-return line due to wind and sea currents on the sea surface;
(2) The effects of sea current generated by the flow of seawater on the skid-and-mud-return line;
(3) The effects of the reverse torque and vibration generated by the operation of the mud lift pump on the skid-and-mud-return line.

For Factor 1, according to "Mooring Equipment Guidelines" released by Oil Companies International Marine Forum (OCIMF) [35], the specification provides wind load calculation formulas for very large oil tankers, as shown in the following equations, where F_{xw} is the longitudinal (bow–stern direction) wind load, F_{yw} is the transverse (port–starboard direction) wind load, C_{xw} is the longitudinal wind resistance coefficient, C_{yw} is the transverse wind resistance coefficient, ρ_w is air density, V_w is wind speed, A_T is the transverse windward area, and A_L is the transverse windward area.

$$\begin{aligned} F_{xw} &= \tfrac{1}{2} C_{xw} \rho_w V_w{}^2 A_T \\ F_{yw} &= \tfrac{1}{2} C_{yw} \rho_w V_w{}^2 A_L \end{aligned} \quad (1)$$

Moreover, the calculation formula of water flow load on ultra-large oil tankers is shown as follows, where F_{xc} is the longitudinal (bow–stern direction) water flow load, F_{yc} is the transverse (port-starboard direction) water flow load, C_{xc} is the longitudinal flow resistance coefficient, C_{yc} is the transverse flow resistance coefficient, ρ_c is seawater density, V_c is flow velocity, L_{BP} is vertical line spacing, T is the average draft depth.

$$\begin{aligned} F_{xc} &= \tfrac{1}{2} C_{xc} \rho_c V_c{}^2 L_{BP} T \\ F_{yc} &= \tfrac{1}{2} C_{yc} \rho_c V_c{}^2 L_{BP} T \end{aligned} \quad (2)$$

When these calculation formulas are used, it is important to refer to the corresponding charts in the OCIMF guidelines. The wind resistance coefficients and flow resistance coefficients for different wind angles and water flow angles can be obtained. Then, these coefficients are substituted into the calculation formulas to obtain the longitudinal wind load (F_{xw}), transverse wind load (F_{yw}), longitudinal water flow load (F_{xc}), and transverse water flow load (F_{yc}). It can be observed that the OCIMF calculation method treats all force structures as a whole.

For Factor 2, calculations can be made based on the wave-induced load theory, which can be divided into the following two methods.

(1) Flow-induced drag load

When a steady and uniform water flow passes around a circular structure, the force on the circular structure in the direction of flow is called flow-induced drag force. Flow-induced drag force is generally composed of two force components: friction drag force and pressure drag force.

Friction drag force is generated due to the viscosity of the fluid, which forms a boundary layer on the structure's surface. The sum of the frictional shear stress is the friction drag force on the circular structure. Pressure drag force is caused by the separation of the boundary layer at a certain point on the structure's surface. The sum of the normal pressure stress is the pressure drag force on the cylinder.

The drag force on a unit length of the structure can be determined using Equation (3), as shown in the following, where f_c is the drag force per unit length of the structural member, ρ is the density of the fluid, C_D is the drag force coefficient, D_C is the hydraulic outside diameter of the structural member, and v is the velocity of the fluid at the point perpendicular to the structural member.

$$f_c = \frac{1}{2} \rho D_C C_D v |v| \quad (3)$$

(2) Flow-induced inertia load

In unsteady flow around a circular structure, the fluid exerts not only drag force but also inertial force due to the fluid's acceleration. The structure disturbs and changes the velocity of fluid particles around it. This leads to a change in the pressure distribution of the original flow field. These changes are most significant near the structure surface and gradually decrease with an increase in the distance from the structure. The decay pattern depends on the shape of the structure's cross-section and the direction of fluid flow. Therefore, the disturbance caused by the structure results in an additional inertial force on the structure from the portion of the surrounding fluid. Consequently, this change in the fluid motion state results in an additional inertial force on the structure. This force is also known as the added mass force.

The true effect of the inertial force of the accelerating fluid on the structure in the direction of flow can be determined using Equation (4), as shown in the following, where f_I is the drag force per unit length of the structural members, and C_M is the inertial force coefficient

$$f_I = \rho C_M \frac{\pi D_C^2}{4} \frac{du}{dt} \quad (4)$$

The lateral load acting on the mud return pipeline is primarily caused by the seawater current's force. In the condition of pure seawater current, the Morison equation is used to calculate the water flow force on the pile column. The wave-induced force per unit length is derived from Equation (5), as shown in the following [36]:

$$f = \frac{1}{2}\rho D_C C_D v|v| + \rho C_M \frac{\pi D_C^2}{4} \frac{du}{dt} \quad (5)$$

For Factor 3, the relationship between the torque and the rotational speed, as well as the one between the amplitude and the rotational speed, can be obtained by carrying out experiments. These are shown in Tables 1 and 2. By using the correlated torque and amplitude data, an analysis can be conducted on the skid-and-mud-return line.

Table 1. Reverse torque at 6 different rotational speeds.

Rotational Speed (rpm)	Reverse Torque (N·m)
0	−320.4
300	224.75
600	669.83
900	1343.63
1200	2328.65
1500	3620.28

Table 2. Analysis data for 6 sets at voltage Un.

Number	Rotational Speed (rpm)	Torque (N·m)	Amplitude (mm)
1	0	4546.54055	2.2229
2	300	5796.57208	2.3618
3	600	7240.85487	2.5285
4	900	8735.65304	2.6466
5	1200	10,590.07644	2.7322
6	1500	0	0.00129

ABAQUS is adopted to analyze the effect of the aforementioned factors on the stability of the skid-and-mud-return line. The mud return pipeline is made of V150 steel with an outer diameter of 168.3 mm and a wall thickness of 12.7 mm. The upper pipeline section is 700 m long, and the lower pipeline section is 1300 m long. The relevant variables for each analysis factor are shown in Table 3.

Table 3. Relevant variables for each analysis factor.

Variables	Wind Recurrence Period	Wind Direction Angle	Sea Recurrence Period	Water Flow Attack Angle	Voltage	Rotational Speed
Value	1 year 5 years 10 years 25 years	0° 45° 90° 135° 180°	1 year 5 years 10 years 25 years	0° 45° 90° 135° 180°	60%Un 80%Un 90%Un Un 110%Un	0 rpm 300 rpm 600 rpm 900 rpm 1200 rpm 1500 rpm

3. Analysis of the Individual Factors Affecting the Stability of the Skid-and-Mud-Return Line

3.1. Effect Analysis of Drilling Ship Motion

The effect of drilling ship motion on the stability of the skid-and-mud-return line is analyzed in consideration of the effect of sea wind and current flow. Based on the measured wind speed, flow velocity, and parameters for calculating drilling ship forces in the work area, as shown in Tables 4 and 5, these five different wind direction angles and the water flow attack angles are taken into consideration. The corresponding wind resistance coefficients and flow resistance coefficients for different wind and water flow attack angles are obtained. They can refer to the relevant charts in the OCIMF guidelines, as shown in Table 6. Using the formulas, the longitudinal and transverse load corresponding to different recurrence periods for various wind and water flow attack angles can be calculated. For this analysis, the attack angles of 0°, 45°, 90°, 135°, and 180° are selected, as shown in Tables 7–10.

Table 4. Measured wind speed in the work area.

Recurrence Period	1 Year	5 Years	10 Years	25 Years
1-h average wind speed V_w (m/s)	13.0	31.7	35.9	40.6
Average flow velocity V_c (m/s)	0.68	1.40	1.49	1.59

Table 5. Parameters required for calculating drilling ship forces.

Name	Value
Air density ρ_w (kg/m^3)	1.28
Sea water density ρ_c (kg/m^3)	1025
Length of the drilling ship L (m)	200
Width of the drilling ship B (m)	35
Longitudinal (beam) wind-exposed area A_L (m^2)	2970
Transverse (frontal) wind-exposed area A_T (m^2)	540
Vertical line spacing L_{BP} (m)	155
Average draft depth T (m)	8.5

Table 6. Wind resistance coefficients and flow resistance coefficients for different attack angles.

Wind and Water Flow Attack Angle (Degrees)	0°	45°	90°	135°	180°
Longitudinal wind resistance coefficient C_{xw}	0.75	0.48	0.05	−0.55	−0.95
Transverse wind resistance coefficient C_{yw}	0	0.60	0.72	0.5	0
Longitudinal flow resistance coefficient C_{xc}	0.04	−0.01	0.01	0	−0.035
Transverse flow resistance coefficient C_{yc}	0	0.4	0.6	0.4	0

Table 7. Drilling ship load corresponding to various attack angles for a one-year recurrence period.

Wind and Water Flow Attack Angle (Degrees)	0°	45°	90°	135°	180°
Longitudinal wind load F_{xw} (kN)	43.80	28.04	2.92	−32.12	−55.49
Transverse wind load F_{yw} (kN)	0	192.74	231.29	160.62	0
Longitudinal water flow load F_{xc} (kN)	12.49	−3.12	3.12	0	−10.93
Transverse water flow load F_{yc} (kN)	0	124.89	187.33	124.89	0
Total longitudinal load F_x (kN)	56.29	24.91	6.04	−32.12	−66.41
Total transverse load F_y (kN)	0	317.63	418.62	285.51	0
Total load F_y (kN)	56.29	318.61	418.67	287.31	66.41

Table 8. Drilling ship load corresponding to various attack angles for a 5-year recurrence period.

Wind and Water Flow Attack Angle (Degrees)	0°	45°	90°	135°	180°
Longitudinal wind load F_{xw} (kN)	260.47	166.70	17.36	−191.01	−329.93
Transverse wind load F_{yw} (kN)	0	1146.06	1375.27	955.05	0
Longitudinal water flow load F_{xc} (kN)	52.94	−13.23	13.23	0	−46.32
Transverse water flow load F_{yc} (kN)	0	529.37	794.06	529.37	0
Total longitudinal load F_x (kN)	313.40	153.46	30.60	−191.01	−376.25
Total transverse load F_y (kN)	0	1675.43	2169.33	1484.42	0
Total load F_y (kN)	313.40	1682.44	2169.54	1496.66	376.25

Table 9. Drilling ship load corresponding to various attack angles for a 10-year recurrence period.

Wind and Water Flow Attack Angle (Degrees)	0°	45°	90°	135°	180°
Longitudinal wind load F_{xw} (kN)	334.06	213.80	22.27	−244.98	−423.14
Transverse wind load F_{yw} (kN)	0	1469.86	1763.83	1224.89	0
Longitudinal water flow load F_{xc} (kN)	59.96	−14.99	14.99	0	−52.47
Transverse water flow load F_{yc} (kN)	0	599.62	899.43	599.62	0
Total longitudinal load F_x (kN)	394.02	198.81	37.26	−244.98	−475.61
Total transverse load F_y (kN)	0	2069.48	2663.27	1824.51	0
Total load F_y (kN)	394.02	2079.01	2663.53	1840.88	475.61

Table 10. Drilling ship load corresponding to various attack angles for a 25-year recurrence period.

Wind and Water Flow Attack Angle (Degrees)	0°	45°	90°	135°	180°
Longitudinal wind load F_{xw} (kN)	427.25	273.44	28.48	−313.32	−541.19
Transverse wind load F_{yw} (kN)	0	1879.92	2255.91	1566.60	0
Longitudinal water flow load F_{xc} (kN)	68.28	−17.07	17.07	0	−59.75
Transverse water flow load F_{yc} (kN)	0	682.81	1024.21	682.81	0
Total longitudinal load F_x (kN)	495.54	256.37	45.55	313.32	−600.94
Total transverse load F_y (kN)	0	2562.73	3280.12	2249.41	0
Total load F_y (kN)	495.54	2575.52	3280.43	2271.13	600.94

By applying the drilling ship's load in ABAQUS, the effects of different wind and water flow angles on the stability of the skid-and-mud-return line for different recurrence periods are analyzed, as shown in Figures 3–6.

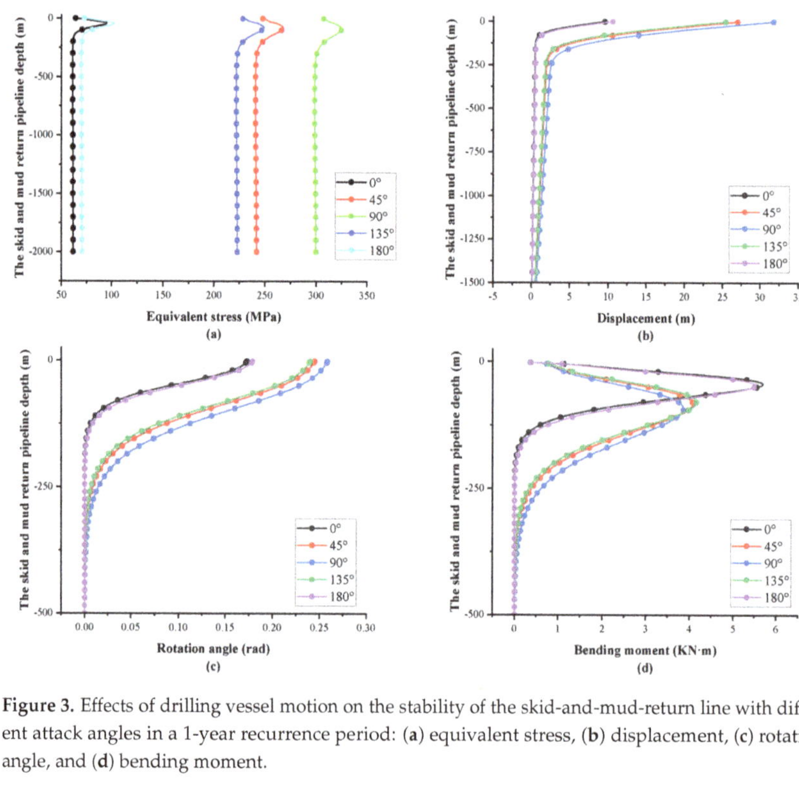

Figure 3. Effects of drilling vessel motion on the stability of the skid-and-mud-return line with different attack angles in a 1-year recurrence period: (**a**) equivalent stress, (**b**) displacement, (**c**) rotation angle, and (**d**) bending moment.

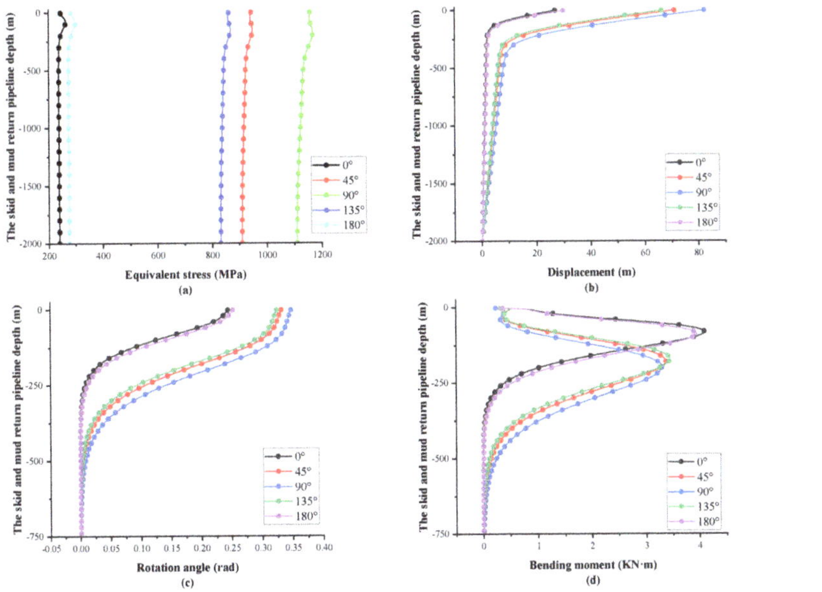

Figure 4. Effects of drilling vessel motion on the stability of the skid-and-mud-return line with different attack angles in a 5-year recurrence period: (**a**) equivalent stress, (**b**) displacement, (**c**) rotation angle, and (**d**) bending moment.

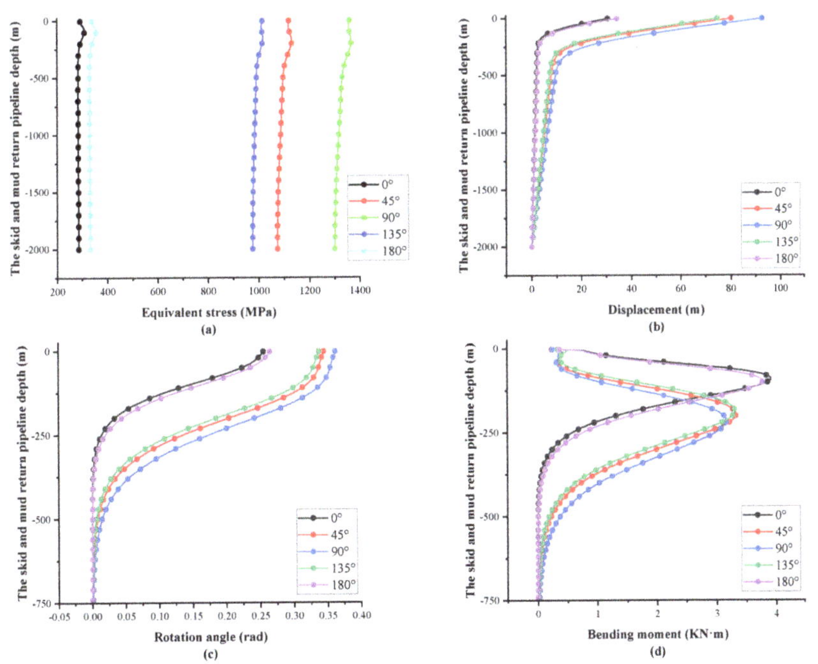

Figure 5. Effects of drilling vessel motion on the stability of the skid-and-mud-return line with different attack angles in a 10-year recurrence period: (**a**) equivalent stress, (**b**) displacement, (**c**) rotation angle, and (**d**) bending moment.

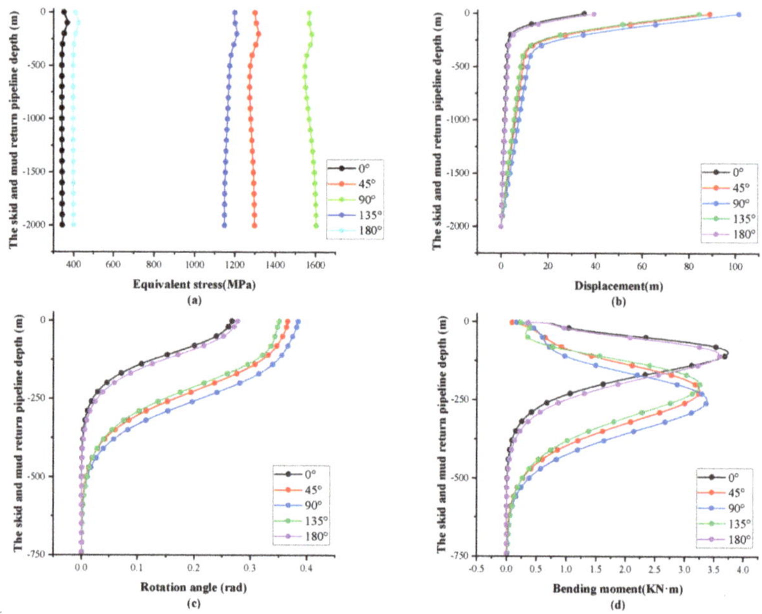

Figure 6. Effects of drilling vessel motion on the stability of the skid-and-mud-return line with different attack angles in a 25-year recurrence period: (**a**) equivalent stress, (**b**) displacement, (**c**) rotation angle, and (**d**) bending moment.

The analysis results indicate that the equivalent stress of the skid-and-mud-return line increases sequentially with the attack angles of 0°, 180°, 135°, 45°, and 90°. The equivalent stress increases less at 0° and 180° and increases more from 45° to 135°. The equivalent stress reaches its peak at a depth of 100 m, as shown in Figure 3a. The displacement of the skid-and-mud-return line also increases sequentially with the attack angles of 0°, 180°, 135°, 45°, and 90°, with a significant increase from 45° to 135°. The displacement increases with a decrease in the depth of the skid-and-mud-return line, with a lesser increase between 2000 m and 250 m or so. It increases sharply between 250 m and the sea surface. The maximum displacement occurs at the sea surface. This indicates that the main displacement of the skid-and-mud-return line occurs between 250 m and the sea surface, as shown in Figure 3b. With attack angles of 0°, 180°, 135°, 45°, and 90°, the rotation angle of the skid-and-mud-return line increases sequentially. The rotation angle remains almost unchanged between 2000 m and 500 m or so and significantly increases from around 500 m. The maximum rotation angle occurs at the sea's surface, as shown in Figure 3c. The maximum bending moment of the skid-and-mud-return line decreases sequentially with the attack angles of 0°, 180°, 135°, 45°, and 90°. The bending moment remains almost constant between about 2000 m and 500 m and shows a peak at 120 m when reaching the water surface, as shown in Figure 3d. Similarly, Figures 4–6 show the effects of different attack angles on the skid-and-mud-return line for the recurrence periods of 5, 10, and 25 years, respectively.

In a 1-year recurrence period, the maximum equivalent stress at an attack angle of 90° is only 300 MPa, which is lower than the yield strength of V150 steel (1115 MPa). Therefore, the operation can be conducted within a 1-year recurrence period. In a 5-year recurrence period, at an attack angle of 0° and 180°, the maximum equivalent stress ranges from 200 MPa to 300 MPa, guaranteeing normal operations. At an attack angle of 45° and 135°, the maximum equivalent stress is about 850 MPa and 900 MPa, respectively. These are lower than the yield strength of V150 steel, guaranteeing the operation. However, at an attack angle of 90°, the maximum equivalent stress is about 1150 MPa. This exceeds the yield strength of V150 steel. In this case, the ship's orientation needs to be adjusted to change the attack angle for operation. In a 10-year recurrence period, at an attack angle of 0° and 180°, the maximum equivalent stress is about 300 MPa, allowing for operations. At an attack angle of 135°, the maximum equivalent stress is about 1000 MPa, guaranteeing the operation. However, at an attack angle of 45° and 90°, the maximum equivalent stress exceeds the yield strength of V150 steel (1115 MPa), preventing the operation. In a 25-year recurrence period, at an attack angle of 0° and 180°, the maximum equivalent stress is close to 400 MPa, allowing for operations. However, at attack angles of 45°, 90°, and 135°, the maximum equivalent stress exceeds the yield strength of V150 steel, preventing the operation.

To analyze the effects of the recurrence period on the stability of the skid-and-mud-return line, the effects of the recurrence period of 1 year, 5 years, 10 years, and 20 years on the stability of the skid-and-mud-return line are selected with an attack angle of 90°. The analysis results are shown in Figure 7. They indicate that the equivalent stress, displacement, and rotation angle of the skid-and-mud-return line increases with an increase in the recurrence period. It is observed that the increase is the most significant from 1 year to 5 years. However, the maximum bending moment of the skid-and-mud-return line decreases with an increase in the recurrence period. The maximum equivalent stress is 300 MPa, 1100 MPa, 1300 MPa, and 1600 MPa, respectively, as shown in Figure 7a. The maximum displacement is 30 m, 80 m, 90 m, and 100 m, respectively, as shown in Figure 7b. The maximum rotation angle is 0.25 rad, 0.325 rad, 0.35 rad, and 0.375 rad, respectively, as shown in Figure 7c. The maximum bending moment is 3.1 kN·m, 3.3 kN·m, 3.5 kN·m, and 3.8 kN·m, respectively, as shown in Figure 7d.

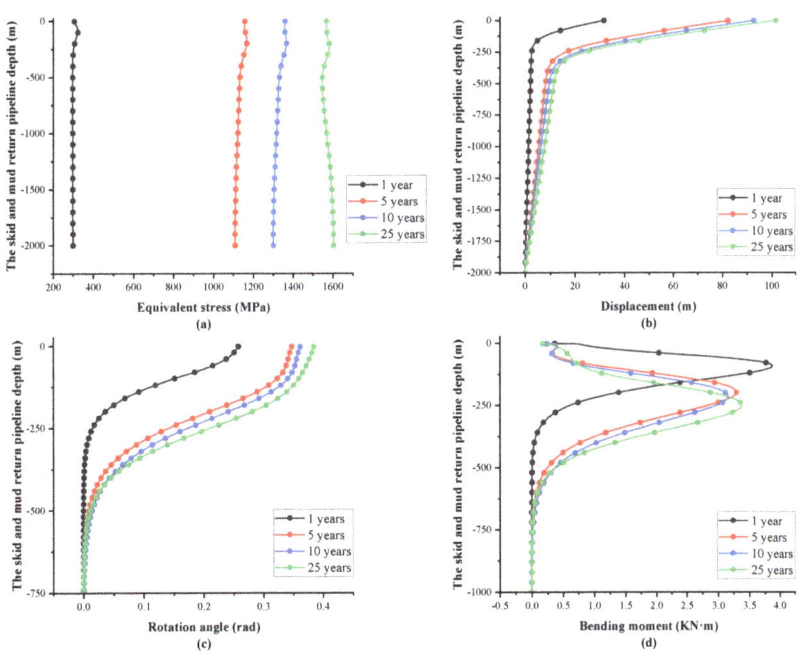

Figure 7. Effects of drilling ship motion on the stability of the skid-and-mud-return line with an attack angle of 90° in different recurrence periods: (**a**) equivalent stress, (**b**) displacement, (**c**) rotation angle, and (**d**) bending moment.

As the water depth decreases for the skid-and-mud-return line, equivalent stress, displacement, and rotation angle all increase. The equivalent stress and the rotation angle start to increase significantly from a depth of about 500 m. The displacement starts to increase significantly from a depth of about 250 m. The bending moment shows an increasing trend followed by a decreasing trend as the water depth decreases. For recurrence periods of 5 years, 10 years, and 25 years, the bending moment starts to increase significantly from a depth of about 500 m, with a peak at 250 m or so. For a recurrence period of 1 year, the bending moment starts to increase significantly from a depth of about 250 m, with the peak occurring at 125 m or so.

If the attack angle is 90°, the maximum equivalent stress is about 300 MPa only when the recurrence period is 1 year, which is lower than the yield strength of V150 steel (1115 MPa), allowing for operation. However, when the recurrence period is 5 years, 10 years, and 25 years, the maximum equivalent stress exceeds the yield strength of V150 steel, indicating that the operation cannot be performed.

3.2. Effect Analysis of Sea Current Disturbances

The effects of sea current disturbance at different depths in the work area in various recurrence periods are shown in Table 11. The ABAQUS/Aqua module, widely used for simulating the characteristics of steady-state sea currents, is extensively applied in numerical simulations for offshore petroleum engineering. It can be used to calculate the drag force, buoyancy, and fluid inertia load caused by steady-state and wave-induced flow. The wind load can be applied to the structures above the free water surface. Therefore, the ABAQUS/Aqua module is utilized to simulate the effects of sea current on the skid-and-mud-return line. The analysis results are presented in Figure 8.

Table 11. Environmental parameters of the regional well.

	Depth \ Recurrence Period	1 Year	5 Years	10 Years	25 Years
Current (m/s)	10 m	0.68	1.40	1.49	1.59
	20 m	0.66	1.39	1.48	1.57
	30 m	0.70	1.36	1.46	1.57
	50 m	0.45	1.35	1.46	1.58
	75 m	0.57	1.29	1.40	1.53
	100 m	0.48	1.21	1.31	1.42
	150 m	0.44	0.99	1.06	1.14
	200 m	0.43	0.81	0.86	0.91
	250 m	0.40	0.75	0.81	0.87
	300 m	0.35	0.73	0.77	0.82
	500 m	0.35	0.56	0.61	0.67
	1000 m	0.30	0.41	0.45	0.49

With an increase in the recurrence period of the sea, the equivalent stress and displacement of the skid-and-mud-return line significantly increase at all depths. The rotation angle and the bending moment only exhibit significant increases near the seabed. As the water depth decreases for the skid-and-mud-return line, the equivalent stress decreases. The displacement sharply increases at the seabed, slightly decreases afterward, and then is stabilized. The rotation angle sharply increases at the seabed, sharply decreases to 0 rad, and then remains stable. The bending moment sharply decreases at the seabed, increases to 0 kN·m, and then remains stable. The maximum equivalent stress is 17 MPa, 60 MPa, 67 MPa, and 75 MPa, as shown in Figure 8a. The maximum displacement is 0.13 m, 0.41 m,

0.45 m, and 0.49 m, as shown in Figure 8b. The maximum rotation angle is 0.006 rad, 0.02 rad, 0.022 rad, and 0.024 rad, as shown in Figure 8c. The maximum bending moment is 2.82 kN·m, 10.06 kN·m, 11.19 kN·m, and 12.51 kN·m, as shown in Figure 8d. It can be observed that the maximum stress, displacement, rotation angle, and bending moment of the skid-and-mud-return line appear near the seabed. This indicates that sea currents primarily affect the stability of the pipelines at the seabed. The effect of sea currents on the stability of the skid-and-mud-return line is relatively minor compared with that of the ship. It mainly affects the stability of the pipelines near the seabed in different recurrence periods.

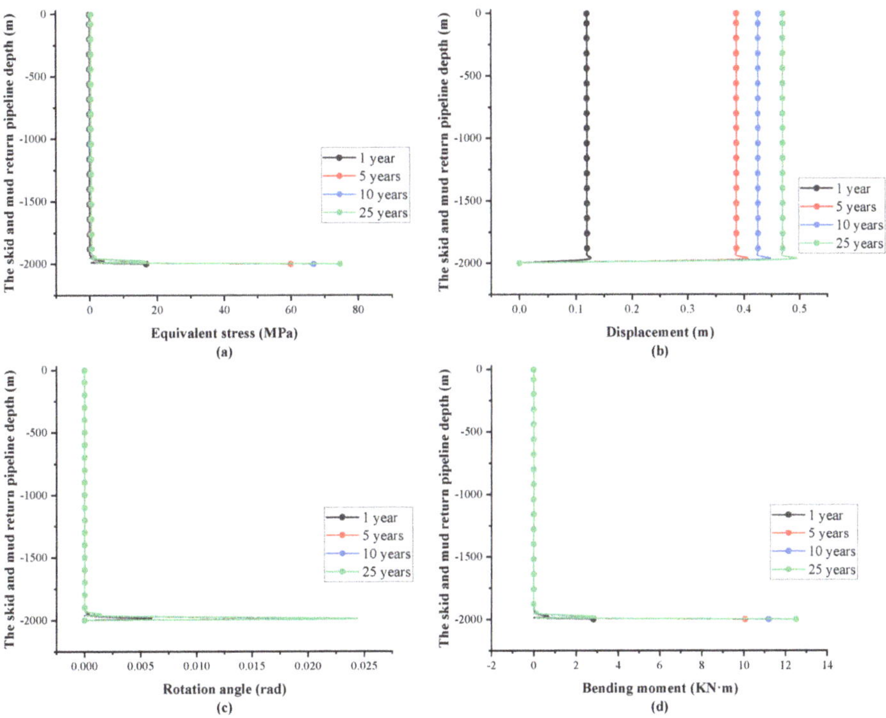

Figure 8. Effects of sea current disturbances on the stability of the skid-and-mud-return line in different recurrence periods: (**a**) equivalent stress, (**b**) displacement, (**c**) rotation angle, and (**d**) bending moment.

3.3. Effect Analysis of Reverse Torque and Vibration on Mud Pumps

At the rated voltage, six sets of data corresponding to reverse torsion and amplitudes at speeds of 0, 300, 600, 900, 1200, and 1500 rpm in Table 2 are selected as the analysis data, as shown in Table 12. They are used to analyze the effects of reverse torque and vibration on the stability of the skid-and-mud-return line. The analysis results are shown in Figure 9.

Table 12. Reverse torque and vibration corresponding to different speeds at the rated voltage.

Rotational Speed (rpm)	Reverse Torque (N·m)	Amplitude (mm)
0	−320.4	0.76665
300	228.27	0.9315
600	667.98	1.1281
900	1332.79	1.2728
1200	2327.27	1.4643
1500	3602.80	0.0977

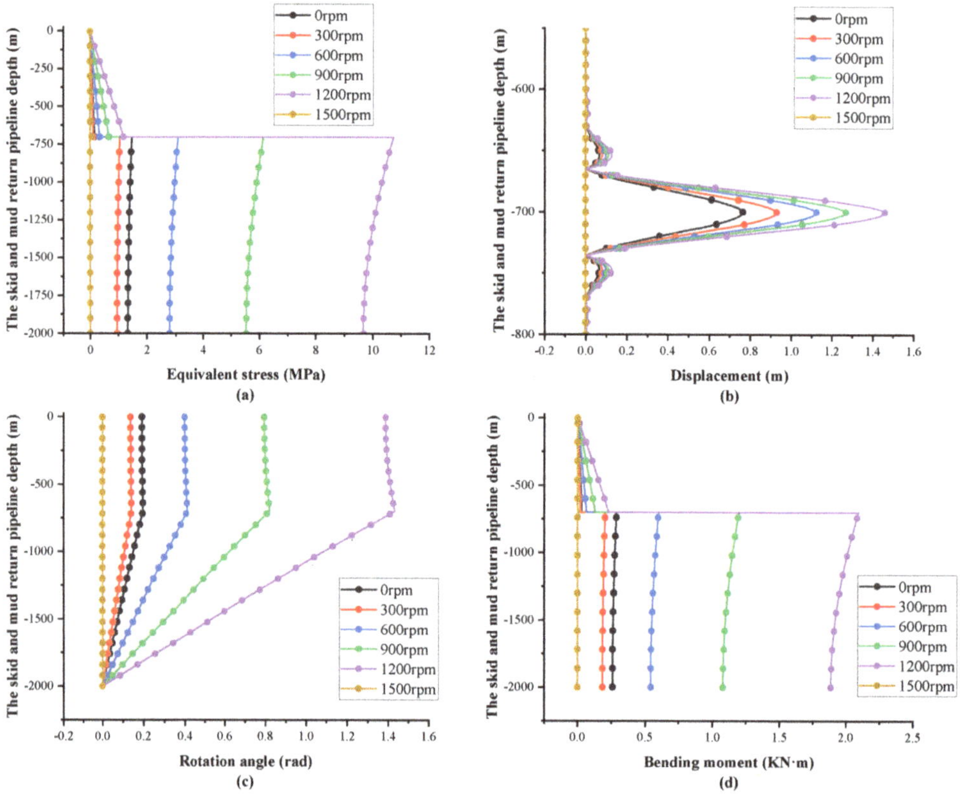

Figure 9. Effects of reverse torque and vibration on the stability of the skid-and-mud-return line (**a**) equivalent stress, (**b**) displacement, (**c**) rotation angle, and (**d**) bending moment.

The analysis results indicate that the stability of the skid-and-mud-return line is most affected by reverse torque and vibration at a speed of 1200 rpm. Due to the effect of reverse torque and vibrations on the mud pump, significant changes in various stability parameters of the skid-and-mud-return line take place at a depth of 1300 m. The stress in the skid-and-mud-return line increases relatively steadily from a depth of 0 m to 1300 m. It decreases sharply at 1300 m and then gradually decreases with an increase in the depth of the mud return pipeline. The displacement of the skid-and-mud-return line is relatively stable overall. But it suddenly increases at 1300 m and then rapidly decreases. The rotation angle of the skid-and-mud-return line increases with the depth from 0 m to 1300 m and then remains stable after 1300 m. The bending moment of the skid-and-mud-return line is relatively stable from 0 to 1300 m, decreases sharply around 1300 m, and then remains stable The maximum equivalent stress is 1.48 MPa, 1.05 MPa, 3.08 MPa, 6.15 MPa, 10.74 MPa, and 0.06 MPa, as shown in Figure 9a. The maximum displacement is 0.77 mm, 0.93 mm, 1.13 mm, 1.27 mm, 1.46 mm, and 0.0005 mm, as shown in Figure 9b. The maximum rotation angle is 0.2 rad, 0.14 rad, 0.41 rad, 0.82 rad, 1.43 rad, and 0.07 rad, as shown in Figure 9c The maximum bending moment is 0.29 kN·m, 0.21 kN·m, 0.60 kN·m, 1.20 kN·m, 2.10 kN·m, and 0.01 kN·m, as shown in Figure 9d. It can be observed that reverse torque and vibration mainly affect the rotation angle of the skid-and-mud-return line. The effects from the equivalent stress, displacement, and bending moment of the skid-and-mud-return line are relatively minor compared with other factors.

4. Quantitative Analysis of the Effect of Various Factors

4.1. Stability Analysis in Extreme Operating Conditions

To further guide practical operations, it is necessary to perform a stability analysis of the skid-and-mud-return line in extreme operating conditions. Based on the individual factor analysis mentioned above, the most critical scenario for each factor is selected to form the current set of extreme operating conditions. The selected conditions for the extreme operating conditions are as follows: a wind condition with a recurrence period of 25 years and a wind direction angle of 90°, a sea condition with a recurrence period of 25 years, a water flow angle of 90°, an electric pump voltage at 110% of the rated voltage (110 Un), and an electric pump speed of 1200 rpm. The analysis results are shown in Figure 10.

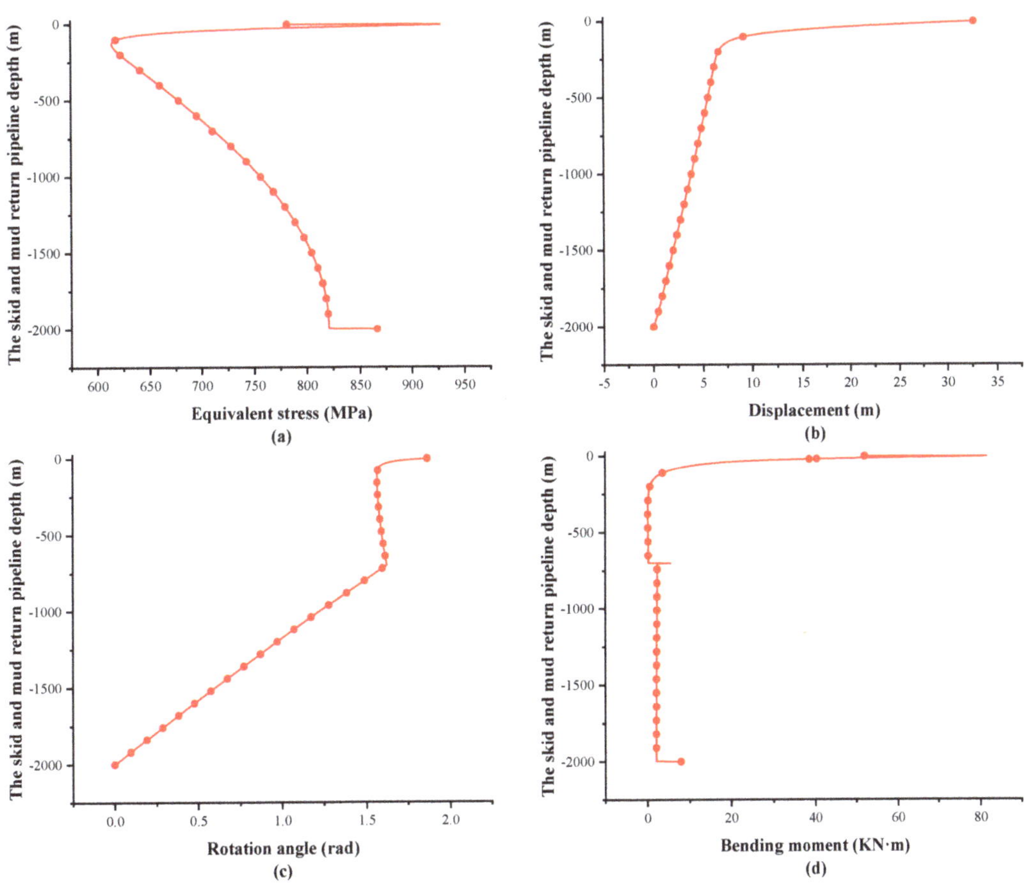

Figure 10. Stability analysis of the skid-and-mud-return line in extreme operating conditions: (**a**) equivalent stress, (**b**) displacement, (**c**) rotation angle, and (**d**) bending moment.

The analysis results indicate that in extreme operating conditions, as the depth of water at the location of the skid-and-mud return line decreases, the stress initially decreases and then increases. The minimum stress value is about 570 MPa at a depth of about 50 m, as shown in Figure 10a. The displacement increases as the depth decreases, and it starts to rapidly increase at a depth of 50 m, reaching a maximum displacement of about 32 m, as shown in Figure 10b. The angular rotation of the pipelines approximates a linear increase at a depth between 2000 m and 700 m. It is stabilized at a depth between 700 m and 50 m and then rapidly increases to a maximum rotation of about 1.8 rad, as shown in Figure 10c. The bending moment of the pipelines is relatively small at a depth between 2000 m and 50 m, but it rapidly increases to about 80 kN-m from 25 m, as shown in Figure 10d. Obviously, the stability of the skid-and-mud-return line is significantly affected in extreme operating conditions. The occurrence of such conditions needs to be avoided.

4.2. SVR-DSWA Parameterization Setting

The SVR-DSWA [37] algorithm is a method used for predicting time series data. It integrates SVR and DSWA techniques. The SVR is utilized to establish a regression model for time series data. The DSWA is applied for data processing. Specifically, the algorithm divides time series data into multiple windows, and each contains a certain number of consecutive time points. Then, within each window, SVR is used to construct a regression model for predicting the value of the next time point. Finally, the predicted results from each window are aggregated to obtain the overall prediction results for the entire time series. The SVR-DSWA algorithm is advantageous in its ability to process nonlinear time series data. It also can adaptively adjust the window size to accommodate the changing patterns within the time series data. This enables the algorithm to function well in many time series prediction tasks.

To analyze the effect weight of various factors on the stability of the skid-and-mud-return line, this paper conducts a weight analysis based on the SVR-DSWA algorithm. The principle of the derivative significance weight analysis (DSWA) algorithm is shown as follows. The derivative significance analysis method based on SVR can be used to analyze the effect weight of input values on the desired output. The decision-making function can only be constructed using training samples corresponding to non-zero coefficients $(a_i - a_i^*)$. Therefore, the decision-making function can be represented as follows [37]:

$$f(x_i) = \sum_{j=1}^{Ns} (a_j - a_j^*) K(x_i, x_j) + b \tag{6}$$

In Equation (6), *Ns* represents the number of support vectors. The sensitivity of the SVR network's output to the input of the feature can be estimated by computing the partial derivative as follows:

$$\begin{aligned} \frac{\partial f(x_i)}{\partial x_{ik}} &= \frac{\partial (\sum\limits_{j=1}^{Ns} (\alpha_j - \alpha_j^*) K(x_i, x_j) + b)}{\partial x_{ik}} = \frac{\partial \sum\limits_{j=1}^{Ns} (\alpha_j - \alpha_j^*) K(x_i, x_j)}{\partial x_{ik}} + \frac{\partial b}{\partial x_{ik}} \\ &= \sum_{j=1}^{N} (\alpha_j - \alpha_j^*) \frac{\partial K(x_i, x_j)}{\partial x_{ik}} \\ &= -2\gamma \sum_{j=1}^{Ns} (\alpha_j - \alpha_j^*) \cdot (x_{ik} - x_{jk}) \exp[-\gamma \sum_{l=1}^{K} x_{il} - x_{jl}{}^2] \end{aligned} \tag{7}$$

In Equation (7), K represents the number of feature inputs. The significance coefficient of the feature input is calculated as the absolute average sensitivity of the output values regarding the input value across all training data in the training set $\{(x_i, y_i), i = 1, 2, \cdots, N\}$. The computed value can be represented as follows:

$$S(k) = \frac{\sum\limits_{i=1}^{N} \left| \frac{\partial f(x_i)}{\partial x_{ik}} \right|}{N} \tag{8}$$

The effect weight of the feature input on the prediction result is translated as follows:

$$C(k) = \frac{S(k)}{\sum\limits_{k=1}^{k} S(k)} \tag{9}$$

Substituting Equations (7) and (8) into Equation (9) yields the weight of the feature input:

$$C(k) = \frac{\sum_{i=1}^{N} \left| -2\gamma \sum_{j=1}^{Ns} (\alpha_j - \alpha_j^*)(x_{ik} - x_{jk}) \exp\left[-\gamma \sum_{l=1}^{N} (x_{il} - x_{jl})^2 \right] \right|}{\sum_{k=1}^{K} \sum_{i=1}^{N} \left| -2\gamma \sum_{j=1}^{Ns} (\alpha_j - \alpha_j^*)(x_{ik} - x_{jk}) \exp\left[-\gamma \sum_{l=1}^{N} (x_{il} - x_{jl})^2 \right] \right|} \tag{10}$$

Based on the SVR-DSWA algorithm, the steps for calculating the weight of each parameter on the maximum displacement are shown in Figure 11.

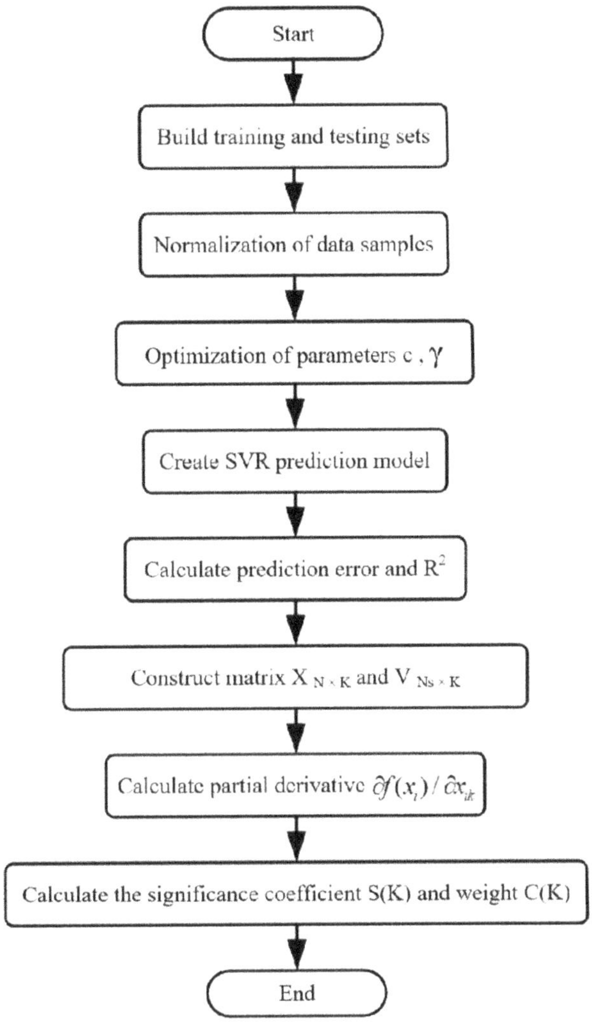

Figure 11. SVR-DSWA algorithm flowchart.

Step 1: The training and testing sets are constructed. From the given n data samples, N data samples are selected as the training set. The remaining (n-N) test samples are used as the testing set.

Step 2: The data samples are normalized. The "mapminmax" function is used to normalize the data samples at a range of $[-1, 1]$ to improve the convergence speed and accuracy of SVR.

Step 3: The parameters are optimized. The Gaussian radial basis kernel function is used for training and prediction. The grid search method combined with ten-fold cross-validation is employed to find the optimal penalty factor c and the kernel parameter γ. The training set is randomly divided into ten subsets, among which nine subsets are used as the training set and one subset as the validation set. The training sets and the validation set are alternately iterated for ten rounds. In each iteration, the "svmtrain" function is used to train the prediction model, and the RMSEe_i of the prediction model is recorded. The average value, $E_i = \frac{1}{10}\sum_{i=1}^{10} e_i$, is calculated. Next, the parameters are changed, the grid is traversed to train the model for m rounds, and the average RMSE of each round of training is obtained the minimum value is obtained, and the optimal parameters of the model are determined

Next, the parameters are modified, and the grid is traversed to train the model for m rounds to obtain the average RMSE for each round of training. The minimum E_m value is obtained, which determines the optimal parameters for the model.

Step 4: The predicted results are then found. The prediction model is trained with the optimal parameters, the model is used to predict the output values for the (n-N) test samples, and the model's output values are obtained.

Step 5: The predicted results are then analyzed. The predicted output data are de-normalized, and it is compared with the experimental data. The MRE, mean absolute error (MAE), mean square error (MSE), RMSE, and squared correlation coefficient (R^2) between the predicted values and the experimental data are calculated. the predicted results are analyzed.

Step 6: The training sample's input matrix and support vector matrix are constructed All training samples corresponding to non-zero coefficients $(a_i - a_i^*)$ are found, which are the support vectors. The training sample input matrix $X_{N\times K}$ and the support vector matrix $V_{Ns\times K}$ are constructed, where N is the number of training samples, K is the number of influencing factors, N_S is the number of support vectors, and $N_S \le N$.

Step 7: the partial derivative of the decision function is computed. Based on the optimal kernel parameter γ of the Gaussian radial basis function and the coefficients $(a_i - a_i^*)$ of the support vectors, the partial derivative $\partial f(x_i)/\partial x_{ik}$ of the decision function for the input variable is calculated.

Step 8: The significance coefficient and weight are calculated. The significance coefficient S(k) is calculated. The average absolute value of the partial derivatives of the decision function for all training samples corresponding to that factor is selected. Weight C(k) r is calculated based on S(k).

The parameter setting is shown in Table 13.

Table 13. SVR-DSWA algorithm parameter settings.

Values of penalty factor C	$[-10, 0.5, 10]$
Values of kernel parameter γ	$[-10, 0.5, 10]$
Number of cross-validation folds v	5
Training set	50
Test set	10

4.3. Validating the Weight Analysis Results Based on the DSWA Algorithm

To validate the accuracy of the DSWA (derivative significance weight analysis), the orthogonal experimental method is employed to verify the weight analysis results based on the DSWA algorithm. In the orthogonal experiment, it selects the following factors: wind recurrence period between 1 year and 25 years, wind direction angle between 0° and 180°, sea recurrence period between 1 year and 25 years, water flow attack angle between 0° and 180°, voltage between 60 and 110, and rotational speed between 0 and 1500 rpm. The levels of the orthogonal experimental factors are presented in Table 14. The input data for the weight analysis based on the SVR-DSWA algorithm are shown in Appendix A.

Table 14. Table of parameter levels.

Wind Recurrence Period	Wind Direction Angle	Sea Recurrence Period	Water Flow Attack Angle	Voltage	Rotational Speed (RPM)
1 year	0°	1 year	0°	60%Un	0
5 years	45°	5 years	45°	80%Un	300
10 years	90°	10 years	90°	90%Un	600
25 years	135°	25 years	135°	100%Un	900
	180°		180°	110%Un	1200
					1500

The displacement prediction results based on the SVR algorithm are shown in Figure 12. The weight calculation results based on the SVR-DSWA algorithm are presented in Table 15.

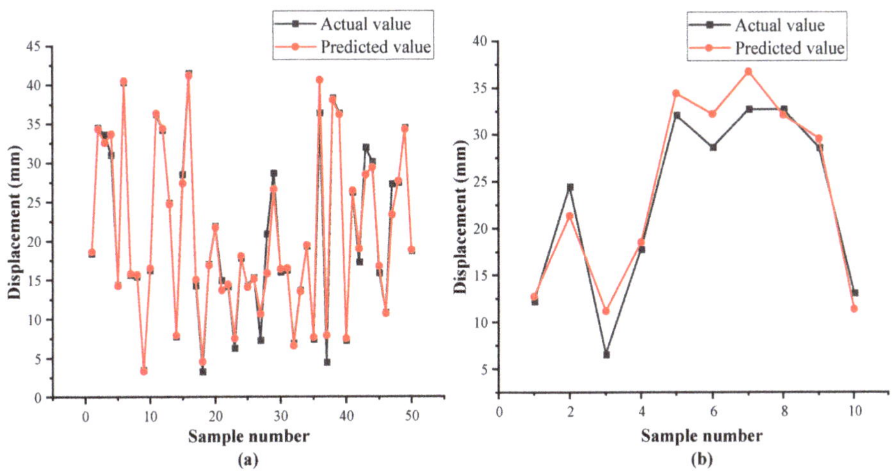

Figure 12. (**a**) Comparison of prediction results for the training set: mse = 0.0064949 and R^2 = 0.97927. (**b**) Comparison of the prediction results for the testing set: mse = 0.019341 and R^2 = 0.93767.

Table 15. Weight calculation results based on the SVR-DSWA algorithm.

	Significance Coefficient	Weight
Wind recurrence period	0.52316	23.29%
Wind direction angle	0.31974	14.23%
Sea recurrence period	0.52316	23.29%
Water flow attack angle	0.31974	14.23%
Voltage	0.32997	14.69%
Rotational speed	0.23033	10.25%

5. Conclusions and Recommendations

Based on the stability analysis of the skid-and-mud-return line conducted in this study, the following conclusions and recommendations can be drawn:

(1) Among these factors on the stability of the skid-and-mud-return line, drilling vessel motion plays a dominant role. In normal operating conditions, the stability of the skid-and-mud-return line meets the requirements for drilling and production operations. However, in extreme operating conditions, the effect on the stability of the skid-and-mud-return line is significant. Therefore, it is necessary to avoid the occurrence of extreme operating conditions.

(2) Due to the dominant effect of drilling vessel action induced by sea wind and sea current flow on the stability of the mud return pipeline, it is crucial to monitor the wind direction and sea current direction when the recurrence interval is large. Adjusting the vessel's orientation based on the wind direction and sea current flow can change the attack angle and reduce the load on the drilling vessel by sea wind and sea current flow. The safety of drilling, production, and operations can be ensured.

(3) According to the weight analysis carried out using the SVR-DSWA algorithm, the relative effect weight on the stability of the skid-and-mud-return line from the highest to the lowest are wind condition recurrence interval, sea condition recurrence interval, wind direction angle, water flow attack angle, voltage, and rotational speed. Both wind condition recurrence interval and sea condition recurrence interval are equal in weight, ranking the highest and accounting for 23.29%, respectively.

A key research direction in the future is the mechanical state monitoring and dynamic analysis of mud upstream and return circulating pipelines. Subsequent research will be compared with other research to obtain environmental and process parameter data suitable for efficient drilling in RMR.

Author Contributions: Conceptualization, G.H. and R.Q.; data curation, H.C. and B.X.; formal analysis, L.W. and J.W.; methodology, C.L. and J.W.; project administration, G.H. and R.Q.; software, G.Y.; writing—original draft, R.Q., C.L. and Q.L. All authors have read and agreed to the published version of the manuscript.

Funding: This research is funded by the National Key Research and Development Program of China (2021YFC2800802; 2021YFC2800803), the Hi-tech Ship Project of Ministry of Industry and Information Technology (CJ05N20), the Marine Economy Development Foundation of Guangdong Province (GDNRC [2022]44), and Technical Support for Stimulation and Testing of Gas Hydrate Reservoirs.

Conflicts of Interest: The authors declare no conflict of interest.

Nomenclature

F_{xw}	Longitudinal (bow–stern direction) wind load, kN
F_{yw}	Transverse (port–starboard direction) wind load, kN
C_{xw}	Longitudinal wind resistance coefficient
C_{yw}	Transverse wind resistance coefficient
ρ_w	Air density, kg/m^3
V_w	Wind speed, m/s
A_T	Transverse windward area, m^2
A_L	Longitudinal windward area, m^2
F_{xc}	Longitudinal (bow-stern direction) water flow load, kN
F_{yc}	Transverse (port-starboard direction) water flow load, kN
C_{xc}	Longitudinal flow resistance coefficient
C_{yc}	Transverse flow resistance coefficient
ρ_c	Sea water density, kg/m^3
V_c	Flow velocity, m/s
LBP	Vertical line spacing, m
T	Average draft depth, m
f_c	Drag force per unit length of the structural member, N
ρ	Density of the fluid, kg/m^3
C_D	Drag force coefficient: 0–150 m below sea level to take 1.2; 150 m below sea level to the sea floor to take 0.7
D_C	Hydraulic outside diameter of the structural member, m
v	Velocity of the fluid at the point perpendicular to the structural member, m/s
f_l	Drag force per unit length of the structural members, N
C_M	Inertia force coefficient; taken as 2.0
$\frac{du}{dt}$	Seawater motion at the calculation point generated by the water quality point acceleration perpendicular to the structural members, m/s^2

Appendix A

Table A1. Input data are analyzed based on the weight of the SVR-DSWA algorithm.

Number	Wind Recurrence Period (Year)	Wind Direction Angle (°)	Sea Recurrence Period (Year)	Water Flow Attack Angle (°)	Voltage (%Un)	Rotational Speed (RPM)	Maximum Displacement (m)
1	25	180	25	180	110	900	18.42
2	25	45	25	45	80	900	34.51
3	5	90	5	90	90	300	33.64
4	10	135	10	135	100	600	31.04
5	5	0	5	0	60	300	14.46
6	25	90	25	90	90	300	40.33
7	5	180	5	180	110	300	15.69
8	1	135	1	135	100	0	15.48
9	1	180	1	180	110	1200	3.568
10	1	45	1	45	80	0	16.31
11	10	45	10	45	80	0	36.18
12	10	135	10	135	100	0	34.22
13	25	135	25	135	100	1500	24.94
14	1	180	1	180	110	0	7.83
15	5	135	5	135	100	300	28.56
16	10	90	10	90	90	0	41.48
17	1	45	1	45	80	600	14.31
18	1	0	1	0	60	1200	3.35
19	25	0	25	0	60	900	17.07
20	5	45	5	45	80	1500	21.95
21	5	180	5	180	110	900	14.99
22	10	0	10	0	60	1200	14.21
23	1	135	1	135	100	1200	6.329
24	25	0	25	0	60	300	17.84
25	25	180	25	180	110	1500	14.27
26	10	180	10	180	110	1200	15.35
27	1	90	1	90	90	1200	7.318
28	5	135	5	135	100	1500	20.94
29	5	45	5	45	80	900	28.68
30	1	90	1	90	90	600	16.03
31	10	0	10	0	60	0	16.28
32	1	0	1	0	60	600	6.965
33	1	135	1	135	100	600	13.71
34	25	180	25	180	110	300	19.38
35	1	180	1	180	110	600	7.456
36	10	90	10	90	90	600	36.36
37	5	180	5	180	110	1500	4.522
38	25	90	25	90	90	900	38.28
39	25	45	25	45	80	300	36.36
40	1	0	1	0	60	0	7.299
41	25	45	25	45	80	1500	26.22
42	10	180	10	180	110	600	17.28
43	5	90	5	90	90	900	31.94
44	5	45	5	45	80	300	30.13
45	10	0	10	0	60	600	15.92
46	5	0	5	0	60	1500	10.88
47	5	135	5	135	100	900	27.28
48	10	135	10	135	100	1200	27.45

Table A1. *Cont.*

Number	Wind Recurrence Period (Year)	Wind Direction Angle (°)	Sea Recurrence Period (Year)	Water Flow Attack Angle (°)	Voltage (%Un)	Rotational Speed (RPM)	Maximum Displacement (m)
49	25	135	25	135	100	300	34.45
50	1	90	1	90	90	0	18.73
51	5	0	5	0	60	900	12.24
52	5	90	5	90	90	1500	24.46
53	1	45	1	45	80	1200	6.588
54	10	180	10	180	110	0	17.78
55	10	90	10	90	90	1200	32.04
56	10	45	10	45	80	1200	28.6
57	10	45	10	45	80	600	32.7
58	25	135	25	135	100	900	32.7
59	25	90	25	90	90	1500	28.55
60	25	0	25	0	60	1500	13.1

References

1. Gao, D.; Sun, T.; Zhang, H.; Tang, H. Displacement and Hydraulic Calculation of the SMD System in Ultra-deepwater Condition *Pet. Sci. Technol.* **2013**, *31*, 1196–1205. [CrossRef]
2. Lindstrom, J. *Ultra-Deep Drilling Cost Reduction; Design and Fabrication of an Ultra-Deep Drilling Simulator (UDS)*; Terratek Incorporated: Salt Lake City, UT, USA, 2010.
3. Das, B.; Samuel, R. Reliability informed drilling: Analysis for a dual-gradient drilling system. In Proceedings of the SPE Annual Technical Conference and Exhibition, Amsterdam, The Netherlands, 27–29 October 2014; OnePetro: Richardson, TX, USA, 2014
4. Carter, G.; Bland, B.; Pinckard, M. Riserless drilling-applications of an innovative drilling method and tools. In Proceedings of the Offshore Technology Conference, Houston, TX, USA, 2–5 May 2005; OnePetro: Richardson, TX, USA, 2005.
5. Aird, P. *Deepwater Drilling*; Gulf Professional Publishing: Houston, TX, USA; Elsevier: Amsterdam, The Netherlands, 2019 Volume 441, p. 475.
6. Liu, J.; Zhao, H.; Yang, S.X.; Liu, Q.; Wang, G. Nonlinear dynamic characteristic analysis of a landing string in deepwater riserless drilling. *Shock Vib.* **2018**, *2018*, 8191526. [CrossRef]
7. Mao, L.; Cai, M.; Liu, Q.; Wang, G. Dynamical Simulation of High-Pressure Gas Kick in Ultra-Deepwater Riserless Drilling. *J. Energy Resour. Technol.* **2021**, *143*, 063001. [CrossRef]
8. Rezk, R. Safe and Clean Marine Drilling with Implementation of Riserless Mud Recovery Technology-RMR. In Proceedings of the SPE Arctic and Extreme Environments Technical Conference and Exhibition, Moscow, Russia, 15–17 October 2013; OnePetro: Richardson, TX, USA, 2013.
9. Zhong, C.; Lu, J.a.; Kang, D. Design and Experimental Research of a Wellhead Overflow Monitoring System for Open-Circuit Drilling of Natural Gas Hydrate. *Energies* **2022**, *15*, 9606. [CrossRef]
10. Hinton, A.J.; Kvalvaag, E.; Jongejan, A.E.; Seim, K.; Becker, G. BP Egypt Uses RMR on a Jack-up to Solve a Top Hole Drilling Problem. In Proceedings of the SPE/IADC Drilling Conference and Exhibition, Amsterdam, The Netherlands, 17–19 March 2009, OnePetro: Richardson, TX, USA, 2009.
11. Hinton, A.; Eikemo, B.; Tilley, V.; Nolan, T. Taming the Grebe Sand–Tophole Drilling Success in the Ichthys Field. In Proceedings of the Asia Pacific Oil and Gas Conference & Exhibition, Jakarta, Indonesia, 4–6 August 2009; OnePetro: Richardson, TX, USA, 2009.
12. Jarvis, S.; Grebe, C.; Lively, R. Use of innovative technology to manage impacts in a sensitive environment. *APPEA J.* **2009**, *49*, 566 [CrossRef]
13. Power, M.R. Seismic Expression of Loss Zones within Carbonates of the Browse Basin. In Proceedings of the International Petroleum Technology Conference, Kuala Lumpur, Malaysia, 3–5 December 2008.
14. Cohen, J.H.; Kleppe, J.; Grønås, T.; Martin, T.B.; Tveit, T.; Gusler, W.; Christian, C.F.; Golden, S. Gulf of Mexico's first application of riserless mud recovery for top-hole drilling-a case study. In Proceedings of the Offshore Technology Conference, Houston, TX, USA, 3–5 May 2010; OnePetro: Richardson, TX, USA, 2010.
15. Ali, T.H.; Mathur, R.; Sharma, N. Build-to-Suit Technologies for Wellbore Construction in Deepwater and Ultradeepwater Gulf of Mexico. In Proceedings of the SPE Deepwater Drilling and Completions Conference, Galveston, TX, USA, 5–6 October 2010, OnePetro: Richardson, TX, USA, 2010.
16. Allan, R.M.; Arain, Z.; Fraser, B.; Short, S.; Davidson, S. First Oil Requirements Drive Simultaneous Drilling and Subsea Construction Operations on UK Central North Sea Development. In Proceedings of the SPE/IADC Drilling Conference, Amsterdam, The Netherlands, 5–7 March 2013; OnePetro: Richardson, TX, USA, 2013.

17. Pessanha, W.; Indio, M.; Miranda, A. Riserless Drilling Technique for Maximizing Bore Hole Stability: Dynamic Kill Drilling Application in Offshore Brazil. In Proceedings of the OTC Brasil, Rio de Janeiro, Brazil, 27–29 October 2015; OnePetro: Richardson, TX, USA, 2015.
18. Goenawan, J.; Goncalves, R.; Dooply, M.; Pasteris, M.; Heu, T.; Chan, L.; Bhaskaran, S.; Hinoul, W. Overcoming shallow hazards in deepwater malikai batch-set top-hole sections with engineered trimodal particle-size distribution cement. In Proceedings of the Offshore Technology Conference Asia, Kuala Lumpur, Malaysia, 22–25 March 2016; OnePetro: Richardson, TX, USA, 2016.
19. Jødestøl, K.; Furuholt, E. Will drill cuttings and drill mud harm cold water corals? In Proceedings of the SPE International Conference on Health, Safety and Environment in Oil and Gas Exploration and Production, Rio de Janeiro, Brazil, 12–14 April 2010; OnePetro: Richardson, TX, USA, 2010.
20. Daniel, M. Use of Riserless Mud Recovery for Protection of Cold Water Corals While Drilling in Norwegian Sea. In Proceedings of the SPE International Conference and Exhibition on Health, Safety, Security, Environment, and Social Responsibility, Stavanger, Norway, 11–13 April 2016; OnePetro: Richardson, TX, USA, 2016.
21. Rye, H.; Furuholt, E. Validation of numerical model for simulation of drilling discharges to sea. In Proceedings of the Abu Dhabi International Petroleum Exhibition and Conference, Abu Dhabi, United Arab Emirates, 1–4 November 2010; OnePetro: Richardson, TX, USA, 2010.
22. Peyton, J.; McPhee, A.; Eikemo, B.; Evans, H.; Utama, B. World first: Drilling with casing and riserless mud recovery. In Proceedings of the IPTC 2013: International Petroleum Technology Conference, Beijing, China, 26–28 March 2013; European Association of Geoscientists & Engineers Publications BV: Utrecht, The Netherlands, 2013; p. cp-350-00269.
23. Ong, G.; Yamaguchi, S.; Imai, R. Successful casing drilling experience with premium connection for production casing application on a subsea well. In Proceedings of the SPE Asia Pacific Oil and Gas Conference and Exhibition, Jakarta, Indonesia, 22–24 October 2013; OnePetro: Richardson, TX, USA, 2013.
24. Odden, D.A.; Lende, G.; Rehman, K.; Lilledal, L.; Smyth, C.; Diesen, M.; Bjørnstad, L.; Laget, M. Use of foam cement to prevent shallow water flow on three wells in Norwegian waters. In Proceedings of the International Petroleum Technology Conference, Dhahran, Saudi Arabia, 13–15 January 2020; OnePetro: Richardson, TX, USA, 2020.
25. Claudey, E.; Maubach, C.; Ferrari, S. Deepest Deployment of Riserless Dual Gradient Mud Recovery System in Drilling Operation in The North Sea. In Proceedings of the SPE Bergen One Day Seminar, Bergen, Norway, 20 April 2016.
26. Rajabi, M.M.; Rohde, B.; Maguire, N.; Stave, R.S.; Tapper, R. Successful implementations of tophole managed pressure cementing and managed pressure drilling in the caspian sea. In Proceedings of the SPE/IADC Managed Pressure Drilling and Underbalanced Operations Conference and Exhibition, Milan, Italy, 20–21 March 2012; OnePetro: Richardson, TX, USA, 2012.
27. Stave, R.; Nordas, P.; Fossli, B.; French, C. Safe and efficient tophole drilling using riserless mud recovery and managed pressure cementing. In Proceedings of the Offshore Technology Conference-Asia, Kuala Lumpur, Malaysia, 25–28 March 2014; OnePetro: Richardson, TX, USA, 2014.
28. Kozicz, J. Managed-Pressure Drilling–Recent Experience, Potential Efficiency Gains, and Future Opportunities. In Proceedings of the IADC/SPE Asia Pacific Drilling Technology Conference and Exhibition, Bangkok, Thailand, 13–15 November 2006; OnePetro: Richardson, TX, USA, 2006.
29. Godhavn, J.; Gaassand, S.; Hansen, K.H.; Morris, R.; Nott, D. Development and First Use of Controlled Mud Level System in US Deepwater GoM. In Proceedings of the SPE/IADC Managed Pressure Drilling and Underbalanced Operations Conference & Exhibition, Dubai, United Arab Emirates, 13–14 April 2015; OnePetro: Richardson, TX, USA, 2015.
30. Myers, G. Ultra-deepwater riserless mud circulation with dual gradient drilling. *Sci. Drill.* **2008**, *6*, 48–51. [CrossRef]
31. Brown, J.D.; Urvant, V.V.; Thorogood, J.L.; Rolland, N.L. Deployment of a riserless mud-recovery system offshore Sakhalin Island. In Proceedings of the SPE/IADC Drilling Conference and Exhibition, Amsterdam, The Netherlands, 20–22 February 2007; SPE: Richardson, TX, USA, 2007; p. SPE-105212-MS.
32. Smith, D.; Winters, W.; Tarr, B.; Ziegler, R.; Riza, I.; Faisal, M. Deepwater riserless mud return system for dual gradient tophole drilling. In Proceedings of the SPE/IADC Managed Pressure Drilling and Underbalanced Operations Conference and Exhibition, Kuala Lumpur, Malaysia, 24–25 February 2010; OnePetro: Richardson, TX, USA, 2010.
33. Claudey, E.; Fossli, B.; Elahifar, B.; Qiang, Z.; Olsen, M.; Mo, J. Experience using managed pressure cementing techniques with riserless mud recovery and controlled mud level in the barents sea. In Proceedings of the SPE Norway One Day Seminar, Bergen, Norway, 18 April 2018; OnePetro: Richardson, TX, USA, 2018.
34. Li, X.; Zhang, J.; Ye, W.; Liu, X.; Sun, X. A new transient simulation model and changing characteristics for circulating fluid temperature in a deepwater riserless mud recovery system. *Ocean Eng.* **2023**, *281*, 114735. [CrossRef]
35. Legieć, W. Position cross-checking on ecdis in view of international regulations requirements and OCIMF recommendations. *TransNav Int. J. Mar. Navig. Saf. Sea Transp.* **2016**, *10*, 105–113. [CrossRef]
36. Morison, J.R.; Johnson, J.W.; Schaaf, S.A. The Force Exerted by Surface Waves on Piles. *J. Pet. Technol.* **1950**, *2*, 149–154. [CrossRef]
37. Yang, Y.; Zhu, W.; Li, Y.; Zhao, B.; Zhang, L.; Song, J.; Deng, Z.; Shi, Y.; Xie, C. Modeling of PEMFC and analysis of multiple influencing factors on output characteristics. *J. Electrochem. Soc.* **2022**, *169*, 034507. [CrossRef]

applied sciences

Article

Event-Triggered Robust Fusion Estimation for Multi-Sensor Time-Delay Systems with Packet Drops

Xiaolei Du [1], Huabo Liu [1,2,*] and Haisheng Yu [1,2]

1 School of Automation, Qingdao University, Qingdao 266071, China; 2021023714@qdu.edu.cn (X.D.); yhsh_qd@qdu.edu.cn (H.Y.)
2 Shandong Key Laboratory of Industrial Control Technology, Qingdao 266071, China
* Correspondence: hbliu@qdu.edu.cn

Abstract: This paper investigates the robust fusion estimation problem for multi-sensor systems with communication constraints, parameter uncertainty, d-step state delays, and deterministic control inputs. The multi-sensor system consists of a fusion center and some sensor nodes with computational capabilities, between which there are random packet drops. The state augmentation method is utilized to transform a time-delay system into a non-time-delay one. The robust state estimation algorithm is derived based on the sensitivity penalty for each sensor node to reduce the impact of modelling errors, and modelling errors here are not limited to a unique form, which implies that the fusion estimator applies to a wide range of situations. An event-triggered transmission strategy has been adopted to effectively alleviate the communication burden from the sensor node to the fusion center. Moreover, the fusion estimator handles packet drops arising from unreliable channels, and the corresponding pseudo-cross-covariance matrix is provided. Some conditions are given to ensure that the estimation error of the robust fusion estimator is uniformly bounded. Two sets of numerical simulations are provided to illustrate the effectiveness of the derived fusion estimator.

Keywords: multi-sensor systems; robust fusion estimation; event-triggered; random packet drops; d-step state delay; deterministic control inputs

Citation: Du, X.; Liu, H.; Yu, H. Event-Triggered Robust Fusion Estimation for Multi-Sensor Time-Delay Systems with Packet Drops. *Appl. Sci.* **2023**, *13*, 8778. https://doi.org/10.3390/app13158778

Academic Editor: Ki-Yong Oh, Sheng Du, Xiongbo Wan, Wei Wang and Hao Fu

Received: 26 June 2023
Revised: 19 July 2023
Accepted: 27 July 2023
Published: 29 July 2023

1. Introduction

In the last decade, sensor systems have been extensively studied in path planning [1], environmental monitoring [2], motor control [3], and trajectory tracking [4,5], and so on. In multi-sensor systems, the accuracy and stability of the system are improved due to the joint data collection by multiple sensors. However, the impact of sensor failures or network attacks in the channel may lead to data transmission time-delay and random packet drops [6,7]. Therefore, the investigation of multi-sensor systems is of great importance.

Data processing in multi-sensor systems is performed in the form of fusion, and basic fusion methods include centralized [8,9] and distributed [10,11]. Centralized is ideally optimal, but when the number of sensors is large, fusion center data processing may be infeasible [12,13]. In contrast, the suboptimal distributed structure is more stable. As research goes further, adding an event-triggered transmission strategy to the system can reduce the energy consumption of sensors and decrease the communication burden. Ref. [14] proposed a distributed event-triggered policy in which the subsystem only broadcasts state information to neighboring nodes when the local state error exceeds a specified threshold. Ref. [15] proposed a data-driven transmission strategy based minimizing the volume of the non-transmission area. Ref. [16] proposed a trigger decision based on the estimated variance, where a copy of the Kalman filter is run at the sensor node, and its measurement is transmitted only when the measurement prediction variance exceeds a certain threshold. The event-triggered transmission strategy in [17] is based on a threshold-based strategy, where the event generator transmits a state measurement only when a signal exceeds a threshold value. A stochastic–deterministic dynamic event-triggered condition is proposed in [18].

At the same time, the treatment of time-delay problems of systems has received much attention [19]. The linear matrix inequality (LMI) [20,21] and partial differential equation (PDE) [22,23] methods are also commonly used in the time-delay treatment of systems. The state augmentation method in [24] converts time-delay systems into non-time-delay systems with excellent results. The method in [24] was used in [25] for a multi-sensor system, but random packet drop was not considered.

State estimation is a pivotal research domain within industrial automation. Consequently, numerous estimation algorithms have been formulated, encompassing the likes of the Kalman filter, Wiener filter, and other notable methodologies. In the system modeling process, modelling errors are inevitable, so the estimator performance must have no sudden changes when the system parameters reasonably deviate from their nominal parameters [26]. Those with this property are called robust state estimators, and many research methods are available [27–30]. A framework based on regularized least squares (RLS) is proposed in [27], but the modelling errors are restricted to a specific form. A filter that compromises the nominal performance and uncertainty robustness is proposed in [28] A robust state estimator based on sensitivity penalty is proposed in [29], which is not limited to structure-specific modelling errors. In addition, a robust state estimator based on the expectation minimization of estimation error is proposed in [30]. The study [31] presents an error estimator, which can be easily implemented in the code. Therefore, it is significant to employ robust state estimators in multi-sensor systems.

In this paper, we investigated the problem of robust fusion estimation for multi-sensor systems with uncertainty, restricted communication, random packet drops, state delay, and deterministic control inputs. A robust state estimator based on state augmentation and sensitivity penalty is used at the local scale. An analytic expression for the robust fusion estimator is derived based on event-triggered, and the pseudo-cross-covariance matrix of the fusion centers is updated. The consistent boundedness of the estimation error is proved Several simulations verify the effectiveness of the fusion estimator.

The rest of this paper is briefly described below. The problem description and a brief description of the event-triggered transmission strategy are given in Section 2. A robust fusion estimator for multi-sensor systems with state delays, deterministic control inputs, random packet drops, and communication constraints is derived in Section 3. The boundedness of the fusion estimator is studied in Section 4. Several sets of simulations are analyzed in Section 5. Section 6 concludes the paper.

2. Problem Formulation and Some Preliminaries

Consider the following discrete-time uncertain linear stochastic system with deterministic inputs and d-steps state delay

$$\begin{cases} x_{k+1} = A_{1,k}(\varepsilon_k)x_k + A_{2,k}(\varepsilon_k)x_{k-d} + B_{1,k}(\varepsilon_k)u_k + B_{2,k}(\varepsilon_k)w_k \\ y_k^i = C_k^i(\varepsilon_k)x_k + g_k^i, 1 \le i \le L, k \ge 0 \end{cases} \tag{1}$$

where k represents the discrete-time and i represents the sensor label. Furthermore, x_k is the state, y_k^i is the measurement, w_k represents the process noise, u_k is the deterministic control input, and g_k^i is the compound effect of measurement and communication errors. The following assumptions need to be made to guarantee the fitness of the state estimation problem.

(a) w_k and g_k^i are normally distributed with white noise, x_0, w_k, and g_k^i are mutually independent random variables.

$$\mathbf{E}(w_k) = 0, \mathbf{E}(g_k^i) = 0,$$
$$\mathbf{E}\left(\begin{bmatrix} x_0 - \mathbf{E}(x_0) \\ w_k \\ g_k^i \end{bmatrix} (*)^T\right) = \begin{bmatrix} \Pi_0 & & \\ & Q_k\delta_{kj} & \\ & & R_k^i\delta_{kj} \end{bmatrix},$$

where Π_0 , Q_k, and R_k^i are known positive definite matrices and δ_{kj} denotes the Kronecker symbolic function.

(b) The elements in the matrices $A_{1,k}(\varepsilon_k)$, $A_{2,k}(\varepsilon_k)$, $B_{1,k}(\varepsilon_k)$, $B_{2,k}(\varepsilon_k)$ and $C_k^i(\varepsilon_k)$ are known differentiable functions of the modelling errors, and the modelling errors ε_k consist of l mutually independent real-valued scalar bounded uncertainties $\varepsilon_{k,j}$, $j = 1, ..., l$.

In the process of transmitting the measurement value Y from the sensor node to the fusion center, the channel may experience packet drops. A random variable r is defined to indicate the success or failure of the communication between the sensor node and the fusion center, taking the value of 1 for the successful transmission and 0 when the communication channel fails.

The aim of this paper is to develop a fusion algorithm based on local estimates from each sensor node for multi-sensor systems with parameter uncertainty, state delay, random packet drops, and communication rate limitations. To balance communication cost and estimation performance, an event-triggered transmission strategy like in [15] is used in this paper.

Consider the following measurement channel

$$Y = H\phi + g$$

where $Y \in \mathbf{R}^m$ is the measurement output, $h \in \mathbf{R}^{m\times n}$ is the measurement matrix of the system, $\phi \in \mathbf{R}^n$ represents the state, and $g \in \mathbf{R}^m$ represents the measurement noise. A binary variable is denoted by t, and when $t = 1$ indicates that the sensor node sends a measurement Y and the other way around. The specific form of the event-triggered transmission strategy is as follows.

$$t_k^i = \begin{cases} 0, Y - \tilde{Y} \in \Xi, \\ 1, \text{ others }, \end{cases}$$

in which $\tilde{Y} \in \mathbf{R}^m$ and $\Xi \in \mathbf{R}^m$ are measurable sets. Generally, the center of mass of Ξ is at the origin, that is, $\int_\Xi \varphi d\varphi = 0$. Note that the decision transmission in the event-triggered transmission strategy is actually when the difference between the measured value and the determined measured value is greater than a threshold value.

The transmission rate, $a^i \in (0,1)$ for each sensor node is derived by $\lim_{T\to\infty} \frac{1}{T}\sum_{k=1}^{T} E\{t_k^i\} = a^i$. In addition, for any given desired transmission rate a^i, a threshold Ξ can be easily determined.

Based on Lemma 1 in [15], a virtual measure $Y = \tilde{Y} = H\varepsilon + g - v$ is now defined, where it is uniformly distributed over, and is independent of, X and g. Suppose, $f_\phi(x) = N(x; \bar{x}, \Omega_x)$, $f_G(g) = N(g; 0, \Omega_g)$, $f_Y(y) = N(y; H\bar{x}, \Omega_y)$ where $\Omega_y = \Omega_g + H\Omega_x H^T$. Thus, the optimal transmission strategy is derived as

$$\|Y - H\bar{x}\|^2_{\Omega^{-1}} \geq \theta,$$

where $\theta = \gamma_m^{-1}(1 - a)$. The random variable $\|Y - H\bar{x}\|^2_{\Omega^{-1}}$ obeys the chi-square distribution with a degree of freedom m where γ_m is the chi-square distribution function with a degree of freedom m.

Remark 1. *The considered multi-sensor system is shown in Figure 1. Each sensor node has state estimation performance with a state delay. Each sensor sends its local state estimate to the fusion center through an unreliable communication channel. All local estimates are fused with data at the fusion center through the best linear unbiased estimation criterion.*

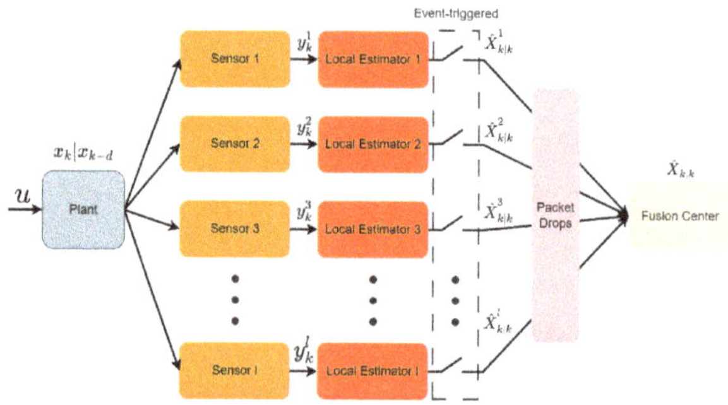

Figure 1. Block diagram of the multi-sensor system with state delay.

3. The Robust Fusion Estimation Procedure

Taking into account the impact of modelling errors on estimation performance, we adopt a robust state estimation algorithm based on sensitivity penalization [29] to obtain local estimates for multi-sensor systems. A design parameter γ_k^i, $0 < \gamma_k^i < 1$, is defined to compromise between nominal estimation performance and performance deterioration due to modelling errors. Derived from the foundation of the Kalman filter, this robust state estimation algorithm utilizes sensitivity penalization of model uncertainty estimation errors. It shares a similar form and comparable computational complexity with the standard Kalman filter. When $\gamma_k^i = 1$, this estimator degenerates to the standard Kalman filter.

By introducing the augmentation matrix X and augmenting the original system (1) with states, the system becomes

$$\begin{cases} X_{k+1} = \bar{A}_k(\varepsilon_k)X_k + \bar{B}_{1,k}(\varepsilon_k)u_k + \bar{B}_{2,k}(\varepsilon_k)w_k, \\ y_k^i = \bar{C}_k^i(\varepsilon_k)X_k + v_k^i, 1 \le i \le L, k \ge 0, \end{cases} \tag{2}$$

in which,

$$\bar{A}_k(\varepsilon_k) = \begin{bmatrix} A_{1,k}(\varepsilon_k) & 0_{n\times n} & \cdots & 0_{n\times n} & A_{2,k}(\varepsilon_k) \\ \mathrm{I}_n & & & & 0_{n\times n} \\ & \mathrm{I}_n & & & 0_{n\times n} \\ & & \ddots & & \vdots \\ & & & \mathrm{I}_n & 0_{n\times n} \end{bmatrix},$$

$$\bar{B}_{1,k}(\varepsilon_k) = \begin{bmatrix} (B_{1,k}(\varepsilon_k))^T & 0_{n\times dn}^T \end{bmatrix}^T,$$

$$\bar{B}_{2,k}(\varepsilon_k) = \begin{bmatrix} (B_{2,k}(\varepsilon_k))^T & 0_{n\times dn}^T \end{bmatrix}^T,$$

$$\bar{C}_k^i(\varepsilon_k) = \begin{bmatrix} C_k^i(\varepsilon_k) & 0_{n\times dn} \end{bmatrix}.$$

As can be seen from the above transformation, the re-modeled system is a discrete linear uncertain system without state delay. Following the transformation of the system model from (1) to (2), it is evident that the system matrix dimension changes from n to $n(d+1)$.

Remark 2. *In this paper, the system is considered only for constant state delays. Based on the state augmentation method, only the system matrix, input (control) matrix, and output matrix of the system need to be changed. The method transforms the original system into a non-time-delay system, but the system dimension will increase from the original n to n(n + d). The state augmentation method is simple and suitable when the delay step is low because the computational burden will* *increase when it is significant. However, the time delay step is generally manageable in practical production so the problem could be more influential.*

To obtain the locally robust state estimate for the $i-$th sensor node, we first define several important matrices S_k^i, $T_{1,K}^i$, and $T_{2,K}^i$, which play a key role in the parameter modification process, as follows:

$$\begin{aligned}
S_k^i &= \left[\left(S_{k,1}^i(0,0)\right)^T, \cdots, \left(S_{k,l}^i(0,0)\right)^T\right]^T, \\
T_{1,k}^i &= \left[\left(T_{1,k,1}^i(0,0)\right)^T, \cdots, \left(T_{1,k,l}^i(0,0)\right)^T\right]^T, \\
T_{2,k}^i &= \left[\left(T_{2,k,1}^i(0,0)\right)^T, \cdots, \left(T_{2,k,l}^i(0,0)\right)^T\right]^T, \\
S_{k,j}^i(\varepsilon_k, \varepsilon_{k+1}) &= \begin{bmatrix} \frac{\partial \bar{C}_{k+1}^i(\varepsilon_{k+1})}{\partial \varepsilon_{k+1,j}} \bar{A}_k(\varepsilon_k) \\ \bar{C}_{k+1}^i(\varepsilon_{k+1}) \frac{\partial \bar{A}_k(\varepsilon_k)}{\partial \varepsilon_{k,j}} \end{bmatrix}, \\
T_{1,k,j}^i(\varepsilon_k, \varepsilon_{k+1}) &= \begin{bmatrix} \frac{\partial \bar{C}_{k+1}^i(\varepsilon_{k+1})}{\partial \varepsilon_{k+1,j}} \bar{B}_{1,k}(\varepsilon_k) \\ \bar{C}_{k+1}^i(\varepsilon_{k+1}) \frac{\partial \bar{B}_{1,k}(\varepsilon_k)}{\partial \varepsilon_{k,j}} \end{bmatrix}, \\
T_{2,k,j}^i(\varepsilon_k, \varepsilon_{k+1}) &= \begin{bmatrix} \frac{\partial \bar{C}_{k+1}^i(\varepsilon_{k+1})}{\partial \varepsilon_{k+1,j}} \bar{B}_{2,k}(\varepsilon_k) \\ \bar{C}_{k+1}^i(\varepsilon_{k+1}) \frac{\partial \bar{B}_{2,k}(\varepsilon_k)}{\partial \varepsilon_{k,j}} \end{bmatrix}, \\
j &= 1, 2, \cdots, l.
\end{aligned}$$

Let $\mu_k^i = \frac{1-\gamma_k^i}{\gamma_k^i}$. The detailed realization of the robust state estimation algorithm based on sensitivity penalty is given in Algorithm 1.

Here $P_{k|k}^i$ and $\hat{P}_{k|k}^i$ are the pseudo-covariance matrices because $\hat{P}_{k|k}^i \neq \mathbf{E}\left\{(X_k - \hat{X}_{k|k}^i)^T (X_k - \hat{X}_{k|k}^i)\right\}$ and $P_{k|k}^i \neq \mathbf{E}\left\{(X_k - \hat{X}_{k|k}^i)^T (X_k - \hat{X}_{k|k}^i)\right\}$.

Based on the event-triggered transmission strategy in the second part, whether each sensor node sends a local state estimate to the fusion center is determined by t_k^i. The transmission strategy mentioned above can be expressed as

$$t_k^i = \begin{cases} 0, \left\| \hat{X}_{k|k}^i - \bar{X}_{k|k}^i \right\|_{\Omega_k^i}^2 \leq \theta^i, \\ 1, \text{ others}. \end{cases} \tag{3}$$

In order to guarantee the transmission rate a^i, the vector $\bar{X}_{k|k}^i$, the positive definite weight coefficient matrix Ω_k^i, and the positive real numbers θ^i must be chosen appropriately. $\hat{X}_{k|k}^i$ is the local state estimate.

Notice that each local state estimate can be interpreted as a measurement y_k^i of the true state X_k collected through the virtual measurement channel defined as

$$Y_k^i = \hat{X}_{k|k}^i = X_k + (\hat{X}_{k|k}^i - X_k) \tag{4}$$

where the estimation error $\hat{X}_{k|k}^i - X_k$ can be regarded as virtual measurement noise.

Algorithm 1: The local robust state estimation based on sensitivity penalty (Appendix A)

Initialization: $P^i_{0|0} = \begin{pmatrix} \left(\bar{C}^i_0(0)\right)^T \left(R^i_0\right)^{-1} \bar{C}^i_0(0) \\ +(\hat{\Pi}^i_0)^{-1} \end{pmatrix}^{-1}$, $\hat{X}^i_{0|0} = P^i_{0|0}\left(\bar{C}^i_0(0)\right)^T \left(R^i_0\right)^{-1} y^i_0$, in which $\hat{\Pi}^i_0 = \left(\Pi_0^{-1} + \mu^i_0 \sum_{j=1}^{l} \left(\frac{\partial\left(\bar{C}^i_0(\varepsilon_0)\right)^T}{\partial \varepsilon_{0,j}}\right)\left(\frac{\partial \bar{C}^i_0(\varepsilon_0)}{\partial \varepsilon_{0,j}}\right)\Bigg|_{\varepsilon_0=0}\right)^{-1}$;

Set design parameters γ^i_k;

for $k = 1 \rightarrow n$ **do**

(a) Replace $T^i_{1,k}, T^i_{2,k}, \bar{A}^i_k(0), \bar{B}^i_{1,k}(0), \bar{B}^i_{2,k}(0), P^i_{k|k}, Q_k$ by:

$$\begin{aligned}
&(\hat{P}^i_{k|k})^{-1} = (P^i_{k|k})^{-1} + \mu^i_k \left(S^i_k\right)^T S^i_k, \\
&\hat{T}^i_{2,k} = T^i_{2,k} - \mu^i_k S^i_k \hat{P}^i_{k|k} \left(S^i_k\right)^T T^i_{2,k}, \\
&\hat{B}^i_{2,k}(0) = \bar{B}_{2,k}(0) - \mu^i_k \bar{A}_k(0) \hat{P}^i_{k|k} \left(S^i_k\right)^T T^i_{2,k}, \\
&(\hat{Q}^i_k)^{-1} = (Q_k)^{-1} + \mu^i_k \left(T^i_{2,k}\right)^T \left(I + \mu^i_k S^i_k P^i_{k|k} \left(S^i_k\right)^T\right)^{-1} T^i_{2,k}, \\
&\hat{A}^i_k(0) = \left(\bar{A}_k(0) - \mu^i_k \hat{B}^i_{2,k}(0) \hat{Q}^i_k \left(T^i_{2,k}\right)^T S^i_k\right)\left(I - \mu^i_k \hat{P}^i_{k|k} \left(S^i_k\right)^T S^i_k\right), \\
&\hat{B}^i_{1,k}(0) = \bar{B}_{1,k}(0) - \mu^i_k \left(\bar{A}_k(0) \hat{P}^i_{k|k} \left(S^i_k\right)^T + \hat{B}^i_{2,k}(0) \hat{Q}^i_k \left(T^i_{2,k}\right)^T\right) T^i_{1,k};
\end{aligned}$$

(b) Update the priori pseudo-covariance and pseudo-covariance matrix:

$$\begin{aligned}
P^i_{k+1|k} &= \bar{A}_k(0) \hat{P}^i_{k|k} \bar{A}^T_k(0) + \hat{B}^i_{2,k}(0) \hat{Q}^i_k (\hat{B}^i_{2,k}(0))^T, \\
P^i_{k+1|k+1} &= P^i_{k+1|k} - P^i_{k+1|k}\left(\bar{C}^i_{k+1}(0)\right)^T \left(R^i_{k+1} + \bar{C}^i_{k+1}(0) P^i_{k+1|k} \left(\bar{C}^i_{k+1}(0)\right)^T\right)^{-1} \\
&\quad \times \bar{C}^i_{k+1}(0) P^i_{k+1|k};
\end{aligned}$$

(c) Update the state of the local estimation:

$$\begin{aligned}
\hat{X}^i_{k+1|k+1} &= \hat{A}^i_k(0) \hat{X}^i_{k|k} + \hat{B}^i_{1,k}(0) u_k + P^i_{k+1|k+1}\left(\bar{C}^i_{k+1}(0)\right)^T \left(R^i_{k+1}\right)^{-1} \\
&\quad \times \left[y^i_{k+1} - \bar{C}^i_{k+1}(0)(\hat{A}^i_k(0) \hat{X}^i_{k|k} + \hat{B}^i_{1,k}(0) u_k)\right].
\end{aligned}$$

Now, considering only the event-triggered transmission strategy, (4) corresponds to the measurements received by the fusion center from sensor node i, that is, $t^i_k = 1$. When sensor data are not transmitted, (4) will be replaced by

$$Y^i_k = \tilde{X}^i_{k|k} = X_k + (\hat{X}^i_{k|k} - X_k) - v^i_k. \tag{5}$$

Here, v^i_k is uniformly distributed within the ellipsoid mentioned in (3) and is not correlated with the estimation error $\hat{X}^i_{k|k} - X_k$.

According to the event-triggered transmission strategy, when there are packet drops in the communication channel from the estimator to the fusion center, the virtual measurement channel can be replaced with

$$Y^i_k = \begin{cases} \hat{X}^i_{k|k}, t^i_k = 1, r^i_k = 1 \\ \hat{X}^i_{k|k-1}, t^i_k = 1, r^i_k = 0 \\ \tilde{X}^i_{k|k}, t^i_k = 0 \end{cases} \tag{6}$$

where r_k^i is explicitly utilized in (6) to indicate whether packet drop occurs in sensor transmission to the fusion center and $r_k^i = \{0,1\}$. The state of the multi-sensor system is shown in Table 1. For simplicity, the event-triggered is abbreviated as ET, and the success of the transmission is simplified as PD. The $\hat{X}_{k|k-1}^i$ in (6) is the predicted values of the i-th sensor node for moment k. η_k^i is the virtual measurement noise of the i-th virtual channel for moment k, which can be derived by

$$\begin{aligned} \hat{X}_{k|k-1}^i &= \hat{A}_{k-1}^i \hat{X}_{k-1|k-1}^i + \hat{B}_{1,k-1}^i u_{k-1}, \\ \eta_k^i &= \begin{cases} \hat{X}_{k|k}^i - X_k, t_k^i = 1, r_k^i = 1, \\ \hat{X}_{k|k-1}^i - X_k, t_k^i = 1, r_k^i = 0, \\ \hat{X}_{k|k}^i - X_k - g_k^i, t_k^i = 0. \end{cases} \end{aligned} \tag{7}$$

Table 1. Multi-sensor system state.

ET / PD	$t_k^i = 0$	$t_k^i = 1$
$r_k^i = 0$	No transmission	Packet drop
$r_k^i = 1$	-	Normal

The fusion estimation with both random packet drops and event-triggered transmission strategies is investigated, and the following matrices are defined as

$$\begin{aligned} Y_k &= \mathrm{col}\left\{ t_k^i \left(r_k^i \hat{X}_{k|k}^i + \left(1 - r_k^i\right) \hat{X}_{k|k-1}^i \right) + \left(1 - t_k^i\right) \hat{X}_{k|k}^i \Big|_{i=1}^{l} \right\}, \\ \eta_k &= \mathrm{col}\left\{ \left(t_k^i r_k^i + \left(1 - t_k^i\right) \right) \left(\hat{X}_{k|k}^i - X_k \right) \Big|_{i=1}^{l} \right\} \\ &+ \mathrm{col}\left\{ t_k^i \left(1 - r_k^i\right) \left(\hat{X}_{k|k-1}^i - X_k \right) \Big|_{i=1}^{l} \right\} + \mathrm{col}\left\{ \left(1 - t_k^i\right) g_k^i \Big|_{i=1}^{l} \right\}, \\ H &= \mathrm{col}\left(I^i \Big|_{i=1}^{l} \right). \end{aligned} \tag{8}$$

The information in the fusion center is obtained from the virtual measurement channel

$$Y_k = HX_k + \eta_k.$$

In accordance with the best linear unbiased criterion (BLUE) in [32], we can obtain the fusion estimate and its error covariance matrix.

$$\begin{aligned} \hat{X}_{k|k} &= \left(H^T \tilde{P}_k^{-1} H \right)^{-1} H^T \tilde{P}_k^{-1} Y_k, \\ P_k &= \left(H^T \tilde{P}_k^{-1} H \right)^{-1}. \end{aligned} \tag{9}$$

In (9), $\tilde{P}_k$ is the covariance matrix of the virtual measurement noise, which is the global error covariance matrix of the estimation error. From η_k in (8), the expression of $\tilde{P}_k$ can be obtained as

$$\tilde{P}_k = \Gamma_k + diag\left\{ \left(1 - t_k^i\right) \tfrac{\theta^i}{n+2} \left(\Omega_k^i\right)^{-1} \Big|_{i=1}^{l} \right\}, \tag{10}$$

in which $\Gamma_k = \Gamma_{k,1} + \Gamma_{k,2} + \Gamma_{k,2}^T + \Gamma_{k,3}$. The matrices $\Gamma_{k,1}$, $\Gamma_{k,2}$, and $\Gamma_{k,3}$ in the formula are equal to

$$\begin{gathered}
\Gamma_{k,1} = \begin{bmatrix} \left(\sigma_{1,k}^1\right)^2 P_{k|k}^{1,1} & \cdots & \sigma_{1,k}^1 \sigma_{1,k}^l P_{k|k}^{1,l} \\ \vdots & \ddots & \vdots \\ \sigma_{1,k}^l \sigma_{1,k}^1 P_{k|k}^{l,1} & \cdots & \left(\sigma_{1,k}^l\right)^2 P_{k|k}^{l,l} \end{bmatrix}, \\
\Gamma_{k,2} = \begin{bmatrix} 0 & \cdots & \sigma_{1,k}^1 \sigma_{2,k}^l \bar{P}_{k|k-1}^{1,l} \\ \vdots & \ddots & \vdots \\ \sigma_{1,k}^l \sigma_{2,k}^1 \bar{P}_{k|k-1}^{l,1} & \cdots & 0 \end{bmatrix}, \\
\Gamma_{k,3} = \begin{bmatrix} \left(\sigma_{2,k}^1\right)^2 P_{k|k-1}^{1,1} & \cdots & \sigma_{2,k}^1 \sigma_{2,k}^l P_{k|k-1}^{1,l} \\ \vdots & \ddots & \vdots \\ \sigma_{2,k}^l \sigma_{2,k}^1 P_{k|k-1}^{l,1} & \cdots & \left(\sigma_{2,k}^l\right)^2 P_{k|k-1}^{l,l} \end{bmatrix}, \\
\sigma_{1,k}^i = \left(t_k^i r_k^i + \left(1 - t_k^i\right)\right), \sigma_{2,k}^i = t_k^i \left(1 - r_k^i\right).
\end{gathered} \tag{11}$$

Then, we consider the state estimation errors of the following dynamic system.

$$\begin{cases} X_{k+1} = \hat{A}_k^i X_k + \hat{B}_{1,k}^i u_k + \hat{B}_{2,k}^i w_k, \\ y_k^i = \bar{C}_k^i X_k + g_k^i, 1 \le i \le l. \end{cases} \tag{12}$$

The following relationships can be easily obtained

$$\begin{aligned}
X_{k+1} - \hat{X}_{k+1|k+1}^i &= \left[I + P_{k+1|k}^i \left(\bar{C}_{k+1}^i\right)^T \left(R_{k+1}^i\right)^{-1} \bar{C}_{k+1}^i\right]^{-1} \\
&\quad \times \left[\hat{A}_k^i \left(X_k - \hat{X}_{k|k}^i\right) + \hat{B}_{2,k}^i w_k\right] \\
&\quad - \left[\left(P_{k+1|k}^i\right)^{-1} + \left(\bar{C}_{k+1}^i\right)^T \left(R_{k+1}^i\right)^{-1} \bar{C}_{k+1}^i\right]^{-1} \\
&\quad \times \left(\bar{C}_{k+1}^i\right)^T \left(R_{k+1}^i\right)^{-1} v_{k+1}^i, \\
X_{k+1} - \hat{X}_{k+1|k}^i &= \hat{A}_k^i \left(X_k - \hat{X}_{k|k}^i\right) + \hat{B}_{2,k}^i w_k .
\end{aligned} \tag{13}$$

According to the above equation, the explicit expressions for the three pseudo mutual covariance matrices $P_{k+1|k+1}^{i,j}$, $\bar{P}_{k+1|k}^{i,j}$, and $P_{k+1|k}^{i,j}$ in (11) can be derived as follows

$$\begin{aligned}
P_{k+1|k+1}^{i,j} &= \begin{bmatrix} I - P_{k+1|k}^i \left(\bar{C}_{k+1}^i\right)^T \\ \times \left(\bar{C}_{k+1}^i P_{k+1|k}^i \left(\bar{C}_{k+1}^i\right)^T + R_{k+1}^i\right)^{-1} \bar{C}_{k+1}^i \end{bmatrix} \\
&\quad \times \left[\hat{A}_k^i P_{k|k}^{i,j} \left(\hat{A}_k^j\right)^T + \hat{B}_{2,k}^i Q_k \left(\hat{B}_{2,k}^j\right)^T\right] \\
&\quad \times \begin{bmatrix} I - P_{k+1|k}^j \left(\bar{C}_{k+1}^j\right)^T \\ \times \left(\bar{C}_{k+1}^j P_{k+1|k}^j \left(\bar{C}_{k+1}^j\right)^T + R_{k+1}^j\right)^{-1} \bar{C}_{k+1}^j \end{bmatrix}^T, (i \neq j)
\end{aligned}$$

$$\begin{aligned}
\bar{P}_{k+1|k}^{i,j} &= \left(I + P_{k+1|k}^i \left(\bar{C}_{k+1}^i\right)^T \left(R_{k+1}^i\right)^{-1} \bar{C}_{k+1}^i\right)^{-1} \\
&\quad \times \left[\hat{A}_k^i P_{k|k}^{i,j} \left(\hat{A}_k^j\right)^T + \hat{B}_{2,k}^i Q_k \left(\hat{B}_{2,k}^j\right)^T\right],
\end{aligned}$$

$$P^{i,j}_{k+1|k} = \hat{A}^i_k P^{i,j}_{k|k} (\hat{A}^j_k)^T + \hat{B}^i_{2,k} Q_k (\hat{B}^j_{2,k})^T,$$

$$\begin{aligned} P^{i,i}_{k+1|k+1} = & \left[\begin{array}{c} I - P^i_{k+1|k}\left(\bar{C}^i_{k+1}\right)^T \\ \times\left(\bar{C}^i_{k+1} P^i_{k+1|k}\left(\bar{C}^i_{k+1}\right)^T + R^i_{k+1}\right)^{-1} \bar{C}^i_{k+1} \end{array} \right] \\ & \times\left[\hat{A}^i_k P^{i,i}_{k|k} (\hat{A}^i_k)^T + \hat{B}^i_{2,k} Q_k (\hat{B}^i_{2,k})^T\right] \\ & \times \left[\begin{array}{c} I - P^i_{k+1|k}\left(\bar{C}^i_{k+1}\right)^T \\ \times\left(\bar{C}^i_{k+1} P^i_{k+1|k}\left(\bar{C}^i_{k+1}\right)^T + R^i_{k+1}\right)^{-1} \bar{C}^i_{k+1} \end{array} \right]^T \\ & + \left(\begin{array}{c} \left(\bar{C}^i_{k+1}\right)^T \left(R^i_{k+1}\right)^{-1} \bar{C}^i_{k+1} \\ +(P^i_{k+1|k})^{-1} \end{array} \right)^{-1} \left(\bar{C}^i_{k+1}\right)^T \left(R^i_{k+1}\right)^{-1} \\ & \times \bar{C}^i_{k+1} \left(\left(\begin{array}{c} \left(\bar{C}^i_{k+1}\right)^T \left(R^i_{k+1}\right)^{-1} \bar{C}^i_{k+1} \\ +(P^i_{k+1|k})^{-1} \end{array} \right)^{-1} \right)^T, \end{aligned}$$

in which $P^{i,i}_{k+1|k} = P^i_{k+1|k}$, $i,j = 1,\cdots,N$. $P^i_{k+1|k+1}$ is a pseudo-covariance matrix in robust state estimation. Thus, there is $P^{i,i}_{k+1|k+1} \neq P^i_{k+1|k+1}$.

4. Some Properties of the Fusion Estimator

This section has the goal of investigating the steady-state properties of event-triggered robust fusion estimators for multi-sensor systems with deterministic inputs, random packet drops, and state delays. Assume that the modelling errors $\varepsilon_{k,j}$ in this section are within the set $\mathcal{E}$, $\mathcal{E} = \left\{\varepsilon \middle| \left|\varepsilon_{k,j}\right| \leq 1, j = 1,2,\cdots,l\right\}$. The matrices $\begin{bmatrix} A_{1,k}(0) & 0_{n\times n(d-1)} & A_{2,k}(0) \\ & I_{nd} & 0_{nd\times n} \end{bmatrix}$, $\begin{bmatrix} B_{2,k}(0) \\ 0_{n\times dn} \end{bmatrix}$, and $\begin{bmatrix} C^i_k(0) & 0_{m\times dn} \end{bmatrix}$ are denoted as M_k, F_k, and O^i_k, respectively. In addition, the following assumptions need to be made.

(A) $A_{1,k}(0)$, $A_{2,k}(0)$, $B_{1,k}(0)$, $B_{2,k}(0)$, $C^i_k(0)$, R^i_k, Q_k, S^i_k, $T^i_{1,k}$, $T^i_{2,k}$, and γ^i_k are time-invariant.

(B) The uncertain linear system of (1) is exponentially stable in the sense of Lyapunov and the matrices $A_{1,k}(\varepsilon_k)$, $A_{2,k}(\varepsilon_k)$, $B_{1,k}(\varepsilon_k)$, $B_{2,k}(\varepsilon_k)$, $C^i_k(\varepsilon_k)$, Π_k, R^i_k, Q_k are bounded whenever $k > 0$ and $\varepsilon_k \in \mathcal{E}$.

(C) For every sensor node, (M_k, N^i_k) is detectable and the following matrix pair is detectable

$$\left(\begin{array}{c} M^T_k - \lambda^i_k (S^i_k)^T \left(I_{n(d+1)} + \lambda^i_k T^i_{2,k} Q_k \left(T^i_{2,k}\right)^T\right)^{-1} T^i_{2,k} Q_k (F_k)^T \\ \left(I_{n(d+1)} + \lambda^i_k Q^{\frac{1}{2}}_k \left(T^i_{2,k}\right)^T T^i_{2,k} Q^{\frac{1}{2}}_k\right)^{-\frac{1}{2}} Q^{\frac{1}{2}}_k (F_k)^T \end{array} \right)^T,$$

where $N^i_k = \begin{bmatrix} (R^i_k)^{-\frac{1}{2}} O^i_k \\ \sqrt{\lambda^i_k} S^i_k \end{bmatrix}$.

Theorem 1 ([15]). *Suppose that Assumptions (A), (B), and (C) hold and that each sensor transmits local estimate $\hat{X}^i_{k|k}$ according to the event-triggered transmission strategy. If the weight matrix Ω^i_k of the sensor node satisfies the condition*

$$\Omega^i_k \geq \omega^i I \quad (14)$$

for some positive real number ω^i*, the estimation error* $X_k - \hat{X}_{k|k}$ *is consistently bounded for any possible choice of* $\left\{\tilde{X}^i_{k|k}, k \in Z_+\right\}$*, which means*

$$\lim_{k\to\infty} \sup E\left\{\left\|X_k - \hat{X}_{k|k}\right\|^2\right\} < +\infty.$$

Proof of Theorem 1. Let $\bar{X}_{k|k}$ be the estimate obtained at time k through $\bar{Y}_k$ instead of Y_k. $\bar{Y}_k = \mathrm{col}\left\{\sigma^i_{1,k}\bar{X}^i_{k|k} + \sigma^i_{2,k}\hat{X}^i_{k|k-1}\Big|_{i=1}^{l}\right\}$, which gives

$$\hat{X}_{k|k} = \bar{X}_{k|k} + \left(H^T\tilde{P}_k^{-1}H\right)^{-1}H^T\tilde{P}_k^{-1}(Y_k - \bar{Y}_k),$$

so we have

$$\begin{aligned} E\left\{\|X_k - \hat{X}_{k|k}\|^2\right\} &\le 2E\left\{\|X_k - \bar{X}_{k|k}\|^2\right\} \\ &+2\left\|\left(H^T\tilde{P}_k^{-1}H\right)^{-1}H^T\tilde{P}_k^{-1}\right\|^2 \\ &\times E\left\{\|Y_k - \bar{Y}_k\|^2\right\}. \end{aligned} \tag{15}$$

Taking into account the first term on the right-hand side in (15), since $\bar{X}_{k|k}$ is based on the vector $\bar{y}_k$, the following inequality can be obtained

$$E\left\{\left\|X_k - \bar{X}_{k|k}\right\|^2\right\} \le tr\left(H^T\tilde{P}_k^{-1}H\right)^{-1}. \tag{16}$$

According to Assumptions (A), (B), and (C), then $P^{i,i}_{k|k}$ is convergent, and $\tilde{P}^{i,j}_{k|k-1}(i \ne j)$ and $P^{i,i}_{k|k-1}$ are also convergent [33]. The estimation error has a bounded covariance matrix at each k. This indicates that Γ_k is converged, and the estimation error covariance matrix is bounded.

From the inequality condition in Theorem 1 and the remainder of $\tilde{P}_k$, we can obtain

$$\begin{aligned} tr\left(\left(1-t^i_k\right)\frac{\theta^i}{n+2}\left(\Omega^i_k\right)^{-1}\right) &= \left(1-t^i_k\right)\frac{\theta^i}{n+2}tr\left(\left(\Omega^i_k\right)^{-1}\right) \\ &\le \left(1-t^i_k\right)\frac{\theta^i}{(n+2)\,\omega^i}. \end{aligned} \tag{17}$$

Hence, the uniform boundedness of $E\left\{\left\|X_k - \bar{X}_{k|k}\right\|^2\right\}$ can be obtained by (16). Now it is only necessary to prove that the second part of the right-hand side of inequality (15) is uniform boundedness. Under the inequality condition in Theorem 1, it can be obtained as

$$\left\|\hat{X}^i_{k|k} - \bar{X}^i_{k|k}\right\|^2_{\Omega^i_k} \ge \omega^i\left\|\hat{X}^i_{k|k} - \bar{X}^i_{k|k}\right\|^2. \tag{18}$$

When $t^i_k = 0$, it means that there is $\|\hat{X}^i_{k|k} - \bar{X}^i_{k|k}\|^2_{\Omega^i_k} \le \theta^i$. Furthermore, it can be obtained that $\|\hat{X}^i_{k|k} - \bar{X}^i_{k|k}\|^2 \le \theta^i / \omega^i$, then $\|Y_k - \bar{Y}_k\|^2 \le \Sigma^l_{i=1}\theta^i / \omega^i$. The proof is done. □

To minimize the volume of the non-transported region, $\bar{X}^i_{k|k}$ and Ω^i_k can be appropriately denoted as

$$\begin{aligned} \bar{X}^i_{k|k} &= \hat{X}^i_{k|k-1} = \hat{A}^i_{k-1}\hat{X}^i_{k-1|k-1}, \\ \Omega^i_k &= \left(\frac{1}{\mathrm{tr}\left(\tilde{P}^i_{k|k-1}\right)}\tilde{P}^i_{k|k-1}\right)^{-1}, \end{aligned} \tag{19}$$

$$\text{in which } \tilde{P}^{i}_{k|k-1} = \bar{A}_{k-1}\begin{bmatrix} \sigma^{i}_{1,k-1}P^{i,i}_{k-1|k-1} \\ +\sigma^{i}_{2,k-1}P^{i}_{k-1|k-2} \\ +\left(1-t^{i}_{k-1}\right)\frac{\theta^{i}}{n+2}\left(\Omega^{i}_{k-1}\right)^{-1} \end{bmatrix}\left(\bar{A}_{k-1}\right)^{T} + \hat{B}^{i}_{2,k-1}\hat{Q}^{i}_{k-1}\left(\hat{B}^{i}_{2,k-1}\right)^{T}.$$

Two methods exist for determining the local prediction of X_k as per (19). The first method utilized in this paper is a local estimation based on sensor nodes. This fusion estimation method does not necessitate broadcasting but requires each sensor node to retain past information. The second method is based on the $k-1$ moment fusion estimation $\hat{X}_{k-1|k-1}$.

5. Numerical Simulations

This section cites the tractor–car system detailed in [34], shown in Figure 2, and extends it to a multi-sensor system for sample simulations. The performance of the derived robust fusion estimator is demonstrated through comparison with the fusion estimator for the Kalman filter based on actual and nominal parameters using the same fusion method across two distinct sets of numerical simulations with modelling errors (fixed or not) and varying transmission rates and packet drop rates. This numerical simulation consists of two sensors. For each set, 500 time experiments were conducted, with 200 moments designated for each set, generating 200 input–output data pairs. In the simulations, the overall average estimated error variance $E\left\|X_k - \hat{X}_{k|k}\right\|^2 \approx \frac{1}{500}\sum_{f=1}^{500}\left\|X_k - \hat{X}^{(f)}_{k|k}\right\|^2$ is computed for each moment, and the implementation of event-triggered and occurrence of packet drops are displayed.

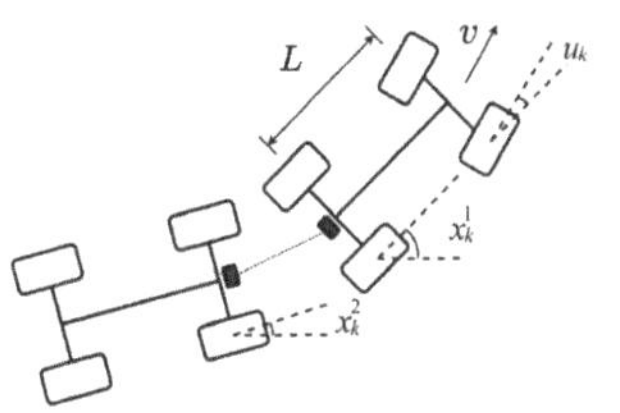

Figure 2. The tractor–car system.

Since the vehicle steering and directional angles in the tractor-car system are nonlinear, they can be linearized and expressed as

$$\begin{cases} x^{1}_{k+1} = \left(1.0000 - \frac{vk}{L}\right)x^{1}_{k} + \left(\frac{vk}{L} - 0.2296\right)x^{1}_{k-d} + \left(0.1764 + \frac{vk}{L}\right)x^{2}_{k} \\ \qquad + \left(0.1764 + \frac{vk}{L}\right)x^{2}_{k-d} + \left(0.9804 + \frac{vk}{L}\right)w^{1}_{k} + \left(0.9804 + \frac{vk}{L}\right)u^{1}_{k}, \\ x^{2}_{k+1} = \left(1.0000 - \frac{vk}{L}\right)x^{2}_{k} + \left(\frac{vk}{L} - 0.2296\right)x^{2}_{k-d} + \left(0.9804 + \frac{vk}{L}\right)w^{2}_{k} \\ \qquad + \left(0.9804 + \frac{vk}{L}\right)u^{2}_{k}, \end{cases} \tag{20}$$

in which x^1_k, x^2_k, u_k, w_k, x^1_{k-d}, and x^2_{k-d} are the direction angle of the tractor, the direction angle of the car, the tractor steering angle, the process noise, d-step time-delay for state 1, and d-step time-delay for state 2, respectively. x_k is the state vector, $x_k = \begin{bmatrix} x^1_k & x^2_k \end{bmatrix}^T$. L, k, and v denote the length of the tractor, the sampling period, and the constant speed, respectively. Considering the system errors at linearization in the form of modelling errors ε_k substituted into the system model, the matrix parameters are obtained as

$$\begin{aligned} A_{1,k}(\varepsilon_k) &= \begin{bmatrix} 1.0000 - \frac{vk}{L} & 0.1764 + \frac{vk}{L} + \varepsilon_k \\ 0.0000 & 1.0000 - \frac{vk}{L} \end{bmatrix}, \\ A_{2,k}(\varepsilon_k) &= \begin{bmatrix} \frac{vk}{L} - 0.2296 & 0.1764 + \frac{vk}{L} + \varepsilon_k \\ 0.0000 & \frac{vk}{L} - 0.2296 \end{bmatrix}, \\ B_{1,k}(\varepsilon_k) &= \begin{bmatrix} 0.9804 + \frac{vk}{L} & 0.0000 \\ 0.0000 & 0.9804 + \frac{vk}{L} \end{bmatrix}, \\ B_{2,k}(\varepsilon_k) &= \begin{bmatrix} 0.9804 + \frac{vk}{L} & 0.0000 \\ 0.0000 & 0.9804 + \frac{vk}{L} \end{bmatrix}. \end{aligned} \tag{21}$$

In the numerical simulation, each parameter is taken as $L = 500$ cm, $k = 0.1$ s, and $v = 98$ cm/s, and a two-step state delay system was used. The matrix parameters are as follows

$$A_{1,k}(\varepsilon_k) = \begin{bmatrix} 0.9804 & 0.196 + 1.99\varepsilon_k \\ 0.0000 & 0.9804 \end{bmatrix}, A_{2,k}(\varepsilon_k) = \begin{bmatrix} -0.2100 & 0.196 + 1.99\varepsilon_k \\ 0.0000 & -0.2100 \end{bmatrix},$$

$$B_{1,k}(\varepsilon_k) = \begin{bmatrix} 1.0000 & 0.0000 \\ 0.0000 & 1.0000 \end{bmatrix}, B_{2,k}(\varepsilon_k) = \begin{bmatrix} 1.0000 & 0.0000 \\ 0.0000 & 1.0000 \end{bmatrix},$$

$$C_k^1(\varepsilon_k) = \begin{bmatrix} 1.0000 & -1.0000 \end{bmatrix}, \; C_k^2(\varepsilon_k) = \begin{bmatrix} 0.4000 & -0.5000 \end{bmatrix},$$

$$R_k^1 = 1.0000, R_k^2 = 1.0000,$$

$$Q_k = \begin{bmatrix} 1.9608 & 0.0195 \\ 0.0195 & 1.9605 \end{bmatrix}, \Pi_0 = \begin{bmatrix} 1.0000 & 0.0000 \\ 0.0000 & 1.0000 \end{bmatrix}, u_k = \begin{bmatrix} 1.0000 \\ 0.1000 \end{bmatrix}.$$

The packet drop process r_k^i is assumed to be a stationary Bernoulli process. A constant value of 0.7300 is assigned to the filter design parameter γ_k^i.

In Case 1, the modeling errors ε_k are assumed to be a fixed value of -0.8508. The transmission and packet drop rates for both sensors are set to 0.8 and 0.2, respectively. Figure 3a illustrates the fusion estimation error over time, demonstrating that the robust fusion estimator proposed in this study outperforms the fusion estimator for the Kalman filter based on nominal parameters by approximately 7.800 dB. Figure 3b,c depict the transmission of the two sensors and the packet drops of the communication channel, respectively. To clearly reflect the execution of the event-triggered, t_k^i is inverted, and r_k^i is treated similarly. Note that the plots of event-triggered realizations and packet drops here are from one of the 500 experiments used.

The modelling errors ε_k are generated randomly and independently, conforming to a normal distribution with a truncation. The mean, standard variance, and truncation values of the normal distribution are set to 0.0000, 1.0000, and 1.0000, respectively. Figure 4a illustrates that the derived estimator surpasses the performance of the Kalman filter based on nominal parameters, and it can be seen from the 200th moment that the estimator derived in this paper is 5.8600 (dB) lower than the nominal parameter-based Kalman filter. Figure 4b,c show the realization of the sensor transmission and the channel packet drop over 200 moments, respectively.

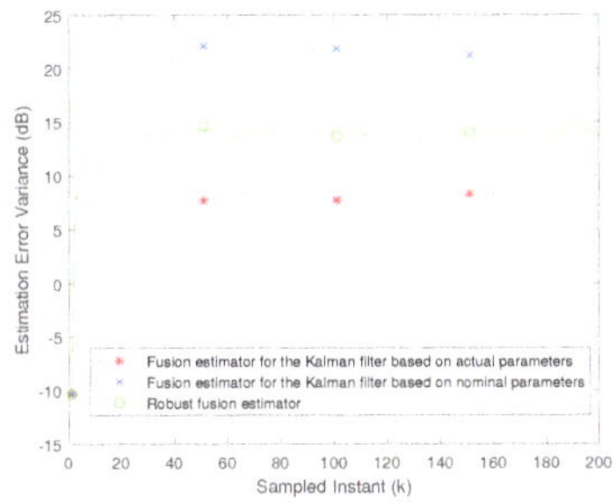

(**a**) Estimation error variance.

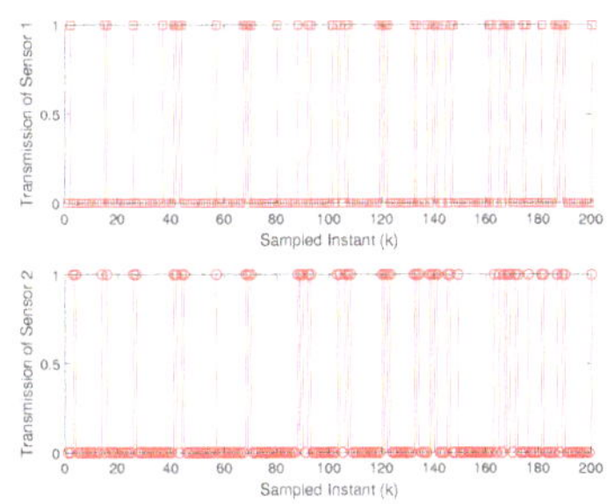

(**b**) Implementation of event-triggered.

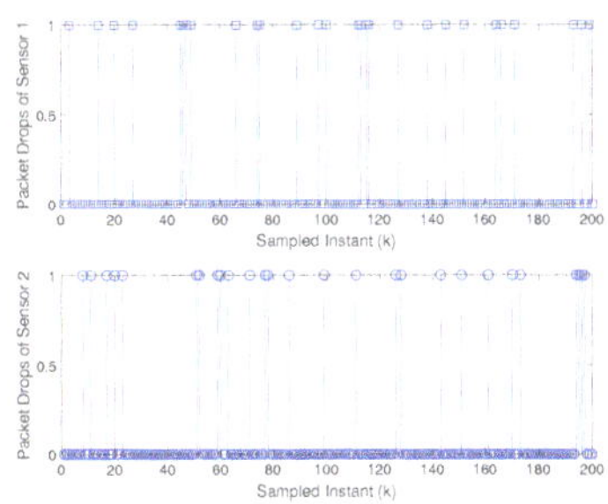

(**c**) Occurrence of packet drop.

Figure 3. ···∗···: the fusion estimator for the Kalman filter based on actual parameters; ···×···: the fusion estimator for the Kalman filter based on nominal parameters; —○—: the method of this paper; □: sensor 1; ○: sensor 2. Data transmission rate: 0.8. Packet drop rate: 0.2. Modelling errors $\varepsilon_k = -0.8508$.

In Case 2, the derived robust fusion estimator is tested using different transmission and packet drop rates. The modelling errors are the same as in Case 1 with a truncated normal distribution. Based on the analysis of Figure 5, it is evident that the derived estimator exhibits effective and reliable operation even under diverse transmission rates generated by the employed event-triggered transmission strategy. However, variations in transmission rates give rise to disparities in estimation performance, a well-studied phenomenon. This can be attributed to the fact that higher transmission rates are associated with improved estimator performance. As the transmission rate increases, the fusion center receives a greater volume of estimation values, thereby leading to more accurate results. A reasonable analysis of Figure 6 demonstrates that the derived estimator effectively maintains its reliability even under diverse packet drop rates. Nonetheless, differing packet drop rates introduce disparities in estimation performance, which is a valid observation. Higher packet drop rates correspond to inferior estimation performance. When compared to Figure 5, it is apparent that the variation in estimation performance is greater for different packet drop rates than for different transmission rates.

As can be seen from the two sets of simulations, the proposed robust fusion estimator exhibits relatively better performance compared to the fusion estimator that ignores uncertainty. The derived robust fusion estimator is still applicable when the selection of modelling errors is not limited to the particular structure. The results show that the method is an effective multi-sensor fusion method in practical applications.

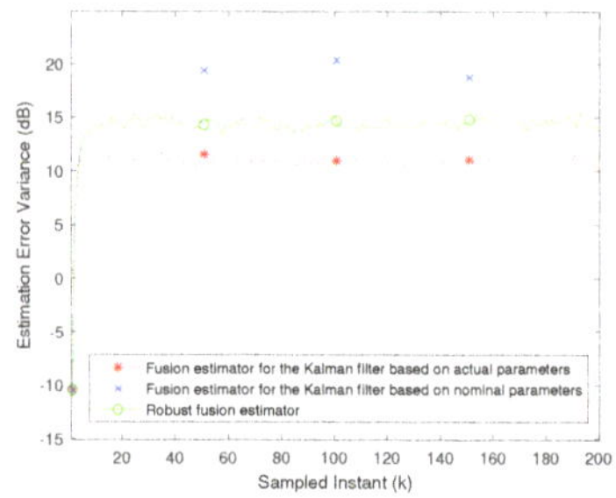

(**a**) Estimation error variance.

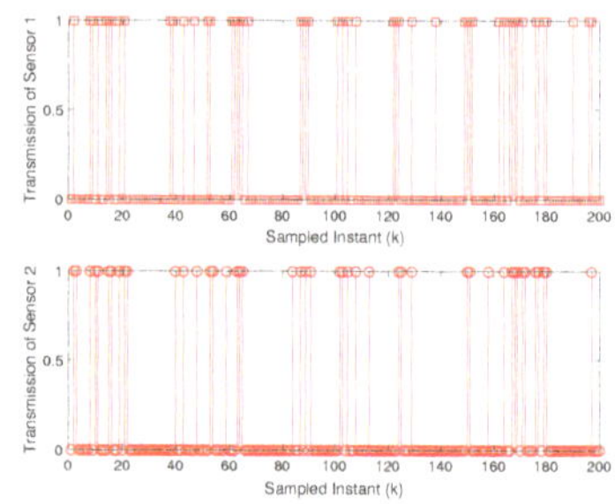

(**b**) Implementation of event-triggered.

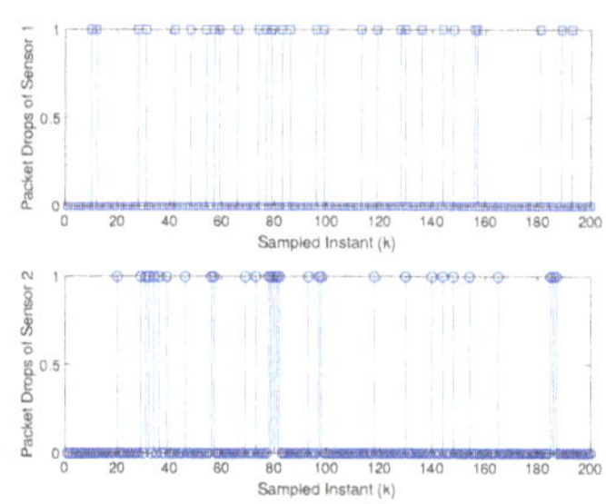

(**c**) Occurrence of packet drop.

Figure 4. ···∗···: the fusion estimator for the Kalman filter based on actual parameters; ···×···: the fusion estimator for the Kalman filter based on nominal parameters; —○—: the method of this paper; □: sensor 1; ○: sensor 2. Transmission rate: 0.8. Packet drop rate: 0.2. The modelling errors ε_k are taken to a normal distribution with truncations.

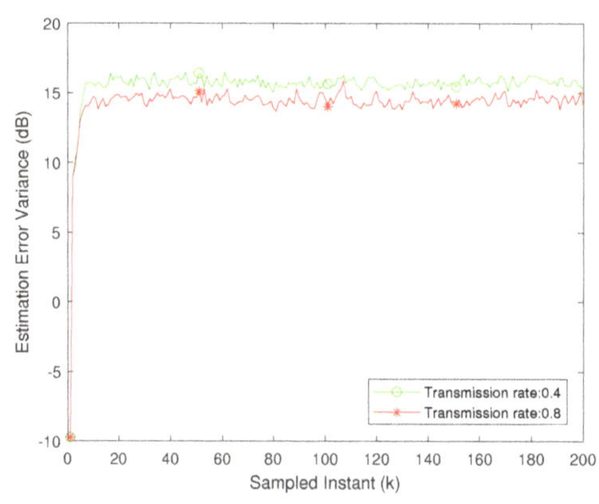

Figure 5. —○—: transmission rate 0.4; —∗—: transmission rate 0.8. Packet drop rate: 0.2. The modelling errors ε_k are taken to a normal distribution with truncations.

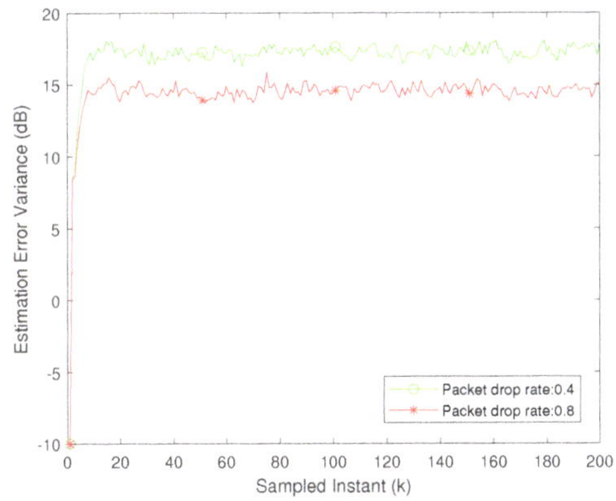

Figure 6. —○—: packet drop rate 0.8; —∗—: packet drop rate 0.4. Transmission rate: 0.2. The modelling errors ε_k are taken to a normal distribution with truncations.

6. Conclusions

In this paper, the effects of deterministic inputs and state delays present in the system are considered based on the study of robust fusion estimators for multi-sensor systems with uncertainty, random packet drops, and transmission constraints. The main contribution of this paper is the derivation of a robust fusion estimator for multi-sensor systems with state delays and external inputs, which penalizes the sensitivity of estimation errors to model uncertainty while minimizing nominal estimation errors and their sensitivity. Model conversion is performed utilizing the state augmentation technique. The event-triggered transmission strategy and the random packet drops generated by channel unreliability are considered. The pseudo-cross-covariance matrix is updated accordingly. This paper delivers robust proof of the fusion estimator of estimation errors being uniformly bounded. Two sets of numerical simulations are executed to illustrate the practical implications of the proposed method, using a tractor–car system as a demonstrative example. The numerical simulation results show that the estimation performance of the updated estimator is better than the fusion estimator for the Kalman filter based on nominal parameters. Since the modelling errors are not restricted to a specific structure, the proposed fusion estimator has a wide range of applicability. In addition, follow-up work on the tractor–car system example is still in progress, and the further stage is to apply the algorithm designed in this investigation to a practical case.

Author Contributions: Conceptualization, X.D. and H.L.; methodology, X.D.; software, X.D.; validation, H.L.; formal analysis, X.D.; investigation, X.D.; resources, H.Y. and H.L.; data curation, H.Y.; writing—original draft preparation, X.D.; writing—review and editing, X.D.; visualization, H.Y. and H.L.; supervision, H.Y. and H.L.; project administration, H.Y. and H.L.; funding acquisition, H.Y. and H.L. All authors have read and agreed to the published version of the manuscript.

Funding: This work is supported by the National Natural Science Foundation of China under grant number 61873138; by the Shandong Provincial Natural Science Foundation under grant number ZR2019MF063 and ZR2020MF064.

Institutional Review Board Statement: Not applicable.

Informed Consent Statement: Not applicable.

Data Availability Statement: Not applicable.

Conflicts of Interest: The authors declare no conflict of interest.

Abbreviations

The following abbreviations are used in this manuscript:

$col[*]$	The stacking vector or matrix
$E[*]$	The mathematical expectation
$f[*]$	The probability density function
M^T	The stacking vector or matrix
$N(\cdot;\sigma,\Omega)$	The notation for the Gaussian probability density function with mean σ and covariance
$tr[*]$	The trace of the matrix

Appendix A. Derivation of Robust State Estimation

In order to reduce the sensitivity of the estimation performance to the modelling error, the following cost function can therefore be minimized

$$J\left(\alpha_k^i\right) = \gamma_k^i\left[\left\|\alpha_k^i\right\|_{\Phi_k^i}^2 + \left\|H_k^i(0,0)\alpha_k^i - \beta_k^i(0,0)\right\|_{\Psi_k^i}^2\right] + \left(1-\gamma_k^i\right)\sum_{j=1}^{l}\left(\left\|\frac{\partial e_k^i(\varepsilon_k,\varepsilon_{k+1})}{\partial \varepsilon_{k,j}}\right\|^2 + \left\|\frac{\partial e_k^i(\varepsilon_k,\varepsilon_{k+1})}{\partial \varepsilon_{k+1,j}}\right\|^2\right)\Bigg|_{\substack{\varepsilon_k=0\\ \varepsilon_{k+1}=0}},$$

in which $\Psi_k^i = \left(R_{k+1}^i\right)^{-1}$, $H_k^i(\varepsilon_k,\varepsilon_{k+1}) = \bar{C}_{k+1}^i(\varepsilon_{k+1})[\bar{A}_k(\varepsilon_k)\quad \bar{B}_{2,k}(\varepsilon_k)]$, $\beta_k^i(\varepsilon_k,\varepsilon_{k+1}) = y_{k+1}^i - \bar{C}_{k+1}^i(\varepsilon_{k+1})\left(\bar{A}_k(\varepsilon_k)\hat{X}_{k|k}^i + \bar{B}_{1,k}(\varepsilon_k)u_k\right)$, $\Phi_k^i = \mathrm{diag}\left\{\left(P_{k|k}^i\right)^{-1}, Q_i^{-1}\right\}$, $\alpha_k^i = col\{X_k - \hat{X}_{k|k}^i, w_k\}$, $e_k^i(\varepsilon_k,\varepsilon_{k+1}) = y_{k+1}^i - \bar{C}_{k+1}^i(\varepsilon_{k+1})\left(\bar{A}_k(\varepsilon_k)\hat{X}_{k|k}^i + \bar{B}_{1,k}(\varepsilon_k)u_k\right) - \bar{C}_{k+1}^i(\varepsilon_{k+1})[\bar{A}_k(\varepsilon_k)\quad \bar{B}_{2,k}(\varepsilon_k)]\alpha_k^i$.

From Φ_k^i and Ψ_k^i, $J(\alpha_k^i)$ is a strictly convex function when $0 < \gamma_k^i \leq 1$. Letting $\frac{\delta J(\alpha_k^i)}{\delta \alpha_k^i} = 0$, the global unique minimum is obtained

$$\left(\Phi_k^i + \left(H_k^i(0,0)\right)^T\Psi_k^i H_k^i(0,0) + \frac{1-\gamma_k^i}{\gamma_k^i}\left[\begin{array}{cc} S_k^i & T_{2,k}^i\end{array}\right]^T\left[\begin{array}{cc} S_k^i & T_{2,k}^i\end{array}\right]\right)\alpha_{k,opt}^i = \left(H_k^i(0,0)\right)^T\Psi_k^i\beta_k^i(0,0) - \frac{1-\gamma_k^i}{\gamma_k^i}\left[\begin{array}{cc} S_k^i & T_{2,k}^i\end{array}\right]^T\left(S_k^i\hat{X}_{k|k}^i + T_{1,k}^i u_k\right). \quad \text{(A1)}$$

The initial state X_0 is estimated such that $e_0^i(\varepsilon_0) = y_0^i - \bar{C}_0^i(\varepsilon_0)X_0$ and the cost function is $J(\alpha_0^i) = \gamma_0^i\left[\|X_0\|_{\Pi_0^{-1}}^2 + \|y_0^i - \bar{C}_0^i(\varepsilon_0)X_0\|_{(R_0^i)^{-1}}^2\right] + (1-\gamma_0^i)\sum_{j=1}^{l}\left(\left\|\frac{\partial e_0^i(\varepsilon_0)}{\partial \varepsilon_{0,j}}\right\|^2\right)\Big|_{\varepsilon_0=0}$.

The following initial state estimate and initial estimation error covariance matrix can be obtained

$$P_{0|0}^i = \left(\begin{array}{c}(\bar{C}_0^i(0))^T(R_0^i)^{-1}\bar{C}_0^i(0)\\ +(\hat{\Pi}_0^i)^{-1}\end{array}\right)^{-1},$$
$$\hat{X}_{0|0}^i = P_{0|0}^i(\bar{C}_0^i(0))^T(R_0^i)^{-1}y_0^i,$$

in which $\hat{\Pi}_0^i = \left(\Pi_0^{-1} + \mu_0^i\sum_{j=1}^{l}\left(\frac{\partial(\bar{C}_0^i(\varepsilon_0))^T}{\partial \varepsilon_{0,j}}\right)\left(\frac{\partial \bar{C}_0^i(\varepsilon_0)}{\partial \varepsilon_{0,j}}\right)\Big|_{\varepsilon_0=0}\right)^{-1}$.

Define $\hat{H}_k^i$, $\hat{T}_{2,k}^i$, $\tilde{X}_{k|k+1}^i$, $\alpha_{k,opt}^i$, $\left(\hat{P}_{k|k}^i\right)^{-1}$, and $(\hat{Q}_k^i)^{-1}$ as $C_{k+1}^i(0)\left[\bar{A}_k(0)\bar{B}_{2,k}^i(0)\right]$, $T_{2,k}^i - \mu_k^i S_k^i\hat{P}_{k|k}^i(S_k^i)^T T_{2,k}^i$, $\hat{X}_{k|k+1}^i + \mu_k^i\hat{P}_{k|k}^i(S_k^i)^T T_{2,k}^i\hat{w}_{k|k+1}$, $col\left\{\hat{X}_{k|k+1}^i - \hat{X}_{k|k}^i, \hat{w}_{k|k+1}\right\}$, and $\left(P_{k|k}^i\right)^{-1} + \mu_k^i(S_k^i)^T S_k^i$, $(Q_k)^{-1} + \mu_k^i\left(T_{2,k}^i\right)^T\left(I + \mu_k^i S_k^i P_{k|k}^i(S_k^i)^T\right)^{-1}T_{2,k}^i$.

It is known by the following algebraic relation

$$\begin{bmatrix} \left(P^i_{k|k}\right)^{-1} & 0 \\ 0 & Q_i^{-1} \end{bmatrix} + \mu^i_k \begin{bmatrix} S^i_k & T^i_{2,k} \end{bmatrix}^T \begin{bmatrix} S^i_k & T^i_{2,k} \end{bmatrix} =$$
$$\begin{bmatrix} I & 0 \\ \mu^i_k \left(T^i_{2,k}\right)^T S^i_k \hat{P}^i_{k|k} & I \end{bmatrix} \begin{bmatrix} \left(\hat{P}^i_{k|k}\right)^{-1} & 0 \\ 0 & (\hat{Q}^i_k)^{-1} \end{bmatrix} \begin{bmatrix} I & \mu^i_k \hat{P}^i_{k|k} (S^i_k)^T T^i_{2,k} \\ 0 & I \end{bmatrix}.$$

Substituting the above equation into (A1) and multiplying equation (A1) left by, it gives

$$\begin{bmatrix} I & 0 \\ \mu^i_k \left(T^i_{2,k}\right)^T S^i_k \hat{P}^i_{k|k} & I \end{bmatrix}^{-1} \left(\begin{bmatrix} \left(P^i_{k|k}\right)^{-1} & 0 \\ 0 & Q_k^{-1} \end{bmatrix} + \left(H^i_k(0,0)\right)^T \Psi^i_k H^i_k(0,0) \right.$$
$$\left. + \mu^i_k \begin{bmatrix} (S^i_k)^T \\ \left(T^i_{2,k}\right)^T \end{bmatrix} \begin{bmatrix} S^i_k & T^i_{2,k} \end{bmatrix} \right) \alpha^i_{k,opt}$$
$$= \begin{bmatrix} I & 0 \\ -\mu^i_k \left(T^i_{2,k}\right)^T S^i_k \hat{P}^i_{k|k} & I \end{bmatrix} \left(\left(H^i_k(0,0)\right)^T \Psi^i_k \beta^i_k(0,0) - \mu^i_k \begin{bmatrix} (S^i_k)^T \\ \left(T^i_{2,k}\right)^T \end{bmatrix} \left(S^i_k \hat{X}^i_{k|k} + T^i_{1,k} u_k \right) \right),$$

$$\begin{aligned} &\left(\begin{bmatrix} \left(\hat{P}^i_{k|k}\right)^{-1} & 0 \\ 0 & (\hat{Q}^i_k)^{-1} \end{bmatrix} + \begin{bmatrix} \bar{A}^T_k(0) \\ -\mu^i_k \left(T^i_{2,k}\right)^T S^i_k \hat{P}^i_{k|k} \bar{A}^T_k(0) + \bar{B}^T_{2,k}(0) \end{bmatrix} \left(C^i_{k+1}(0)\right)^T \Psi^i_k \right. \\ &\left. \times C^i_{k+1}(0) \begin{bmatrix} \bar{A}_k(0) & -\mu^i_k \bar{A}_k(0) \hat{P}^i_{k|k} (S^i_k)^T T^i_{2,k} + \bar{B}_{2,k}(0) \end{bmatrix} \right) \\ &\times \begin{bmatrix} \hat{X}^i_{k|k+1} + \mu^i_k \hat{P}^i_{k|k} (S^i_k)^T \left(T^i_{2,k}\right)^T \hat{w}_{k|k+1} - \hat{X}^i_{k|k} \\ \hat{w}_{k|k+1} \end{bmatrix} = \\ &\begin{bmatrix} \bar{A}^T_k(0) \\ -\mu^i_k \left(T^i_{2,k}\right)^T S^i_k \hat{P}^i_{k|k} \bar{A}^T_k(0) + \bar{B}^T_{2,k}(0) \end{bmatrix} \left(C^i_{k+1}(0)\right)^T \Psi^i_k \left[y^i_{k+1} - C^i_{k+1}(0) \left(\bar{A}_k(0) \hat{X}^i_{k|k} + \bar{B}_{1,k}(0) u_k \right) \right] \\ &-\mu^i_k \begin{bmatrix} (S^i_k)^T \\ -\mu^i_k \left(T^i_{2,k}\right)^T S^i_k \hat{P}^i_{k|k} (S^i_k)^T + \left(T^i_{2,k}\right)^T \end{bmatrix} \left(S^i_k \hat{X}^i_{k|k} + T^i_{1,k} u_k \right). \end{aligned} \tag{A2}$$

Defining $\hat{B}^T_{2,k}(0)$,$\hat{T}^i_{2,k}$, and $\tilde{X}^i_{k|k}$ as $\hat{B}^i_{2,k}(0) = \bar{B}_{2,k}(0) - \mu^i_k \bar{A}_k(0) \hat{P}^i_{k|k} (S^i_k)^T T^i_{2,k}$, $\hat{T}^i_{2,k} = T^i_{2,k} - \mu^i_k S^i_k \hat{P}^i_{k|k} (S^i_k)^T T^i_{2,k}$, and $\hat{X}^i_{k|k+1} + \mu^i_k \hat{P}^i_{k|k} (S^i_k)^T T^i_{2,k} \hat{w}_{k|k+1}$, respectively, to simplify (A2), we can obtain

$$\begin{aligned} &\left(\begin{bmatrix} \left(\hat{P}^i_{k|k}\right)^{-1} & 0 \\ 0 & (\hat{Q}^i_k)^{-1} \end{bmatrix} + \left(\hat{H}^i_k\right)^T \Psi^i_k \hat{H}^i_k \right) \begin{bmatrix} \tilde{X}^i_{k|k+1} - \hat{X}^i_{k|k} \\ \hat{w}_{k|k+1} \end{bmatrix} = \\ &\left(\hat{H}^i_k\right)^T \Psi^i_k \left[y^i_{k+1} - C^i_{k+1}(0) \left(\bar{A}_k(0) \hat{X}^i_{k|k} + \bar{B}_{1,k}(0) u_k \right) \right] \\ &- \mu^i_k \begin{bmatrix} (S^i_k)^T \\ \left(T^i_{2,k}\right)^T \end{bmatrix} \left(S^i_k \hat{X}^i_{k|k} + T^i_{1,k} u_k \right). \end{aligned} \tag{A3}$$

From (A3), we have

$$\begin{cases} \tilde{X}^i_{k,k+1} = \hat{X}^i_{k|k} + \hat{P}^i_{k|k}\bar{A}^T_k(0)\left(C^i_{k+1}(0)\right)^T\left(R^i_{k+1}\right)^{-1} \\ \qquad \times \left[y^i_{k+1} - C^i_{k+1}(0)\left(\bar{B}_{1,k}(0)u_k + \bar{A}_k(0)\tilde{X}^i_{k|k+1} + \hat{B}^i_{2,k}(0)\hat{w}_{k|k+1}\right)\right] \\ \qquad - \mu^i_k\hat{P}^i_{k|k}\left(S^i_k\right)^T\left(S^i_k\hat{X}^i_{k|k} + T^i_{1,k}u_k\right) \\ \hat{w}_{k|k+1} = \hat{Q}^i_k\left(\hat{B}^i_{2,k}(0)\right)^T\left(C^i_{k+1}(0)\right)^T\left(R^i_{k+1}\right)^{-1} \\ \qquad \times \left[y^i_{k+1} - C^i_{k+1}(0)\left(\bar{B}_{1,k}(0)u_k + \bar{A}_k(0)\tilde{X}^i_{k|k+1} + \hat{B}^i_{2,k}(0)\hat{w}_{k|k+1}\right)\right] \\ \qquad - \mu^i_k\hat{Q}^i_k\left(\hat{T}^i_{2,k}\right)^T\left(S^i_k\hat{X}^i_{k|k} + T^i_{1,k}u_k\right) \end{cases}. \tag{A4}$$

Define the variable $\tilde{X}^i_{k+1|k+1} = \bar{A}_k(0)\tilde{X}^i_{k|k+1} + \hat{B}^i_{2,k}(0)\hat{w}_{k|k+1} + \bar{B}_{1,k}(0)u_k$. Bringing (A4) into $\tilde{X}^i_{k+1|k+1}$, we obtain

$$\begin{aligned} \tilde{X}^i_{k+1|k+1} = {} & \bar{B}_{1,k}(0)u_k + \bar{A}_k(0)\tilde{X}^i_{k|k} + \left(\bar{A}_k(0)\hat{P}^i_{k|k}\bar{A}^T_k(0) + \hat{B}^i_{2,k}(0)\hat{Q}^i_k\left(\hat{B}^i_{2,k}(0)\right)^T\right) \\ & \times \left(C^i_{k+1}(0)\right)^T\left(R^i_{k+1}\right)^{-1}\left[y^i_{k+1} - C^i_{k+1}(0)\hat{X}^i_{k+1|k+1}\right] \\ & - \mu^i_k\left[\bar{A}_k(0)\hat{P}^i_{k|k}\left(S^i_k\right)^T + \hat{B}^i_{2,k}(0)\hat{Q}^i_k\left(T^i_{2,k}\right)^T\right]\left(S^i_k\hat{X}^i_{k|k} + T^i_{1,k}u_k\right). \end{aligned} \tag{A5}$$

Letting $P^i_{k+1|k} = \bar{A}_k(0)\hat{P}^i_{k|k}\bar{A}^T_k(0) + \hat{B}^i_{2,k}(0)\hat{Q}^i_k(\hat{B}^i_{2,k}(0))^T$, (A5) can be transformed into the form of (A6)

$$\begin{aligned} & \left(I + P^i_{k+1|k}\left(C^i_{k+1}(0)\right)^T\left(R^i_{k+1}\right)^{-1}C^i_{k+1}(0)\right)\tilde{X}^i_{k+1|k+1} = \bar{B}_{1,k}(0)u_k + \bar{A}_k(0)\hat{X}^i_{k|k} \\ & + P^i_{k+1|k}\left(C^i_{k+1}(0)\right)^T\left(R^i_{k+1}\right)^{-1}y^i_{k+1} - \mu^i_k\left[\bar{A}_k(0)\hat{P}^i_{k|k}\left(S^i_k\right)^T + \hat{B}^i_{2,k}(0)\hat{Q}^i_k\left(T^i_{2,k}\right)^T\right]\left(S^i_k\hat{X}^i_{k|k} + T^i_{1,k}u_k\right). \end{aligned} \tag{A6}$$

According to the matrix inverse lemma $(A+BCD)^{-1} = A^{-1} - A^{-1}B(DA^{-1}B + C^{-1})^{-1}DA^{-1}$, the following procedure can be obtained

$$P^i_{k+1|k+1} = P^i_{k+1|k} - P^i_{k+1|k}\left(\bar{C}^i_{k+1}(0)\right)^T\left(R^i_{k+1} + \bar{C}^i_{k+1}(0)P^i_{k+1|k}\left(\bar{C}^i_{k+1}(0)\right)^T\right)^{-1}\bar{C}^i_{k+1}(0)P^i_{k+1|k}.$$

(A6) can be changed to

$$\begin{aligned} & \left(I + P^i_{k+1|k}\left(C^i_{k+1}(0)\right)^T\left(R^i_{k+1}\right)^{-1}C^i_{k+1}(0)\right)\tilde{X}^i_{k+1|k+1} \\ & = \left[\bar{B}_{1,k}(0) - \mu^i_k\left[\bar{A}_k(0)\hat{P}^i_{k|k}(S^i_k)^T + \hat{B}^i_{2,k}(0)\hat{Q}^i_k\left(T^i_{2,k}\right)^T\right]T^i_{1,k}\right]u_k \\ & + \left[\bar{A}_k(0) - \mu^i_k\left[\bar{A}_k(0)\hat{P}^i_{k|k}(S^i_k)^T + \hat{B}^i_{2,k}(0)\hat{Q}^i_k\left(T^i_{2,k}\right)^T\right]S^i_k\right]\hat{X}^i_{k|k} \\ & + P^i_{k+1|k}\left(C^i_{k+1}(0)\right)^T\left(R^i_{k+1}\right)^{-1}y^i_{k+1}. \end{aligned}$$

Thus, the matrices $\hat{A}^i_k(0)$ and $\hat{B}^i_{1,k}(0)$ can be defined as

$$\hat{A}^i_k(0) = \left(\bar{A}_k(0) - \mu^i_k\hat{B}^i_{2,k}(0)\hat{Q}^i_k\left(T^i_{2,k}\right)^TS^i_k\right)\left(I - \mu^i_k\hat{P}^i_{k|k}(S^i_k)^TS^i_k\right),$$

$$\hat{B}^i_{1,k}(0) = \bar{B}_{1,k}(0) - \mu^i_k\left(\bar{A}_k(0)\hat{P}^i_{k|k}(S^i_k)^T + \hat{B}^i_{2,k}(0)\hat{Q}^i_k\left(T^i_{2,k}\right)^T\right)T^i_{1,k}.$$

Thus (A5) can be simplified as

$$\begin{aligned}\tilde{X}^i_{k+1|k+1} &= \hat{A}^i_k(0)\hat{X}^i_{k|k} + \hat{B}^i_{1,k}(0)u_k + P^i_{k+1|k+1}\left(\bar{C}^i_{k+1}(0)\right)^T\left(R^i_{k+1}\right)^{-1} \\ &\times\left[y^i_{k+1} - \bar{C}^i_{k+1}(0)\left(\hat{A}^i_k(0)\hat{X}^i_{k|k} + \hat{B}^i_{1,k}(0)u_k\right)\right].\end{aligned} \tag{A7}$$

(A7) is similar to the form described in [29], so that $\tilde{X}^i_{k+1|k+1}$ can be specified as $\hat{X}^i_{k+1|k+1}$.

The derivation is complete.

References

1. Li, A.; Cao, J.; Li, S.; Huang, Z.; Wang, J.; Liu, G. Map Construction and Path Planning Method for a Mobile Robot Based on Multi-Sensor Information Fusion. *Appl. Sci.* **2022**, *12*, 2913. [CrossRef]
2. Han, Y.; Sun, H.; Lu, Y.; Zhong, R.; Ji, C.; Xie, S. 3D Point Cloud Generation Based on Multi-Sensor Fusion. *Appl. Sci.* **2022**, *12*, 9433. [CrossRef]
3. Bai, H.; Yu, B.; Gu, W. Research on position sensorless control of rdt motor based on improved smo with continuous hyperbolic tangent function and improved feedforward PLL. *J. Mar. Sci. Eng.* **2023**, *11*, 642. [CrossRef]
4. Qin, J.; Du, J.; Li, J. Adaptive finite-time trajectory tracking event-triggered control scheme for underactuated surface vessels subject to input saturation. *IEEE Trans. Intell. Transp. Syst.* **2023** , 1–13. [CrossRef]
5. Qin, J.; Du, J. Minimum-learning-parameter-based adaptive finite-time trajectory tracking event-triggered control for underactuated surface vessels with parametric uncertainties. *Ocean Eng.* **2023**, *271*, 113634. [CrossRef]
6. Sun, S.; Ma, J. Linear estimation for networked control systems with random transmission delays and packet dropouts. *Inf. Sci.* **2014**, *269*, 349–365. [CrossRef]
7. Wei, Y.; Sun, S. Recursive distributed fusion estimation for multi-sensor systems with missing measurements, multiple random transmission delays and packet losses. *Signal Process.* **2023**, *204*, 108829. [CrossRef]
8. Ma, J.; Sun, S. Centralized Fusion Estimators for Multisensor Systems with Random Sensor Delays, Multiple Packet Dropouts, and Uncertain Observations. *IEEE Sens. J.* **2013**, *13*, 1228–1235. [CrossRef]
9. Shen, Q.; Liu, J.; Zhou, X.; Qin, W.; Wang, L.; Wang, Q. Centralized Fusion Methods for Multi-Sensor System with Bounded Disturbances. *IEEE Access* **2019**, *7*, 141612–141626. [CrossRef]
10. Sun, S. Distributed Optimal Linear Fusion Predictors and Filters for Systems with Random Parameter Matrices and Correlated Noises. *IEEE Trans. Signal Process.* **2020**, *68*, 1064–1074. [CrossRef]
11. Wang, R.; Sun, S. Distributed Matrix-Weighted Fusion Consensus Filtering with Two-Stage Filtering for Sensor Networks. *IEEE Sens. J.* **2023**, *23*, 5003–5013. [CrossRef]
12. Caballero-Águila, R.; Hermoso-Carazo, A.; Linares-Pérez, J. Networked distributed fusion estimation under uncertain outputs with random transmission delays, packet losses and multi-packet processing. *Signal Process.* **2019**, *156*, 71–83. [CrossRef]
13. Li, L.; Fan, M.; Xia, Y.; Geng, Q. Dynamic Event-Triggered Feedback Fusion Estimation for Nonlinear Multi-Sensor Systems with Auto/Cross-Correlated Noises. *IEEE Trans. Signal Inf. Process. Over Netw.* **2022**, *8*, 868–882. [CrossRef]
14. Wang, X.; Lemmon, M.D. Event-Triggering in Distributed Networked Control Systems. *IEEE Trans. Automat. Contr.* **2011**, *56*, 586–601. [CrossRef]
15. Battistelli, G.; Benavoli, A.; Chisci, L. Data-driven communication for state estimation with sensor networks. *Automatica* **2012**, *48*, 926–935. [CrossRef]
16. Trimpe, S.; D'Andrea, R. Event-based state estimation with variance-based triggering. In Proceedings of the 2012 IEEE 51st IEEE Conference on Decision and Control (CDC), Maui, HI, USA, 10–13 December 2012; pp. 6583–6590. [CrossRef]
17. Maity, D.; Baras, J.S. Optimal Event-Triggered Control of Nondeterministic Linear Systems. *IEEE Trans. Automat. Control* **2020**, *65*, 604–619. [CrossRef]
18. Yu, H.; Chen, T.; Hao, F. A New Event-Triggered Control Scheme for Stochastic Systems. *IEEE Trans. Automat. Control* **2023**, *68*, 1463–1478. [CrossRef]
19. Meng, F.; Gao, J.; Liu, H. Robust State Estimation for Time-Delay Linear Systems With External Inputs. *IEEE Access* **2021**, *9*, 106540–106549. [CrossRef]
20. Ren, J. LMI-Based Fault Detection Filter Design for A Class of Neutral System with Time Delay in States. In Proceedings of the 2006 6th World Congress on Intelligent Control and Automation, Dalian, China, 21–23 June 2006; Volume 2, pp. 5581–5585. [CrossRef]
21. Wang, Y.; Liu, H.; Li, X. A Novel Method for Stability Analysis of Time-Varying Delay Systems. *IEEE Trans. Automat. Control* **2021**, *66*, 1422–1428. [CrossRef]
22. Zhang, H.; Zhang, D.; Xie, L. An innovation approach to H∞ prediction for continuous-time systems with application to systems with delayed measurements. *Automatica* **2004**, *40*, 1253–1261. [CrossRef]
23. Wu, Z.; Shi, P.; Su, H.; Chu, J. State estimation for discrete-time neural networks with time-varying delay. *Int. J. Syst. Sci.* **2012**, *43*, 647–655. [CrossRef]

24. Frezzatto, L.; Oliveira, R.C.; Peres, P.L. H∞ non-minimal filter design in finite frequency ranges for discrete-time Takagi–Sugeno fuzzy systems with time-varying delays. *J. Franklin Inst.* **2020**, *357*, 622–634. [CrossRef]
25. Wang, J.; Mao, Y.; Li, Z.; Gao, J.; Liu, H. Robust Fusion Estimation for Multisensor Uncertain Systems with State Delay Based on Data-Driven Communication Strategy. *IEEE Access* **2020**, *8*, 151888–151897. [CrossRef]
26. Liu, H.; Yu, H. Event-triggered robust state estimation for wireless sensor networks. *Asian J. Control* **2020**, *22*, 1649–1658. [CrossRef]
27. Sayed, A.H. A framework for state-space estimation with uncertain models. *IEEE Trans. Automat. Control* **2001**, *46*, 998–1013. [CrossRef]
28. Xu, H.; Mannor, S. A Kalman Filter Design Based on the Performance/Robustness Tradeoff. *IEEE Trans. Automat. Control* **2009**, *54*, 1171–1175. [CrossRef]
29. Zhou, T. Sensitivity Penalization Based Robust State Estimation for Uncertain Linear Systems. *IEEE Trans. Automat. Control* **2010**, *55*, 1018–1024. [CrossRef]
30. Liu, H.; Zhou, T. Robust state estimation for uncertain linear systems with random parametric uncertainties. *Sci. China Inf. Sci.* **2016**, *60*, 157–169. [CrossRef]
31. Zienkiewicz, O.C.; Zhu, J.Z. A simple error estimator and adaptive procedure for practical engineerng analysis. *Int. J. Numer. Methods Eng.* **1987**, *24*, 337–357. [CrossRef]
32. Li, X.; Zhu, Y.; Wang, J.; Han, C. Optimal linear estimation fusion .I. Unified fusion rules. *IEEE Trans. Inf. Theory* **2003**, *49*, 2192–2208. [CrossRef]
33. Zhou, T.; Liang, H. On asymptotic behaviors of a sensitivity penalization based robust state estimator. *Syst. Control Lett.* **2011**, *60*, 174–180. [CrossRef]
34. Zhang, Z.; Yao, M.; Gao, J.; Liu, H. State Estimation of Discrete-Time T–S Fuzzy Systems Based on Robustness Ideas. *Int. J. Fuzzy Syst.* **2023**, *25*, 2007–2019. [CrossRef]

applied sciences

MDPI

Article

Comparative Analysis of Machine Learning Models for Predicting Crack Propagation under Coupled Load and Temperature

Intisar Omar *, Muhammad Khan * and Andrew Starr

School of Aerospace, Transport and Manufacturing, Cranfield University, Cranfield MK43 0AL, UK; a.starr@cranfield.ac.uk
* Correspondence: i.omar@cranfield.ac.uk (I.O.); muhammad.a.khan@cranfield.ac.uk (M.K.)

Abstract: Crack propagation in materials is a complex phenomenon that is influenced by various factors, including dynamic load and temperature. In this study, we investigated the performance of different machine learning models for predicting crack propagation in three types of materials: composite, metal, and polymer. For composite materials, we used Random Forest Regressor, Support Vector Regression, and Gradient Boosting Regressor models, while for polymer and metal materials, we used Ridge, Lasso, and K-Nearest Neighbors models. We trained and tested these models using experimental data obtained from crack propagation tests performed under varying load and temperature conditions. We evaluated the performance of each model using the mean squared error (MSE) metric. Our results showed that the best-performing model for composite materials was Gradient Boosting Regressor, while for polymer and metal materials, Ridge and K-Nearest Neighbors models outperformed the other models. We also validated the models using additional experimental data and found that they could accurately predict crack propagation in all three materials with high accuracy. The study's findings provide valuable insights into crack propagation behavior in different materials and offer practical applications in the design, construction, maintenance, and inspection of structures. By leveraging this knowledge, engineers and designers can make informed decisions to enhance the strength, reliability, and durability of structures, ensuring their long-term performance and safety.

Keywords: crack propagation; machine learning; dynamic load; Random Forest Regressor; Support Vector Regression; Gradient Boosting Regressor; Ridge; Lasso; K-Nearest Neighbors

Citation: Omar, I.; Khan, M.; Starr, A. Comparative Analysis of Machine Learning Models for Predicting Crack Propagation under Coupled Load and Temperature. *Appl. Sci.* **2023**, *13*, 7212. https://doi.org/10.3390/app13127212

Academic Editors: Sheng Du, Xiongbo Wan, Wei Wang and Hao Fu

Received: 30 May 2023
Revised: 12 June 2023
Accepted: 15 June 2023
Published: 16 June 2023

1. Introduction

Crack propagation in materials is a critical phenomenon that can lead to structural failure and compromise the integrity and safety of various engineering applications. Understanding and predicting crack propagation is crucial for the design and maintenance of durable and reliable structures. Traditional approaches for predicting crack propagation involve complex mathematical models and empirical equations, which often have limitations in capturing the intricate nature of crack behavior under different conditions [1,2]. Crack propagation in materials involves complex and nonlinear behavior influenced by factors, such as material properties, loading conditions, and environmental elements. Traditional methods struggle to capture the complex relationships and dynamics involved in crack growth. Conversely, machine learning models excel at identifying complex patterns and nonlinear relationships within data, making them well-suited for modeling crack propagation processes.

Traditional methods often rely on simplified assumptions and limited representations of the factors influencing crack propagation, leading to inaccuracies. In contrast, machine learning models can incorporate a broader range of factors, including material properties, loading conditions, temperature variations, and more. By training on extensive datasets that encompass diverse conditions and parameters, machine learning models can learn the intricate relationships between these factors and crack propagation, enabling more accurate predictions.

Additionally, traditional methods lack generalization capabilities, meaning they struggle to predict crack propagation in materials or loading conditions that differ from those used during model development. On the other hand, machine learning models can generalize well to new and unseen scenarios, provided they are trained on diverse and representative datasets. This ability enables machine learning models to make accurate predictions for various materials, loading conditions, and temperature ranges, enhancing their applicability and reliability [3,4].

The utilization of machine learning models empowers researchers to surpass the limitations of traditional methods when it comes to predicting crack propagation. These models provide improved capabilities in capturing intricate behaviors, integrating a wider array of influencing factors, leveraging extensive and diverse datasets, and adapting to new scenarios [1,2,5–10]. As a result, machine learning-based approaches possess significant potential for advancing our comprehension of crack propagation and enhancing the design and dependability of structures and materials. Previous studies have utilized machine learning techniques for crack prediction in different materials. For instance, Ref. [11] employed Random Forest Regressor to predict crack propagation in composite materials under various loading conditions. The study demonstrated that the model accurately captured crack behavior and showed improved performance compared to traditional analytical methods. Support Vector Regression (SVR) has also been applied for crack propagation prediction. Ref. [12] utilized SVR to predict crack growth in metallic materials subjected to cyclic loading. Their results indicated that the SVR model achieved good accuracy in predicting the crack growth rate and exhibited better performance than conventional regression methods. Gradient Boosting Regressor has shown promise in crack propagation prediction as well. Ref. [13] employed this model to predict crack growth in composite material under different environmental temperatures. Their findings indicated that the Gradient Boosting Regressor achieved high prediction accuracy and outperformed other regression models, such as Random Forest and Support Vector Regression. Ridge, Lasso, and K-Nearest Neighbors are also commonly used models in machine learning applications. In the context of crack propagation prediction, these models have demonstrated effectiveness in capturing complex relationships between crack behavior and influencing factors. For instance, Ref. [1] employed Ridge regression and K-Nearest Neighbors to predict crack growth in polymer materials. The models exhibited high accuracy and provided valuable insights into the crack propagation process.

Several studies have demonstrated the effectiveness of machine learning models for crack propagation prediction, even with small and moderate datasets. For instance, Ref. [14] utilized Random Forest and Support Vector Regression models for predicting crack propagation in concrete structures with a small dataset. Their results showed that both models achieved high prediction accuracy and outperformed traditional analytical methods. Similarly, Ref. [15] employed a Gradient Boosting Machine model for predicting crack growth in metallic materials with a moderate dataset. Their findings showed that the model achieved good accuracy and provided valuable insights into the underlying crack behavior. These studies suggest that machine learning models can effectively predict crack propagation behavior, even with small or moderate datasets. By identifying the most important factors influencing crack behavior and accurately predicting crack growth, these models can aid in the design and development of more reliable and durable structures. In this study, we build upon the existing research by comparing the performance of these machine-learning models for predicting crack propagation in composite, metal, and polymer materials. We utilize experimental data obtained from crack propagation tests performed under varying load and temperature conditions. The performance of each model is evaluated to identify the best-performing models for each material type. The proposed study expands upon previously conducted research and significantly contributes to the current state of the art in crack propagation prediction in several means. The study focuses on crack propagation prediction in three distinct types of materials: composite, metal, and polymer. While previous research has often focused on individual material types, this study provides a comprehensive comparative analysis across multiple materials. By examining crack propagation in different materials, the study offers valuable insights into the varying behaviors, influencing factors, and predictive models specific to each material type. This expanded scope enhances our understanding of crack propagation across a wider range of materials and aids in the development of more versatile and accurate prediction models. The study investigates crack propagation under dynamic load and temperature conditions, which are critical factors influencing crack growth in real-world scenarios. While previous research has often focused on static load conditions, this study expands the understanding of crack propagation by incorporating the effects of dynamic loading and temperature variations. By considering these realistic operational conditions, the study enhances the applicability and relevance of crack propagation predictions to practical engineering scenarios. The proposed study expands upon previous research by conducting a comparative analysis across multiple materials, evaluating various machine learning models, considering dynamic load and temperature effects, validating with experimental data, and emphasizing practical implications for structural design. These contributions enhance our understanding of crack propagation prediction and offer practical tools and insights to improve the reliability and durability of structures in various material systems. The study's findings also could have respective practical implications and applications such as:

1. Risk Assessment and Maintenance Strategies: The study's findings enable better risk assessment and the development of proactive maintenance strategies. This information helps identify critical areas prone to crack initiation and propagation, allowing engineers to focus their efforts on preventive measures and inspections. By integrating machine learning models into structural health monitoring systems, early detection of crack propagation can be achieved, enabling timely maintenance and repair actions. This proactive approach enhances the dependability and longevity of structures by preventing failures and minimizing downtime.
2. Cost and Resource Optimization: Accurate crack propagation prediction enables optimized allocation of resources and cost-effective maintenance strategies. By identifying critical areas and accurately predicting crack propagation rates, engineers can prioritize inspection and maintenance efforts, allocating resources where they are most needed. This targeted approach optimizes resource utilization, reduces unnecessary maintenance activities, and lowers overall costs while ensuring the long-term dependability of structures.

Our study aims to demonstrate the potential of machine learning models as effective tools for predicting crack propagation in materials.

The remaining sections of this paper are structured as follows: Section 2 provides an overview of the materials used in this study, along with the machine learning models employed for crack propagation prediction under coupled load and temperature. In Section 3, we present the results obtained by applying the selected algorithms to three distinct material datasets. A comprehensive analysis of these results is provided, highlighting the performance and effectiveness of the various models. Finally, the concluding section summarizes the key findings of our research, shedding light on the comparative analysis of the machine learning models for crack propagation prediction under coupled load and temperature. It consolidates the main outcomes and implications derived from the study, potentially paving the way for further advancements in this field.

2. Materials and Methods

2.1. Specimen Parameters and Experimental Data Collection

The study involved three different materials: aluminum 2024-T3, control mix concrete, and steel fiber-reinforced concrete (SFRC). Figure 1 displays the geometry of the various specimens manufactured, while the experimental data were obtained from prior experiments. To conduct the experiments, the specimen was affixed onto the shaker and heated to multiple temperatures. The shaker applied mechanical loads, while impact tests were performed to determine the fundamental frequency of the specimen. Measurements were taken using a laser vibrometer. The vibration test was then conducted at the fundamental frequency. If crack propagation occurred, the beam tip's displacement amplitude was reduced, and the shaker was halted to record the new frequency. Further impact tests were executed to determine the new fundamental frequency, which was then set on the shaker. This process was reiterated until the specimen failed catastrophically due to crack propagation. In addition, Fused Deposition Modelling (FDM) 3D printed ABS was also used as representative material and tested in a similar manner.

ABS, concrete, and aluminum are widely used materials in industries, such as automotive, construction, and aerospace. Understanding crack propagation in these materials is of practical significance for ensuring the reliability and durability of structures made from them [16–23]. By evaluating the performance of machine learning models on these materials, the study can provide insights and guidance for real-world applications, aiding in the design and maintenance of structures involving ABS, concrete, and aluminum. While ABS, concrete, and aluminum differ in nature, the comparative analysis allows researchers to gain a comprehensive understanding of crack propagation across materials with diverse properties. This broader evaluation provides insights into the strengths and weaknesses of the machine learning models, facilitating the development of more robust and accurate prediction methods applicable to a range of materials. It is important to acknowledge the differences between these materials and interpret the results accordingly, considering the unique characteristics of each material type [1,2].

The collected data (as shown in Table 1) were plotted using scatter, pair, heatmap, and grid plots to understand the relationships between the features and the predicted crack depth. The data showed non-linearity and overlapping nature, and the problem was a regression problem with small and moderate datasets.

Table 1. The experimental conditions and parameters.

Material	Temperature °C	Crack Location mm	Structural Response
Composite	20		
Control Mix	40	5	
SFRC	60		
Polymer	20		
	50	5	
ABS	60	15	
	70	25	
Metal			
	20–25	4.5	
aluminum 2024-T3	50–100	5	
	150–200	10	

An empirical model had been developed earlier to relate crack depth/location and structural dynamic response. However, this model contained a large number of coefficients and was of a high order, making it challenging to interpret. To overcome this limitation, a more concise and suitable model was necessary that could accurately predict crack propagation and provide physical insight into the coefficients' meaning. Accordingly, selected models were proposed that met these requirements.

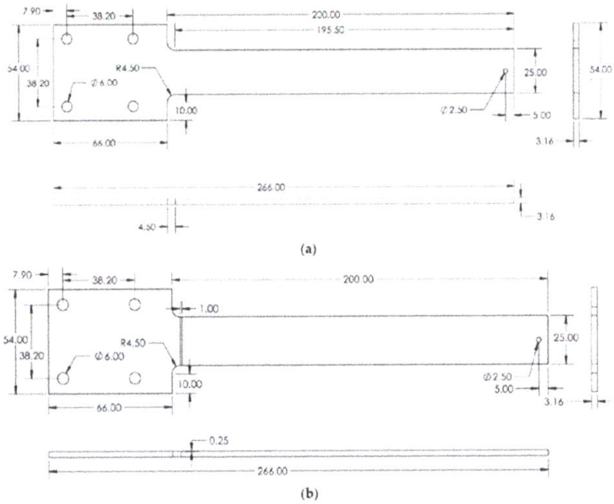

Figure 1. Specimen's geometry dimensions in mm (**a**) without crack (**b**) with crack [19].

2.2. *Data Analysis and Machine Learning Models*

2.2.1. Data Analysis

In this study, Jupyter Notebook was utilized as a tool for creating and sharing documents that contain live code, equations, visualizations, and narrative text. It is commonly used by practitioners and researchers in the fields of data analysis, machine learning, and scientific computing. Before using a dataset for machine learning, data analysis, and feature extraction techniques (as shown in Figure 2) are employed to select the most relevant and informative features for the task at hand. Techniques, such as visualizing the correlations between the features and the target variable, are useful, and visualization libraries, such as Pandas' scatter matrix and Altair, which is a declarative visualization library for Python. It supports a wide range of chart types, including bar charts, line charts, scatter plots, and heat maps. It also provides support for faceting and layering, which allows users to create more complex visualizations.

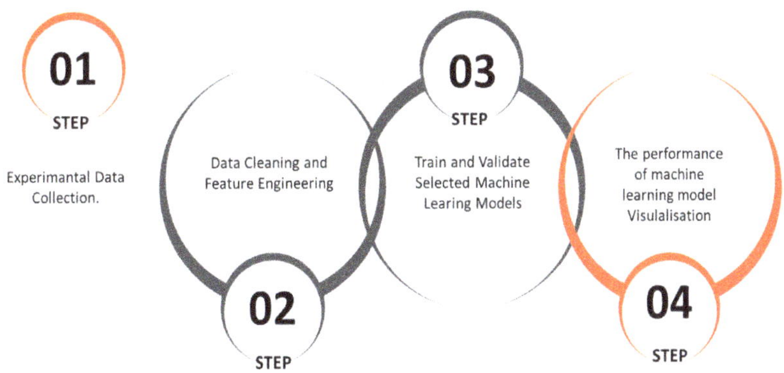

Figure 2. The flow chart of analysis steps.

These tools can aid in understanding data and selecting a suitable machine-learning model. However, it is important to consider the nature of the different machine learning models and how they handle different types of problems. The research utilized experimental data from three materials collected from previous studies [20,21,23]. The experimental data consists of four features: temperature °C, crack location mm, amplitude mm, natural frequency Hz, and a predicted value: crack depth mm. The data were plotted using a scatter matrix, pair plot, and grid plot to investigate the relationship between the features and the predicted value, which indicated that the task was a regression problem. Pre-processing of the data was conducted using libraries, such as Pandas and NumPy, to clean and format the data, perform basic statistical analysis, and use techniques, such as Variance Threshold, to drop features with constant or near-constant values. The data were also scaled using libraries, such as Standard Scaler, to improve the training time of the model and ensure that the features were at a comparable scale.

After cleaning and preparing the data, they are typically split into a training set and a validation set. The training set is used to train the model, while the validation set is used to evaluate the performance of the model on unseen data, ensuring that the model generalizes well to new, unseen data.

2.2.2. Machine Learning Models

Suitable machine learning models for ABS and aluminum 2024-T3 datasets are Ridge Regression, Lasso Regression, and K-Nearest Neighbors (KNN). Ridge Regression, a linear regression model with L2 regularization, was employed to prevent overfitting and handle multicollinearity in the data. It is computationally efficient and provides interpretable coefficients. Lasso Regression, similar to Ridge Regression, employs L1 regularization, encouraging sparsity in the model coefficients. It automatically performs feature selection, identifying the most relevant features for crack propagation prediction. However, both Ridge Regression and Lasso Regression assume a linear relationship and may struggle to capture complex nonlinear patterns. They can also encounter challenges with multicollinearity when highly correlated features are present.

KNN algorithm, a non-parametric method, was used as well. It makes predictions based on the proximity of training samples in the feature space, accommodating complex relationships. KNN is relatively easy to understand and implement. However, it can be sensitive to the choice of the number of neighbors (K) and the distance metric. During prediction, it can be computationally expensive, particularly for large datasets. These models are relatively simple, easy to understand, and do not require a lot of computational power, making them a good choice for tasks that need to be run quickly or on resource-constrained systems [1,2,13,24].

In another experiment using composite data, the crack depth of each concrete type was clustered based on their respective elastic modulus, which varied with changes in temperature. The researcher selected Support Vector Regressor, Random Forest Regressor, and Gradient Boosting Regressor to predict crack depth for composite data based on their ability to handle complex datasets with non-linear relationships between the features and the target variable and many hyperparameters that can be fine-tuned to improve performance. The Random Forest Regressor, an ensemble learning method that combines multiple decision trees, was utilized. It handles high-dimensional data effectively, mitigates overfitting, and captures complex relationships between input features and crack propagation. However, Random Forest Regressor can be computationally expensive, especially for large datasets. It may struggle to identify subtle patterns and can be challenging to interpret due to its ensemble nature.

Support Vector Regression (SVR) was also utilized, as it is effective in handling both linear and nonlinear relationships. It can manage high-dimensional data and, by selecting appropriate kernel functions, capture complex patterns. However, SVR can be sensitive to the choice of hyperparameters and kernel functions. It may also face limitations in dealing with very large datasets due to computational requirements.

Gradient Boosting Regressor, which constructs an ensemble of weak prediction models to sequentially correct errors, was another model used. It handles complex relationships well, performs admirably with high-dimensional data, and typically exhibits high prediction accuracy. However, Gradient Boosting Regressors can be computationally intensive and may require careful hyperparameter tuning. Without proper regularization, it is also prone to overfitting [11,15].

It is important to note that the benefits and drawbacks mentioned above are general considerations for the respective machine learning models. The specific performance of these models in predicting crack propagation would depend on factors, such as the quality and size of the dataset, appropriate feature selection, and careful hyperparameter tuning [1,2].

2.2.3. Training and Validation of Machine Learning Models

Top of Form

To enable proper comparison of the coefficients and account for different material behavior, three identical but independent regression models were trained—one for Aluminium, one for concrete, and one for ABS. The data were randomly and blindly split into training and test sets with a 70/30 split. The choice of a 70% training data split ensures a sufficiently large dataset for training the machine learning models. Having a substantial amount of data for training helps the models capture the underlying patterns and variations in crack propagation accurately. It provides a robust foundation for the models to learn from and develop predictive capabilities. Allocating 30% of the dataset to the validation set allows for a substantial portion of data to evaluate the trained models' performance. Sizable test data helps assess the generalization ability of the models, determining how well they perform on unseen data. This split ensures a meaningful evaluation of the model's predictive accuracy and provides statistical confidence in their performance metrics. The 70/30 split strikes a balance between mitigating overfitting and underfitting issues. If the training set were too small (e.g., 50/50 split), the models might not have enough data to learn effectively, potentially resulting in poor generalization and overfitting. Conversely, if the training set were too large, the risk of underfitting increases as the models may not adequately capture the complexity of crack propagation patterns. The 70/30 split attempts to optimize this balance [7,25–28].

Model hyperparameters were adjusted to minimize mean squared error and achieve satisfactory performance. This approach ensured that the model was always evaluated on previously unseen data, making the evaluated performance more representative of a real-world predictive problem. To prevent any compromise to the model's performance due to breaking the "test on unseen data" principle during training, a K-fold cross-validation methodology was utilized. This methodology is illustrated in Figure 3 and ensured that the model was evaluated using unseen data during training.

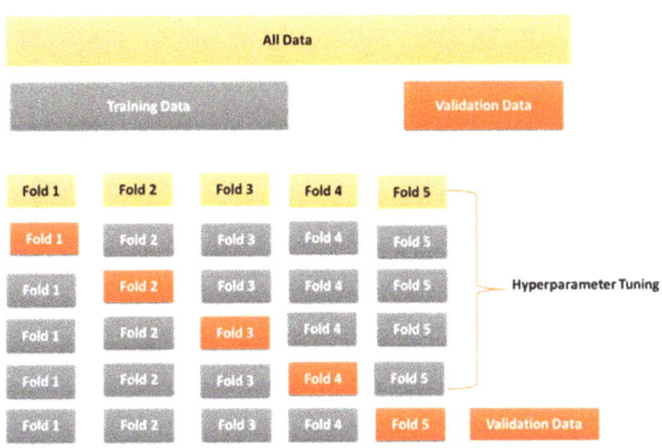

Figure 3. K-Fold cross-validation schematic.

3. Results and Discussion

The study analyses experimental data on natural frequency, structural amplitude, temperature, and crack position. According to [29], Spearman's correlation coefficient is alternative measure of correlation that can capture non-linear relationships and are suitable for ranking or ordinal data. Spearman's correlation coefficient (ρ) is calculated based on the ranks of the variables. Let us denote the ranks of X and Y as R(X) and R(Y), respectively. The Spearman correlation coefficient is given by:

$$\rho = 1 - [(6 \times \Sigma(D^2))/(n \times (n^2 - 1))] \ldots \ldots \quad (1)$$

where:

Σ denotes the sum of the respective values.

D represents the difference between ranks (R(X) − R(Y)).

n is the number of data points.

The value of ρ ranges between −1 and 1, where:

ρ = 1 indicates a perfect monotonic increasing relationship.

ρ = −1 indicates a perfect monotonic decreasing relationship.

ρ = 0 indicates no monotonic relationship.

Results show that ABS, concrete, and aluminum exhibit similar patterns in natural frequency, with frequency decreasing as crack depth increases (as shown in Figure 4) in a nonlinear manner. Additionally, the impact of temperature on natural frequency is more pronounced in aluminum than in concrete and ABS. This is due to the higher elastic modulus and greater consistency of isotropic sheet metal properties relative to additive layer manufactured ABS and concrete [1,5].

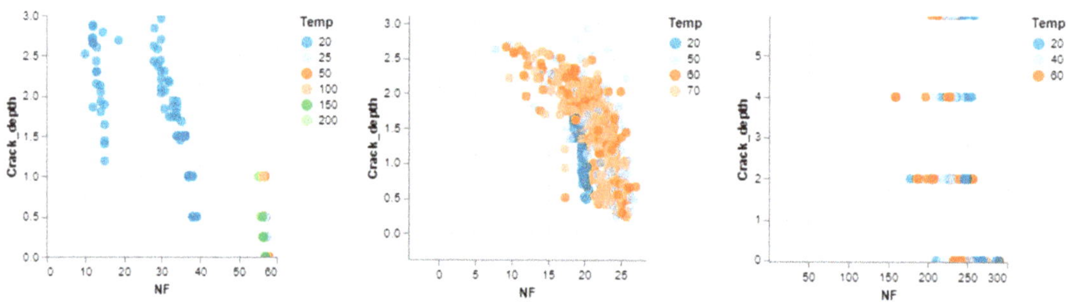

(a) (b) (c)

Figure 4. The correlation between crack depth (mm) and natural frequency (Hz) for (**a**)—Aluminum, (**b**)—ABS, (**c**)—Concrete.

As the crack depth and temperature increase in aluminum specimens, the amplitude also increases, consistent with the expected effect of reduced natural frequency resulting from a crack in the specimen. In contrast, ABS exhibits the opposite behavior: as crack depth increases, the natural frequency drops similarly to aluminum, but amplitude decreases instead of increasing, as shown in Figure 5. This discrepancy can be explained by the different temperature conditions during the experiments. The maximum temperature for aluminum specimens is 200 °C, well below the temperature threshold for any significant material transformations during the short test duration. In contrast, the maximum temperature for ABS samples is 70 °C, which is close to the glass transition temperature.

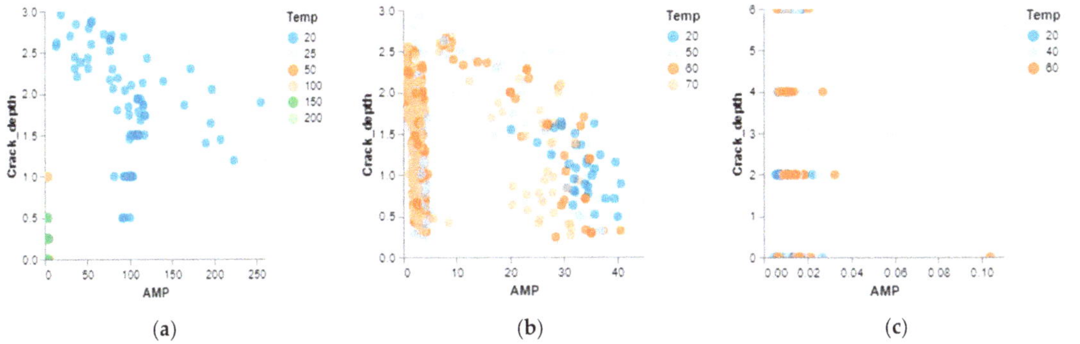

Figure 5. The correlation between Crack depth (mm) and amplitude (mm) for (**a**)—Aluminum, (**b**)—ABS, (**c**)—Concrete.

Figure 6 provides a more comprehensive view of the frequency-amplitude relationship in the experimental data, revealing an inverse relationship between frequency and amplitude in aluminum and concrete, while ABS shows a positive relationship. No clear relationship is evident between crack depth and position in ABS and concrete data, but in aluminum specimens, increasing crack position results in increased amplitude and decreased frequency.

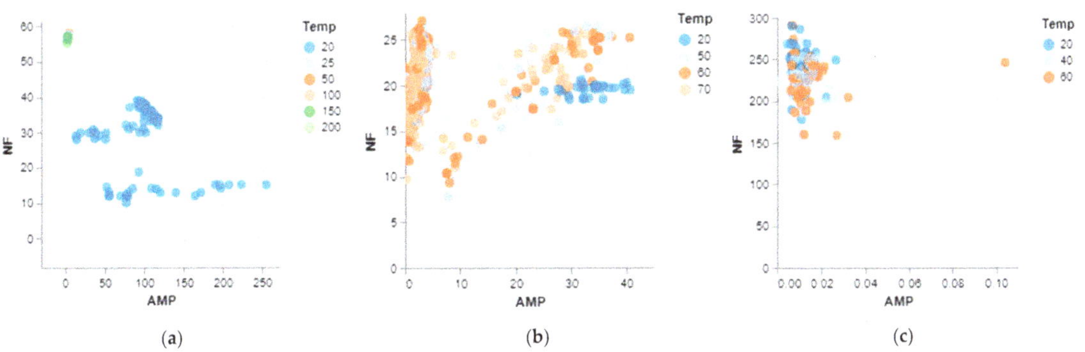

Figure 6. The correlation between natural frequency (Hz) and amplitude (mm) in for (**a**)—Aluminum, (**b**)—ABS, (**c**)—Concrete.

To enable a proper comparison and account for the diverse material behavior discussed earlier, three identical but independent models were trained and validated with unseen data. The proposed models were evaluated for training and validation accuracy using a mean squared error (MSE) metric. MSE is a commonly used statistical metric for evaluating the performance of regression models. MSE measures the average squared difference between the predicted values and the actual values of the target variable. For each data point in the test set, the model generates a predicted value based on the input features. The squared difference between the predicted value and the corresponding actual value is calculated. These squared differences are then averaged across all data points in the test set. MSE provides an estimate of the average squared error between the predicted values and the true values. It gives higher weight to larger errors due to the squaring operation. The value of MSE is always non-negative, with lower values indicating better performance. A value of 0 indicates a perfect fit, where the predicted values match the actual values exactly. In the context of the study, MSE was used as an evaluation metric to assess the performance of the machine learning models in predicting crack propagation.

Lower MSE values indicate that the model's predicted crack propagation values closely match the actual values, suggesting better predictive accuracy. It is worth noting that while MSE provides a useful measure of model performance, it is not the only metric available. Depending on the specific objectives of the study, other metrics, such as mean absolute error (MAE), R-squared (coefficient of determination), or domain-specific metrics, could also be considered to evaluate and compare the performance of the models. Essentially, these metrics represent the average predictive error of the models [1]. Table 2 summarizes the performance metrics for training and validation predictions for the three materials.

Table 2. The evaluation of training and validation for selected models.

Material	Learning Model	Training MSE	Validation MSE
Aluminum	K-Neighbors Regressor	0.189	0.079
	Lasso	0.180	0.15
	Ridge	0.170	0.14
ABS	K-Neighbors Regressor	0.195	0.086
	Lasso	0.195	0.182
	Ridge	0.194	0.182
Concrete	Gradient Boosting Regressor	4.100	0.597
	SVR	4.401	3.114
	Random Forest Regressor	3.903	0.509

ABS Models: The results show that the K-Neighbors Regressor model has a Training MSE of 0.195, indicating that, on average, the model's predictions are off by 0.195 squared units from the actual values in the training set. The Validation MSE of the same model is 0.086, indicating that the model's predictions are off by 0.086 squared units from the actual values in the test set. These values indicate that the K Neighbors Regressor model performs well on both the training and test sets, suggesting that it has not overfit the training data. The Lasso model's Training MSE is 0.195, and the Validation MSE is 0.182. Both values are lower than the MSE of the baseline model (0.437), indicating that the Lasso model outperforms the baseline. The results for the Ridge model show that it has a Training MSE of 0.194 and a Validation MSE of 0.182. These values are also lower than the MSE of the baseline model (0.437), indicating that the Ridge model outperforms the baseline. Overall, all three models (K Neighbors Regressor, Lasso, and Ridge) outperform the baseline model, suggesting that they have learned meaningful patterns in the data and can make reasonably accurate predictions. Figure 7 shows the actual against predicted ABS crack depth for selected machine learning models.

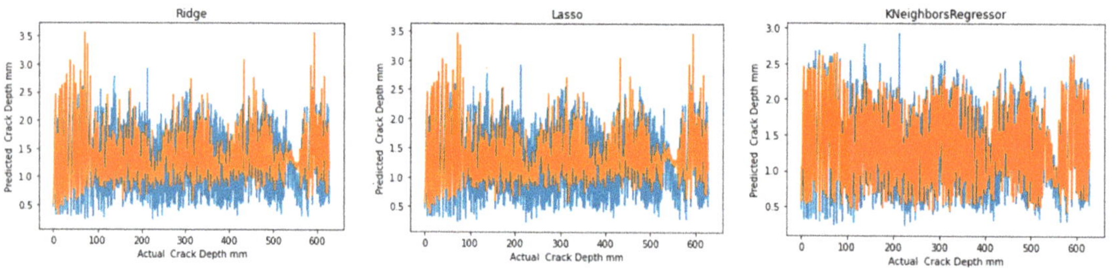

Figure 7. The actual against predicted ABS crack depth.

Aluminum Models: The results show that the K Neighbors Regressor model has a Training MSE score of 0.189 and a Validation MSE score of 0.079, which is lower than the MSE of the baseline model (0.8907), indicating that the K Neighbors Regressor model outperforms the baseline. The lower the MSE score, the better the performance of the model. The Ridge model has a Training MSE score of 0.170 and a Validation MSE score of 0.149, both of which are lower than the MSE of the baseline model. The Lasso model has a Training MSE score of 0.180 and a Validation MSE score of 0.159, which are also lower than the MSE of the baseline model. In all cases, the Validation MSE score is lower than the MSE of the baseline model, suggesting that the models have learned meaningful patterns in the data and can make reasonably accurate predictions. Figure 8 shows the actual against predicted Aluminum crack depth for selected machine learning models.

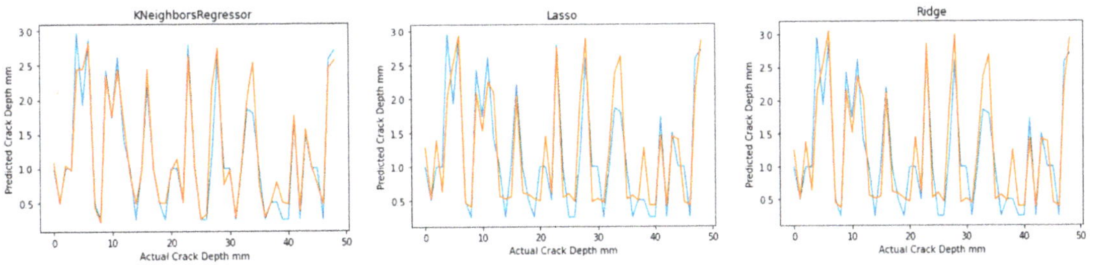

Figure 8. The actual against predicted Aluminum crack depth for selected machine learning models.

Composite Models: For all models (Gradient Boosting Regressor, SVR, and Random Forest Regressor), the Mean Squared Error (MSE) is used to evaluate their performance, which measures the average squared difference between the predicted and actual values. The lower the MSE, the better the performance of the model. The results show that the Gradient Boosting Regressor model has a Training MSE of 0.208 and a Validation MSE of 0.014, indicating that the model performs well on both the training and test sets and has not overfit the training data. The SVR model has a Training MSE of 0.182 and a Validation MSE of 0.094, indicating that the model's predictions are off by 0.182 and 0.094 squared units from the actual values in the training and test sets, respectively. The Random Forest Regressor model has a Training MSE of 1.649 and a Test MSE of 0.189. Although the Test MSE is lower than the baseline model, it is higher than the Test MSE of the other two models, suggesting that the model may have to overfit the training data. Overall, the Gradient Boosting Regressor and SVR models seem to be the best performers based on these results. They have the lowest Test MSE values and are less likely to overfit compared to the Random Forest Regressor model. Figure 9 shows the actual against predicted Concrete crack depth for selected machine learning models.

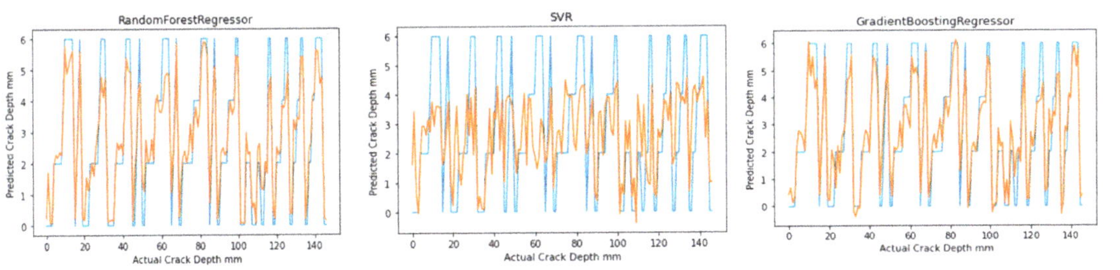

Figure 9. The actual against predicted Concrete crack depth for selected machine learning models.

The most important predictive features for aluminium are natural frequency, crack position, and temperature. For ABS and concrete, natural frequency and amplitude dominate, while the position of the crack appears to be of little importance. This may seem counterintuitive, as theory suggests that the closer the crack is to the fixed position of the specimen, the more pronounced the effect on dynamic response. However, the effect of crack position is captured by the model under the natural frequency and amplitude terms, resulting in a small and relatively unimportant crack position coefficient. Using all of the features as predictors yields the lowest mean squared error (MSE) but removing crack location does not significantly increase the error in ABS and concrete models. Removing amplitude has a much more significant effect on the error, increasing it by approximately 60% relative to the all-features baseline. The most significant effect is seen by removing natural frequency from the feature pool, which increases the error significantly. This suggests that natural frequency is the dominant feature in the prediction of crack depth, and its effect on the model error is greater than the rest of the features combined.

According to [30,31], the formal verification techniques mentioned further down provide a general framework for assessing the correctness, reliability, and robustness of machine learning models used in predicting crack propagation under coupled load and temperature.

1. Formal Specification: The first step in applying formal verification methods is to establish a formal specification that defines the desired properties or requirements for the crack propagation prediction models.
2. Model Checking: Model checking can be used to verify if the machine learning models satisfy the specified properties. In this context, it would involve exhaustively exploring the model's behavior under various coupled load and temperature conditions to check if the predicted crack propagation adheres to known physical laws or expected behavior.
3. Sensitivity Analysis: Formal verification methods can also include sensitivity analysis, which involves analyzing the impact of changes in input parameters or model assumptions on the predicted crack propagation. This analysis can provide insights into the robustness of the models and their sensitivity to variations in load, temperature, or other factors.
4. Statistical Testing: Statistical testing methods can be employed to evaluate the performance of different machine learning models. This involves comparing the predicted crack propagation results from different models using appropriate statistical tests to determine if there are significant differences in their performance or accuracy.

It is important to note that the application of formal verification methods to this specific paper would depend on the details of the research and the methodology employed.

The computational time and complexity associated with training algorithms exhibit variations contingent upon several factors, including dataset size and algorithmic implementation. When dealing with small to moderate datasets, linear regression techniques, such as Ridge and Lasso, demonstrate efficient training times, characterized by a complexity of $O(p^3)$, where p corresponds to the number of features. In contrast, the kNN algorithm, which is non-parametric in nature, requires minimal training time due to its reliance on storing data points. However, the computational expense arises during prediction when kNN necessitates distance calculations, resulting in a complexity of O (n d), with n referring to the number of training instances and d representing the number of features. Support Vector Regression (SVR), an algorithm based on support vector machines, generally exhibits reasonable training time for small to moderate datasets. Nevertheless, the process of solving the quadratic programming problem can become time-consuming for larger datasets. Random Forest Regressor, an ensemble algorithm, exhibits efficient training for small to moderate datasets with a complexity of $O(M \times n \times \log(n))$, where M denotes the number of trees and n represents the number of training instances. On the other hand, Gradient Boosting Regressor, another ensemble algorithm, typically entails higher computational requirements, especially when considering larger numbers of iterations. Consequently, its complexity generally surpasses that of Random Forests. The selected algorithms can typically achieve training within an acceptable timeframe for small and moderate datasets. However, the actual time and complexity involved are contingent upon specific dataset characteristics, implementation nuances, and the availability of computational resources [1,21–24].

4. Conclusions

In recent years, machine learning has emerged as a powerful tool for analyzing and predicting outcomes in various domains. One such application is the prediction of damage severity in materials, which holds great importance in structural health monitoring and maintenance. This paper presents a comprehensive study that harnesses machine learning techniques to predict crack depth in different materials, namely ABS, aluminum, and concrete.

The study utilizes experimental data obtained from previous research and employs a range of regression models, including K Neighbors Regressor, Lasso, Ridge, Gradient Boosting Regressor, and SVR, to forecast crack depth. Model performance is evaluated using Mean Squared Error (MSE) on both training and test sets. The findings demonstrate that all models surpass the baseline model, indicating their ability to capture meaningful patterns within the data and make reasonably accurate predictions.

Furthermore, the study identifies the most influential predictive features for each material. For instance, the impact of temperature on natural frequency exhibits a more pronounced effect in aluminum compared to concrete and ABS. This discrepancy can be attributed to the higher elastic modulus and greater consistency of isotropic sheet metal properties in aluminum relative to additive layer manufactured ABS and concrete. Moreover, the study reveals that all models outperform the baseline, implying their capability to learn significant patterns and facilitate accurate predictions.

The study offers valuable insights into predictive features and model performance in crack depth prediction, which have practical implications for specialists involved in structural health monitoring and maintenance. By employing machine learning algorithms, it becomes possible to reduce the time and cost associated with traditional testing methods while enhancing the accuracy and reliability of damage severity predictions.

It is worth noting that the training times for the discussed algorithms, such as Ridge, Lasso, kNN, SVR, Random Forest Regressor, and Gradient Boosting Regressor, are generally reasonable for small to moderate datasets. However, the actual time and complexity required for training may vary depending on dataset characteristics, implementation intricacies, and computational resources. It is essential to consider these factors when selecting and applying these algorithms to ensure efficient and effective model training.

Future studies could explore the adoption of more advanced machine learning algorithms or ensemble techniques to further improve the accuracy of crack depth prediction. Additionally, investigating the influence of additional features, such as material composition or environmental factors, could provide valuable insights into the mechanisms of crack formation and propagation. Furthermore, evaluating the models on different datasets and in real-world scenarios would contribute to assessing the generalizability of these models beyond the specific materials and conditions used in this study.

This study underscores the potential of machine learning techniques in predicting damage severity in materials. The results highlight the superiority of machine learning models in providing accurate and reliable predictions, which can significantly benefit structural health monitoring and maintenance practices. Further research is warranted to explore advanced techniques and the impact of additional features on predictive accuracy. Ultimately, the continued development and application of machine learning in this field could lead to remarkable advancements in structural engineering and maintenance.

Author Contributions: Writing—review & editing, I.O.; Supervision, M.K. and A.S. All authors have read and agreed to the published version of the manuscript.

Funding: This research received no external funding.

Conflicts of Interest: The authors declare no conflict of interest.

References

1. Omar, I.; Khan, M.; Starr, A. Suitability Analysis of Machine Learning Algorithms for Crack Growth Prediction Based on Dynamic Response Data. *Sensors* **2023**, *23*, 1074. [CrossRef] [PubMed]
2. Omar, I.; Khan, M.; Starr, A. Compatibility and challenges in machine learning approach for structural crack assessment. *Struct. Health Monit.* **2022**, *21*, 2481–2502. [CrossRef]
3. Avci, O.; Abdeljaber, O.; Kiranyaz, S.; Hussein, M.; Gabbouj, M.; Inman, D.J. A review of vibration-based damage detection in civil structures: From traditional methods to Machine Learning and Deep Learning applications. *Mech. Syst. Signal Process.* **2021**, *147*, 107077. [CrossRef]
4. Worden, K.; Manson, G. The application of machine learning to structural health monitoring. *Philos. Trans. R. Soc. A Math. Phys. Eng. Sci.* **2007**, *365*, 515–537. [CrossRef] [PubMed]
5. Fleet, T.; Kamei, K.; He, F.; Khan, M.A.; Khan, K.A.; Starr, A. A Machine Learning Approach to Model Interdependencies between Dynamic Response and Crack Propagation. *Sensors* **2020**, *20*, 6847. [CrossRef]
6. Hsieh, Y.-A.; Tsai, Y.J. Machine Learning for Crack Detection: Review and Model Performance Comparison. *J. Comput. Civ. Eng.* **2020**, *34*, 04020038. [CrossRef]
7. Khan, A.; Kim, N.; Shin, J.K.; Kim, H.S.; Youn, B.D. Damage assessment of smart composite structures via machine learning: A review. *JMST Adv.* **2019**, *1*, 107–124. [CrossRef]
8. Flah, M.; Nunez, I.; Chaabene, W.B.; Nehdi, M.L. Machine Learning Algorithms in Civil Structural Health Monitoring: A Systematic Review. *Arch. Comput. Method. Eng.* **2020**, *28*, 2621–2643. [CrossRef]
9. Yang, Z.; Yabansu, Y.C.; Al-Bahrani, R.; Liao, W.-K.; Choudhary, A.N.; Kalidindi, S.R.; Agrawal, A. Deep learning approaches for mining structure-property linkages in high contrast composites from simulation datasets. *Comput. Mater. Sci.* **2018**, *151*, 278–287. [CrossRef]
10. Khan, A.; Kim, H.S. Assessment of delaminated smart composite laminates via system identification and supervised learning. *Compos. Struct.* **2018**, *206*, 354–362. [CrossRef]
11. Wang, Z.; Chegdani, F.; Yalamarti, N.; Takabi, B.; Tai, B.; El Mansori, M.; Bukkapatnam, S.T. Acoustic Emission Characterization of Natural Fiber Reinforced Plastic Composite Machining Using a Random Forest Machine Learning Model. *J. Manuf. Sci. Eng. Trans. ASME* **2020**, *142*, 031003. [CrossRef]
12. Zhang, M.; Sun, C.-N.; Zhang, X.; Goh, P.C.; Wei, J.; Hardacre, D.; Li, H. High cycle fatigue life prediction of laser additive manufactured stainless steel: A machine learning approach. *Int. J. Fatigue* **2019**, *128*, 105194. [CrossRef]
13. Liu, H.; Liu, S.; Liu, Z.; Mrad, N.; Milani, A.S. Data-Driven Approaches for Characterization of Delamination Damage in Composite Materials. *IEEE Trans. Ind. Electron.* **2021**, *68*, 2532–2542. [CrossRef]
14. Ghassemi, P.; Rajabi, H.; Toufigh, V. Fatigue performance of polymer and ordinary cement concrete under corrosive conditions: A comparative study. *Eng. Fail. Anal.* **2020**, *111*, 104493. [CrossRef]
15. Liu, Y.; Zhao, K.; Hou, F.; Hao, X. Crack Extension Life and Critical Crack Length Prediction Based on XGBoost. 2022. Available online: https://ceur-ws.org/Vol-3304/paper11.pdf (accessed on 29 May 2023).
16. Isametova, M.E.; Nussipali, R.; Martyushev, N.V.; Malozyomov, B.V.; Efremenkov, E.A.; Isametov, A. Mathematical Modeling of the Reliability of Polymer Composite Materials. *Mathematics* **2022**, *10*, 3978. [CrossRef]
17. Yelemessov, K.; Sabirova, L.B.; Martyushev, N.V.; Malozyomov, B.V.; Bakhmagambetova, G.B.; Atanova, O.V. Modeling and Model Verification of the Stress-Strain State of Reinforced Polymer Concrete. *Materials* **2023**, *16*, 3494. [CrossRef]
18. Alshammari, Y.L.A.; He, F.; Khan, M.A. Modelling and Investigation of Crack Growth for 3D-Printed Acrylonitrile Butadiene Styrene (ABS) with Various Printing Parameters and Ambient Temperatures. *Polymers* **2021**, *13*, 3737. [CrossRef]
19. Kamei, K.; Khan, M.A.; Khan, K.A. Characterising Modal Behaviour of a Cantilever Beam at Different Heating Rates for Isothermal Conditions. *Appl. Sci.* **2021**, *11*, 4375. [CrossRef]
20. Kamei, K.; Khan, M.A. Investigating the Structural Dynamics and Crack Propagation Behavior under Uniform and Non-Uniform Temperature Conditions. *Materials* **2021**, *14*, 7071. [CrossRef]
21. He, F.; Ning, H.; Khan, M. Effect of 3D Printing Process Parameters on Damping Characteristic of Cantilever Beams Fabricated Using Material Extrusion. *Polymers* **2023**, *15*, 257. [CrossRef]
22. He, F.; Khan, M.; Aldosari, S. Interdependencies between Dynamic Response and Crack Growth in a 3D-Printed Acrylonitrile Butadiene Styrene (ABS) Cantilever Beam under Thermo-Mechanical Loads. *Polymers* **2022**, *14*, 982. [CrossRef] [PubMed]
23. Khalel, H.H.Z.; Khan, M.; Starr, A. Dynamic response-based crack resistance analysis of fibre reinforced concrete specimens under different temperatures and crack depths. *J. Build. Eng.* **2023**, *66*, 105865. [CrossRef]
24. Song, J.; Zhao, J.; Dong, F.; Zhao, J.; Qian, Z.; Zhang, Q. A Novel Regression Modeling Method for PMSLM Structural Design Optimization Using a Distance-Weighted KNN Algorithm. *IEEE Trans. Ind. Appl.* **2018**, *54*, 4198–4206. [CrossRef]
25. Chun, P.J.; Yamane, T.; Izumi, S.; Kuramoto, N. Development of a Machine Learning-Based Damage. *Sensors* **2020**, *20*, 2780. [CrossRef]
26. Bao, Y.; Tang, Z.; Li, H. Compressive-sensing data reconstruction for structural health monitoring: A machine-learning approach. *Struct. Health Monit.* **2020**, *19*, 293–304. [CrossRef]
27. Rafiei, M.H.; Adeli, H. A novel machine learning-based algorithm to detect damage in high-rise building structures. *Struct. Des. Tall Spéc. Build.* **2017**, *26*, e1400. [CrossRef]

28. Catbas, F.N.; Malekzadeh, M. A machine learning-based algorithm for processing massive data collected from the mechanical components of movable bridges. *Autom. Constr.* **2016**, *72*, 269–278. [CrossRef]
29. Chatterjee, S. A New Coefficient of Correlation. *J. Am. Stat. Assoc.* **2021**, *116*, 2009–2022. [CrossRef]
30. Krichen, M.; Mihoub, A.; Alzahrani, M.Y.; Adoni, W.Y.H.; Nahhal, T. Are Formal Methods Applicable to Machine Learning and Artificial Intelligence? In Proceedings of the 2022 2nd International Conference of Smart Systems and Emerging Technologies (SMARTTECH), Riyadh, Saudi Arabia, 22–24 May 2022; pp. 48–53.
31. Kang, E.-Y.; Huang, L. Formal Specification & Analysis of Autonomous Systems in PrCCSL/Simulink Design Verifier. 2018. Available online: http://arxiv.org/abs/1806.07702 (accessed on 12 June 2023).

applied sciences

MDPI

Article

Fault Type Diagnosis of the WWTP Dissolved Oxygen Sensor Based on Fisher Discriminant Analysis and Assessment of Associated Environmental and Economic Impact

Alexandra-Veronica Luca, Melinda Simon-Várhelyi, Norbert-Botond Mihály and Vasile-Mircea Cristea *

Department of Chemical Engineering, Faculty of Chemistry and Chemical Engineering, Babes-Bolyai University of Cluj-Napoca, 11 Arany János Street, 400028 Cluj-Napoca, Romania

* Correspondence: mircea.cristea@ubbcluj.ro

Featured Application: For automatically controlled WWTPs, the prompt DO sensor faults identification is essential, and the efficiency of the straightforward proposed methodology is shown. Barely revealed by the literature, the presented results also reveal the combined assessment of the impact on the environment and costs of different sensor failures. They are of great interest to researchers and practitioners seeking safe and optimal WWTP operation and provide a robust quantitative impact assessment methodology aimed at improving plant sustainability.

Abstract: Sensor failures are common events in wastewater treatment plant (WWTP) operations, resulting in ineffective monitoring and inappropriate plant management. Efficient aeration control is typically achieved by the dissolved oxygen (DO) control, and its associated sensor becomes critical to the whole WWTP's reliable and economical operation. This study presents the Fisher discriminant analysis (FDA) used for fault diagnosis of the DO sensor of a currently operating municipal WWTP. Identification of the bias, drift, wrong gain, loss of accuracy, fixed value, complete failure minimum and maximum types of DO sensor fault was investigated. The FDA-proposed methodology proved efficiency and promptitude in obtaining the diagnosis decision. The consolidated fault identification showed an accuracy of 87.5% correct identification of the seven faulty and normal considered classes. Depending on the fault type, the results of the diagnosing time varied from 2.5 h to 16.5 h during the very first day of the fault appearance and were only based on observation data not included in the training data set. The latter aspect reveals the potential of the methodology to learn from incomplete data describing the faults. The rank of the fault type detection promptitude was: bias, fixed value, complete failure minimum, complete failure maximum, drift, wrong gain and loss of accuracy. Greenhouse gases (GHGs) such as nitrous oxide (N_2O) and carbon dioxide (CO_2) emitted during wastewater treatment, electrical energy quantity in association with costs spent in the WWTP water line and clean water effluent quality were ranked and assessed for the normal operation and for each of the DO sensor faulty regimes. Both for CO_2 and N_2O, the on-site emissions showed the most significant GHG contribution, accounting for about three-quarters of the total emissions. The complete failure maximum, fixed value and loss of accuracy were the DO sensor faults with the highest detrimental impact on GHG-released emissions. The environmental and economic study reveals the incentives of the proposed DO sensor faults identification for the WWTP efficient and environmentally friendly operation.

Keywords: fault identification; Fisher discriminant analysis; dissolved oxygen sensor; energy costs assessment; GHG emissions assessment

Citation: Luca, A.-V.; Simon-Várhelyi, M.; Mihály, N.-B.; Cristea, V.-M. Fault Type Diagnosis of the WWTP Dissolved Oxygen Sensor Based on Fisher Discriminant Analysis and Assessment of Associated Environmental and Economic Impact. *Appl. Sci.* **2023**, *13*, 2554. https://doi.org/10.3390/app13042554

Academic Editors: Xiongbo Wan, Sheng Du, Wei Wang and Hao Fu

Received: 19 December 2022
Revised: 11 February 2023
Accepted: 14 February 2023
Published: 16 February 2023

1. Introduction

Wastewater treatment plants (WWTPs) are essential for eliminating pollutants from wastewater and converting it into clean effluent discharged in rivers or reused for different applications, including irrigation [1]. As treatment regulations on effluent quality are continuously tightening, the values of primary variables, such as organic matter, ammonia, nitrates, phosphorus and suspended solids, have to strictly conform to the challenging water quality requirements. The operation of WWTPs has proven to be tough and challenging. A detailed description of wastewater treatment processes, as well as the models associated with them, are extremely complicated, exhibit nonlinear behavior and are characterized by a large number of intensely changing variables. The most difficult challenges for WWTP management and control of the operation, aside from establishing the appropriate control system structure and determining optimal values for the operated variables, are the daily, weekly or seasonal influent composition fluctuations. Comparing control systems' performance in different plants is difficult and necessitates the creation of simulation standards and benchmarks [2]. Modeling water treatment processes has become a very useful tool for the design, optimization and automatic control of the WWTP, as the importance of the treatment plants has grown today within the circular economy concept, and they are presently considered water resources recovery facilities.

The International Water Association (IWA) is well-known for its vast contributions to discovering solutions and fighting against global water problems. IWA experts have created advanced phenomenon-based models to support the construction and control of sewerage treatment processes as part of the envisioned goal of improving standards for sustainable water management [3]. Activated Sludge Model #1 (ASM1) is the most common of these. It was created by the IWA (formerly IAWQ) to standardize nomenclature and set a milestone in wastewater treatment plant modeling [4]. ASM2, ASM2d and ASM3 have been developed over time with the scope of increasing the capability of revealing the intrinsic behavior of processes that occur within the activated sludge water processing [5]; they are currently being employed for sewage treatment plant design and control and are widely used for estimation or optimization [6]. Benchmark Simulation Model #1 (BSM1) was created to enlarge the ASM1 modeling with a defined plant configuration. It has become a standard tool and method for evaluating the performance of wastewater treatment plants. BSM1 has been expanded to Benchmark Simulation Model #2 (BSM2) as a more comprehensive plant description for long-term evaluation, coupled with extensions to primary sewage treatment plant subunits and processing of sludge by digestion in anaerobic conditions [7]. Both benchmarks have become powerful tools to support the development of various control strategies and the evaluation of their performance [8].

Chemical process monitoring is important for evaluating process performance and improving process efficiency and wastewater quality in wastewater treatment plants. The advantage of using mechanical or statistical mathematical models is to estimate the behavior of main key process parameters in different operating circumstances and, in association with real-time measurements, to use them for process design or operational improvement. Calibrated models are a necessary prerequisite for model-driven control solutions due to their high usage in advanced process automation. Supervision and control of water plant process variables are attained to varying extents and complexity depending on the wastewater inflow, the specificity of the bioreactor configuration, and the actual equipment available at each facility. Successful, accurate monitoring and control of processes requires reliable information on the composition, flow rate and temperature for both influents, recycle flows or effluents. Despite the system's high level of complexity, process monitoring and simple univariate fault detection methods proved critical to ensuring that the control system has access to consistent data [9]. Consequently, efficient process control is strongly relying on the capability of detecting sensor faults before the extension of their undesired effects [10]. Failures that are detected early on can help avert further induced undesired plant performance consequences and breakdowns. In general, process monitoring addresses four different tasks: (1) fault detection, which indicates that something is wrong in the process operation; (2) fault identification (or diagnosis), which establishes the cause of the issue; (3) fault estimation, which determines the magnitude of the defect; and (4) fault reconstruction, which calculates the fault-free process variable values aimed to perform the operation in the presence of faults [11]. Traditional fault detection and isolation methods use a mathematical model of the system to detect and isolate faults. In order to reveal discrepancies between the fault-affected and fault-free circumstances, these methods employ state estimation, parameter identification techniques, and parity relations [12–14]. However, developing precise mathematical models that characterize all physical and biochemical phenomena that occur in industrial processes is often challenging and time costly. When analytical models are not available, knowledge-based approaches such as expert systems might be considered as an alternative or supplement to analytical model-based approaches [15]. However, the development of these knowledge-based systems also necessitates a significant amount of time and work [16].

Online field-installed probes and offline laboratory analyses are used in traditional monitoring of the process variables. Real-time monitoring is difficult to be implemented, expensive and time-consuming. Field instruments require frequent and qualified maintenance, and the lack of satisfying these demands often makes the field measurements unreliable. On the other hand, time implied by laboratory measurements may range from a few minutes to several days to build a trustful assessment of the most standard wastewater monitoring metrics [17]. Both hardware and soft sensors can manifest a series of problems. While hardware sensors require a long time for maintenance and calibration, they exhibit insufficient accuracy and high noise levels. Their accuracy can quickly deteriorate in time and propose the support of soft sensors that can solve some of these issues. The soft sensors may also be subjected to faults as they rely on some sensor measurements and might depend on insufficiently accurate models [18]. As a result, detecting fault-affected measurements among the recorded data is critical for obtaining high WWTP effluent quality results. Sensor deficiencies and techniques utilized for detecting, amending and identifying faulty information or broken sensors were studied in a few works using different multivariate statistical process control (MSPC) methods, such as principal component analysis (PCA) [10,19–21], independent component analysis (ICA) [22,23], partial least squares (PLS) [24–26] or control charts [27,28]. Dynamic multiblock partial least squares (DMBPLS) was implemented to detect chemical oxygen demand (COD) bias and pH drifting sensor faults at a Chinese papermaking wastewater treatment plant [29], while a combination of t-distribution stochastic neighbor embedding with a Gaussian mixture model (t-SNE-GMM) was proposed for detecting bias, drift and complete failure sensor faults in a similar plant [30]. Because of their natural identification limit, MSPC techniques show high potential and are proficient in observing the sensor deficiencies occurring in time-varying, poorly characterized and nonlinear behavior framework of measurements in wastewater treatment plants [31–33]. MSPC methodologies are most frequently used and are superior to statistical process control (SPC) methods since they straightforwardly consider and use idle factors, prompting effective issue identification [34]. Another study combined sub-period division strategies with multiway principal component analysis for the fault diagnosis on a sequence batch reactor of the wastewater treatment process in a paper mill [35].

FDA is a widely used pattern classification technique [36], and its application to chemical process data analysis has continuously increased in the last two decades [37–40]. FDA application to faulty sensors was studied for air handling units [41]. The target of the FDA method is to determine the Fisher optimal discriminant vector that maximizes the Fisher criterion function. Fisher discriminant analysis uses a linear combination of features to distinguish between two or more classes in an optimal way. It is an empirical method based on observed characteristics over a large number of cases. FDA for fault identification provides the best lower dimensionality representation in terms of a discriminant between data classes, where each class corresponds to data acquired during a specific and known fault. Unlike PCA, which is looking for directions that are effective for representation, FDA is looking for directions that are effective for discrimination. From a theoretical standpoint, FDA has advantages for fault visualization and diagnostics [39].

In general, the economic efficiency and environmental friendliness of sewage treatment plants are primarily based on the process of removing nitrogen from pollutants by biologically catalyzed oxidation using aeration control. This control is primarily sustained by measuring the content of dissolved oxygen (DO) with dedicated sensors. The effectiveness of an aeration control system can be significantly impacted by wrong information that emerges from the dissolved oxygen sensors, resulting in unexpected degradation of the system functionality and problems throughout the wastewater treatment process. Negative consequences in terms of deterioration consist of a decrease in the quality of the effluent, an increase in energy consumption, a decrease in environmentally sustainable performance or even a temporary shutdown of the plant. Few studies have addressed the problem of DO sensor failure. In typical research, the PCA approach was used to detect the set of three categories of faults [42]. Another study examined single-type DO and level sensor failures [28]. The wrong output signal from the sensor was analyzed in another work [19], and a study was devoted to detecting the clogging bias of the sensor based on PCA [20]. The variety of the detection tools was extended to element recognition, neural networks with radial transfer functions [43], binary classification instruments [44], approaches based on impulse response [45] or deep dropout neural networks (MC-DDNN) to identify incipient faults of sensors installed in wastewater treatment plants [46]. In a previous investigation, the current authors developed a study based on the PCA-based methodology that proposed detection solutions for determining the presence of inappropriate functioning of the dissolved oxygen sensor [47].

The motivation and contributions of the present work originate from several reasons. The reported fault identification studies referring to the DO sensor defects were typically focused on a single or on a limited number of fault types. Extending the set of fault types embedded in a single diagnosing tool hinders the efficiency of the fault categorization but makes it most appreciated. The DO sensor fault diagnosing in a controlled A^2O configured WWTP, where the sensor is implied in the automatic operation associated with the nitrites and nitrates concentration control loop, was not specifically reflected in the literature. The literature presenting the impact of different DO sensor fault types on the WWTP energy, effluent water quality and GHG emissions performance is still lacking, although they have become of very high interest. To the best of the authors' knowledge, no WWTP single study of fault detection has been conducted on six different sensor fault types combined with the assessment of associated environmental and economic impact.

The goal of the present paper is to use multivariate statistical methods to construct a fault diagnosis method for the wastewater treatment facility. This research was intended to explore efficient diagnostic ways for specific defects caused by faulty DO sensors and to assess the environmental and cost impact of the faults. Six different fault types were considered: bias, drift, wrong gain, loss of accuracy, fixed value or complete failure. The original contribution of the paper addresses the comprehensive and comparative disclosure of the DO sensor fault types and evaluation of the fault identification benefits for the automatically controlled treatment plant. Furthermore, for the municipal WWTP case study, the energy costs and environmental impact on the treatment plant performance were assessed for the various defects of the DO sensor.

The structure of the present work presents (i) the dynamic model of the WWTP used for simulations, (ii) the basic theory underpinning the FDA fault-diagnosis methodology for the identification of the sensors' faults and (iii) the presentation of the equations used to compute the GHGs released by the fault affected WWTP process; then, the (iv) results and discussions of the performance are obtained by the FDA approach and the environmental and economic study, and the paper ends with (v) the conclusions of the research for improving the performance of the case study WWTP.

2. Materials and Methods

2.1. Process Model

The sewage treatment plant considered in this work has an anaerobic–anoxic–aerobic (A^2O) arrangement. The sewage enters the treatment plant with a flow rate of about 115,000 m^3/day, and it undergoes a series of physical treatments: mechanical filtration, separation of sand and grease, and the first sedimentation step. Water exits the first clarification step and enters the bioreactor tank. Here, activated sludge technology significantly reduces the concentration of carbon, nitrogen and phosphorus compounds. Three different zones characterize the biodegradation basins. The first one is anaerobic and has a capacity of about 9000 m^3; the second is anoxic and has a volume of around 12,800 m^3; the third is aerobic and comprises a total volume of approximately 33,000 m^3. Biological phosphorous removal is accomplished in the first bioreactor, and the transformation of nitrates and nitrites (NO) produced in the aerobic reactor occurs in the second bioreactor [48]. Physical separation takes place in a secondary clarifier. The clean water is discharged to the emissary river as purified water. A small fraction of 0.5% of the secondary clarifier bottom sludge flow rate is directed to the plant unit for digestion. The bulk is returned to the anaerobic degradation tank as external recycling. A second recycle flow, called internal recycling, recycles nitrate from the aerated bioreactor to the anoxic bioreactor for denitrification [49,50].

The WWTP model constructed and developed in this study was built on the foundations of BSM1 and ASM1. To comply with the A^2O configuration, size and operation characteristics of the case study targeted municipal WWTP, the appropriate modifications were made to the original BSM1. Each of the basic structural parts of the WWTP was described by a set of differential and algebraic equations: primary settler [51], anaerobic bioreactor, anoxic bioreactor, the set of 3 aerated bioreactors [51] and secondary settling unit [52]. Design and dry weather operation data were collected and reconciled from municipal WWTP measurements. The plant data were further used to update and calibrate the modified BSM1 model [53]. The model parameters were calibrated using optimization techniques. This previously calibrated model of the municipal sewage treatment plant was used in the current work to carry out the scenarios of simulations for the normal and faulty sensors. The model was implemented in MATLAB version 8.5 (MathWorks, Natick, MA, USA) and Simulink^{TM} version 8.5 software (MathWorks, Natick, MA, USA). In order to speed up the simulation of Simulink S-functions and save computational resources, the mathematical model was written in the C++ programming language and compiled as a MATLAB executable.

Automatic control is widely regarded as critical for the efficient and safe operation of the WWTP, and the prevention of pollution spread into freshwater ecosystems [54,55]. When associated with optimization, control generates a highly valuable synergy [56,57]. This study employed two typical automatic control structures: one is the dissolved oxygen control loop, and the other one is devoted to the control of nitrates and nitrites concentration. Both control loops used proportional–integral (PI) controllers. The DO concentration in the aerated bioreactor was controlled by the first loop, and the nitrate and nitrite concentrations in the anoxic bioreactor were controlled by the second loop. For the DO concentration control loop, the DO reference of 2 mg O_2/L was enforced in the aerobic bioreactor in order to prevent excessive aeration [58]. The DO controller manipulates the air control valves to supply the necessary oxygen flow rate into the aerated reactors. In the sequence of three aerobic reactors (No. 3, 4 and 5) of the modified BSM1 plant configuration, the airflow calculated by the oxygen controller was unevenly distributed. The following weighting factors for the DO controller generated airflow rate were used: 0.9 for the first aerobic (reactor no. 3), 0.5 for the second aerobic (reactor no. 4) and 0.3 for the third aerated bioreactor (reactor no. 5). The oxygen requirements for the nitrification bioreactors were used to set this distribution strategy, according to the commonly used practice emerged approach. The concentrations of nitrate and nitrite in the anoxic reactor (reactor #2) were controlled via a second control loop by adjusting the recycle flow of nitrates and nitrites. This control loop had a reference of 0.01 mg N/L. The overall configuration of the WWTP reactors, settlers and the main control loops are shown in Figure 1.

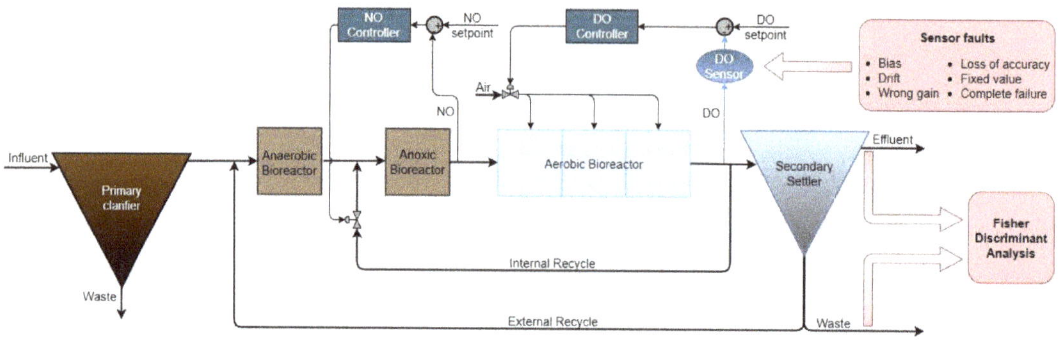

Figure 1. WWTP general structure.

The main parameters of the plant are presented in Table 1. They correspond to the typical structure and actual operating configuration of plant equipment and influent wastewater of the municipal WWTP of Cluj-Napoca, Romania.

Table 1. Municipal WWTP main equipment parameters.

Equipment	Parameter	Value	Measurement Unit
Primary settler	Area	2125	m^2
	Height	3.5	m
Anaerobic bioreactor	Volume	9015	m^3
Anoxic bioreactor	Volume	12,678	m^3
Aerated bioreactors	Volume	33,066	m^3
	Area	6012	m^2
Secondary settler	Area	11,304	m^2
	Height	3	m

Data for both the normal state and faulty state of operation were generated by simulation for each of the specifically designed cases (types) of DO sensor failure. Corresponding data to every individual type of fault and of normal operation were separated into classes and subjected to global Fisher discriminant analysis in order to obtain a distinct class representation of the high-dimensional data.

2.2. Sensor Faults

Lack of sensor functioning can be caused by a variety of factors, including a signal of doubtful quality [59], a bad electrical connection or sensor failure [60]. The diagnosis of different DO sensor malfunction types is of critical importance for the overall WWTP operation. The sensor defects investigated in the present work are:

- Bias—caused by the miscalibrated sensor, is a change in the sensor-generated signal, such as the supplied value being altered typically by a constant value in addition to the actual sensor signal [59].
- Drift—characterized by the continuous time-varying divergence of the sensor signal from the genuine one [59].
- Wrong gain—happens when the sensor slope is affected by an inappropriate gain factor, frequently determined incorrectly during the calibration process [61].
- Loss of accuracy—occurs when the signal of the sensor changes randomly, and its value is imprecise when it is compared to the genuine value [59].
- Fixed value—the sensor displays the same value all the time [61].
- Complete failure (with two cases, minimum and maximum)—characterized by a measured value that is equal or close to the sensor minimum calibration limit or is equal or close to the maximum calibration value [61].

The majority of the simulated fault types originate from the WWTP operating practice, and they were chosen to follow the typical behavioral pattern. For the loss of accuracy type of fault, the random scenario used for the faulty sensor signal was chosen to roughly reflect an irregular change in the true signal. For other magnitudes of the bias, drift, wrong gain and loss of accuracy types of faults, where fault samples are not directly available from the actual WWTP measurements, the use of the calibrated dynamic model can be used for generating data to be embedded in the FDA fault identification methodology. As new faults and magnitudes show up, the database of faults behavior can be completed by the plant measurements, and the diagnosing methodology becomes adaptive.

2.3. Simulation Methodology of the Faults

Specific software modules were created to simulate the behavior of the 6 DO concentration sensor errors. These are integrated into the dynamic WWTP simulator. For each error type, the error signal produced by the sensor is given to the proportional–integral oxygen controller as the measured oxygen process variable. The DO control loop adjusts the airflow accordingly, returning back the controlled DO to the desired reference of 2 mg O_2/L without offset. For each failure, the scenario was run as follows. The simulation was carried out without any fault for 139 days. Then, each fault was implemented starting from the 140th day of the simulations and lasted for a period of 28 days. The process variables taken into consideration for the fault identification were considered from the first day of faulty operation, i.e., day 140 [47]. The same type period of measurements for 28 days was considered for collecting data that describe the normal operation.

The DO sensor uses a galvanic measurement technique. The considered scenarios for the faults were generated, such as to imitate the real DO sensor's steady state and dynamic behavior.

For the bias type of fault, the DO sensor signal was considered to have a bias of +1.5 mg O_2/L added to the actual DO process value. It simulates a constant difference between the true DO value and the DO signal produced by the defective air blast assembly of the sensor.

To mimic the drift, a constant value of 0.05 mg O_2/L was integrated in time, and it supplemented the DO genuine value. This fault simulates a defect in the electronic circuits of the transmitter, produced by a time-increasing parasite electric capacitance.

The sensor's wrong gain type of fault was considered by an incorrect gain factor of 1.4, which consists of an incorrect relationship between the actual DO process value and the sensor output. This fault type is potentially generated by a wrong calibration of the sensor or by calibration sudden change due to internal membrane deterioration. The first order filter with a time constant of 0.3 days was used to smoothly introduce the faulty gain over time for passing from normal to fault-affected operation.

The defective measured process variable for the case of loss of accuracy type of fault was generated by adding to the DO true value a random signal value from the interval −2.5, 2.5. Such sensor fault can originate from the partial dislocation of the anode of the cathode electrodes of the sensor. Each random sample value had a duration of 0.1 days.

For the fixed value type of fault, the constant of 2.2 mg O_2/L was used. This sensor defect can be produced due to the leak of the sensor filling solution.

For the complete failure minimum and maximum types of faults, the two very low and very high values of 0.1 mg O_2/L and 6 mg O_2/L were considered. They correspond to the practical circumstances when the electrical supply or parts of the transmitter electronic circuits are malfunctioning or due to the defective self-cleaning assembly, which leads to sensor clogging.

2.4. Fisher Discriminant Analysis

Fisher discriminant analysis is a pattern categorization approach that features a very efficient classification potential. FDA's main goal is to determine the Fisher optimal discriminant vector that maximizes the Fisher criterion function. The higher-dimensional feature space of process measurements can be projected onto the obtained optimal discriminant vector space for constructing a lower-dimensional feature space. Let $X \in R^{n\times m}$ be the matrix that contains the training data for all classes. The total set of n observations for the m measured variables that build the X matrix contains the submatrix X_i as the subset of measurements consisting of n_i rows and corresponding to the class i of samples. $\overline{x_i}$ denotes the m-dimensional sample mean vector for the class i and is given by:

$$\overline{x_i} = \frac{1}{n_i} \sum_{x_j \in X_i} x_j \tag{1}$$

with x_j the set of vectors that belong to class j, then the within-class scatter matrix is defined by:

$$S_w = \sum_{i=1}^{c} S_i \tag{2}$$

where c is the number of classes, and

$$S_i = \sum_{x_j \in X_i} (x_j - \overline{x_i})(x_j - \overline{x_i})^T \tag{3}$$

is the within-scatter matrix for class i.

The between-class scatter matrix is then defined by:

$$S_b = \sum_{i=1}^{c} n_i \, (\overline{x_i} - \overline{x})(\overline{x_i} - \overline{x})^T \tag{4}$$

where $\overline{x}$ is the total mean vector of all means of the columns of X.

The optimal discriminant direction is found by maximizing the Fisher criterion:

$$J(\varphi) = \frac{\varphi^T S_b \varphi}{\varphi^T S_w \varphi} \tag{5}$$

where the maximizer φ is the Fisher optimal discriminant direction that maximizes the ratio of the between-class scatter to the within-class scatter. It may be shown that a vector φ that maximizes $J(\cdot)$ must satisfy the equation:

$$S_b \varphi = \lambda S_w \varphi \tag{6}$$

for some constants, λ indicates the separability between classes. If S_w is nonsingular, it is obtained as a conventional eigenvalue problem, described by the following expression:

$$S_w^{-1} S_b \varphi = \lambda \varphi \tag{7}$$

The total-scatter matrix is given by the sum of S_b and S_w:

$$S_t = S_b + S_w \tag{8}$$

If data vector x_j from the m-dimensional space is reduced to the a-dimensional space of the FDA vectors, then its linear transformation is given by:

$$z_i = W_a^T x_j \tag{9}$$

where W_a^T has the a FDA vectors as columns and $z_i \in R^a$.

To diagnose the faults, FDA examines observed data collected under various faults and uses a discriminant function that assesses the similarity between the current data and the data belonging to each class. When the maximum discriminant function value, g_i, satisfies the following conditions, the observation is allocated to the class i:

$$g_i(x) > g_j(x),\ \forall j \neq i \tag{10}$$

$g_i(x)$ is the discriminant function given by a measured vector x for class i, and $g_j(x)$ is the discriminant function given by the measured vector x for class j. The discriminant function can be calculated, for each class i, with the following equations:

$$g_i(x) = -\frac{1}{2}(x - \overline{x_i})^T W_a \left(\frac{1}{n_i - 1} W_a^T S_i W_a\right)^{-1} W_a^T (x - \overline{x_i}) + lnP_i - \frac{1}{2}\, ln\left[det\left(\frac{1}{n_i - 1} W_a^T S_i W_a\right)\right] \tag{11}$$

where P_i is the a posteriori probability of x to belong to class i, and it is computed by the ratio between the number of observations from a class and the total number of observations for all classes. FDA was implemented using the Statistics and Machine Learning Toolbox version 10.0 of Matlab.

2.5. Assessment of the GHGs Impact

Conventional wastewater treatment facilities purify wastewater and reduce water pollution, but they also discharge GHGs into the air through direct emissions, and as they require a considerable amount of energy to process the influent, they also indirectly contribute to the release of GHGs in the atmosphere. Biological wastewater treatment using activated sludge technology removes organic matter and N and P nutrients from wastewater in an effective manner. Comprehensive environmental implications of the WWTP's contribution to GHG emissions have been highlighted in different studies. Greenhouse gas emissions have been identified as a key negative impact of the WWTP operation and have been studied in several works [62–64]. Other studies offer a complex image of the total environmental impact of a wastewater treatment plant through life cycle assessments [65–67]. However, no studies have looked at the environmental impact of the wastewater treatment plant when anomalies occur in the process due to sensor fault inappropriate operation. This study emphasizes the differences, with respect to both the environmental impact and the energy costs, between the plant performance when it is operated normally and when its operation is affected by faulty operation of the dissolved oxygen sensor.

Overall impact assessment of the WWTP disturbed operation on plant performance was performed using a cumulative performance index composed of three different performance indices: aeration energy (AE), pumping energy (PE) and effluent quality (EQ) [50]. The oxygen mass transfer coefficient of the aerobic bioreactors (K_La), which is directly related to the airflow rate, is used to compute the aeration energy index, as stated in Equation (12).

$$AE = \frac{SO_{sat}}{T \cdot 1.8 \cdot 1000} \cdot \int_0^T \sum_{aerated\ reactors} V_{bioreactor} \cdot K_La_i(t)dt \tag{12}$$

where SO_{sat} is the oxygen saturation concentration (mg O_2/L), T is the time in days of faulty operation, $V_{bioreactor}$ represents the volume of the bioreactor (m^3) and K_La_i is the mass transfer coefficient in the aerated bioreactor i.

The pumping energy index is calculated using the flow rates of nitrate recirculation, return-activated sludge recycling and waste, as shown in Equation (13). The energy used for aeration and pumping is computed in kWh per day. The effluent quality index is determined by a weighted sum of total suspended solids (*TSS*), chemical oxygen demand, biochemical oxygen demand (*BOD*), total Kjeldahl nitrogen (*TKN*), and nitrates and nitrites concentrations in the effluent flow stream, as indicated by Equation (14). The effluent quality is expressed in kilograms of pollutant units per day.

$$PE = \frac{1}{T} \cdot \int_0^T [0.004 \cdot Q_{NR}(t) + 0.08 \cdot Q_{RAS}(t) + 0.05 \cdot Q_{waste}(t)]dt \tag{13}$$

where Q_{NR} is the flow rate of the nitrate recirculation (m^3/day), Q_{RAS} is the flow rate of the return-activated sludge (m^3/day) and Q_{waste} is the flow rate of waste from the secondary settler (m^3/day).

$$EQ = \frac{1}{T \cdot 1000} \cdot \int_0^T [PU_{TSS}(t) + PU_{COD}(t) + PU_{BOD}(t) + PU_{TKN}(t) + PU_{NO}(t)] \cdot Q_{effluent}(t)dt \tag{14}$$

where PU_{TSS} denotes total suspended solids, PU_{COD} refers to the chemical oxygen demand and PU_{BOD} to the biochemical oxygen demand, PU_{TKN} considers the Total Kjeldahl Nitrogen, PU_{NO} accounts for the nitrates and nitrites, and $Q_{effluent}$ is the effluent flow rate.

Total GHG emissions generated by the water line of the municipal wastewater treatment plant are composed of both on-site and off-site emissions and consist of CO_2 and N_2O gases. Besides CO_2, N_2O is also considered an important contributor to the GHGs, as it has a global warming potential (GWP) of about 265–298 times higher than CO_2, with an average residence time of 100 years [68].

Off-site CO_2 emissions (kg CO_2/day) include indirect CO_2 emissions from the electrical power generation plant that are associated with the electrical energy consumed at the WWTP. They are described by:

$$P_{CO_2,\ off-site} = k_{PG} \cdot e_D \tag{15}$$

where k_{PG} is the site-specific emission factor per unit of energy generated, considered with a value of 0.19 kg CO_{2e}/kWh, and e_D is the total energy demand, calculated as the sum of the aeration energy and pumping energy [64,69].

Off-site N_2O emissions include N_2O that results from biological degradation in the effluent (downstream) of the wastewater treatment plant [64,70]:

$$P_{N_2O,\ off-site} = N_{effluent} \cdot EF_{effluent} \tag{16}$$

where $N_{effluent}$ is the nitrogen load in the effluent discharged into aquatic environments and $EF_{effluent} = 0.005 \cdot 44/28$ kg N_2O/kg N [71] is the emission factor for N_2O emissions from the discharged wastewater.

The on-site CO_2 emissions emerging from the water line of the aerobic biological processes are computed by the following expression:

$$P_{CO_2,on-site} = Q_{influent} \cdot 0.99 \cdot (1 - Y_H) \cdot \eta_{ASP} \cdot bCOD + Q_{influent} \cdot 1.03 \cdot Y_H \cdot \eta_{ASP} \cdot bCOD \cdot \frac{k_{d,H} \cdot MCRT}{1 + k_{d,H} \cdot MCRT} \tag{17}$$

where $Q_{influent}$ is the plant influent flow rate (m^3/day); 0.99 kg CO_{2e}/kg COD is the emission factor related to organic compounds; Y_H is the heterotrophic biomass yield (massVSS/massCOD) [2]; η_{ASP} is the biodegradable COD (bCOD) removal in the activated sludge reactors; 1.03 kg CO_{2e}/kg COD is the emission factor related to activated sludge biomass; $k_{d,H}$ is the decay rate of heterotrophic biomass and has a value of 0.3 day^{-1} [2]; and MCRT is the mean cell retention time, which is 15 days for this case [64,72].

The on-site N_2O emissions from the water line can be estimated using the following relationship:

$$P_{N_2O,on-site} = Q_{influent} \cdot (TN_{in} - TN_{out}) \cdot r_{N_2O} \tag{18}$$

where TN_{in} represents the total nitrogen from the influent (kg N/m^3), TN_{out} is the total nitrogen in the effluent (kg N/m^3) [73] and r_{N_2O} is the emission rate of N_2O (kg N_2O/kg N) [74].

3. Results and Discussion

3.1. Normal and Abnormal Operation Data Sets

The FDA methodology relied on 17 WWTP process variables. They were the bottom effluent sludge concentrations (10 variables), secondary settler clean effluent concentrations (six variables) and temperature. The set of these variables consisted of total nitrogen (N_{total}); total Kjeldahl nitrogen; chemical oxygen demand; free and saline ammonia (S_{NH}); nitrate and nitrite nitrogen (S_{NO}); total suspended solids (TSS); slowly biodegradable substrate (X_S); heterotrophic biomass ($X_{B,H}$); autotrophic biomass ($X_{B,A}$); inert particulate products (X_P); particulate biodegradable organic nitrogen (X_{ND}); soluble, biodegradable organic nitrogen (S_{ND}); readily biodegradable substrate (S_S); alkalinity (S_{alk}); and temperature (T). The first six of them characterized the secondary settler clean water, while the last eleven variables described the bottom effluent.

Eight separate scenarios were created, and simulations were performed to generate the data sets for the different fault classes, one for normal functioning and six for malfunctioning of the DO sensor. The DO and NO-controlled WWTP were simulated for 168 days of operation. The starting set of 139 days of nominal (faults lacking) functioning was considered to bring the plant to a quasi-steady state. Sensor faults were applied from day 140 in the simulation scenario. Operational data of the simulated wastewater treatment plant were collected with a sampling time of 15 min. Data generated in the time period from the 141st to the 145th day of DO sensor normal and faulty operation were used for training the FDA model. Each set of 480 observations is considered to be a class. They formed the training matrix of observations (3840 lines and 17 columns). The fault diagnosis performance of the trained FDA model was tested using data corresponding to the 140th day, i.e., the first day of faulty sensor operation. The testing data set contained 96 measurements for each fault. This testing approach was designed in order to investigate the capability of the FDA diagnosis method to identify the type of fault in the very first hours following the fault appearance.

The chosen scenarios and the emerged dimensions of the data sets used for training and validation of the fault detection methodology were considered to make a fair trade-off between diagnosis accuracy and required computational resources.

3.2. Fault Diagnosis

The values obtained for the discriminant function of each class, g_i, were compared in order to diagnose the flawed sensor operation. The discriminant function with the highest value indicated the class of the faulty sensor, and consequently, it diagnosed the fault. The values of the discriminant functions g_i were computed for each of the 15 min time-sampled measurements of the testing day no. 140, affected by the different types of faults (seven classes) and for measurements corresponding to normal operation (one class). They are presented in Figures 2–9.

Each of these figures presents the values $g_i(x)$, $i = 1$ to 8, i.e., the values of the FDA discriminant functions associated with the eight classes of the normal and of the seven fault types. For every sampling time moment of the 140th day (considered as multiples of the 15 min sample time), the eight discriminant function values were computed and presented with a particular line and marker. The vector of observations x, as the independent variable of the discriminant functions $g_i(x)$, consists of the set of values of the features at a particular sampling time moment. Separately, each of the Figures 2–9 shows the values of the discriminant functions at all considered validation sampling time moments and the corresponding vectors of observations obtained for only one of the normal or of the seven investigated fault types cases. In each figure, the discriminant function with the permanent highest value identifies the type of fault indicated by its associated class.

In order to reveal the moment of time within the first day after the fault appearance, when the fault is firmly and permanently diagnosed by the maximum discriminant function, a magnified representation of the time interval in the vicinity of this moment is presented in the special inserted detailed graphical representation.

As may be observed in Figure 2, all of the g_i values confirmed the normal operation (faulty-free values) diagnosis after 16.5 h of the fault appearance.

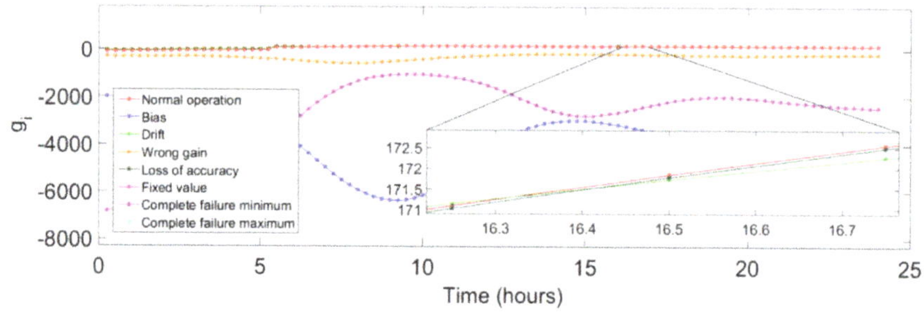

Figure 2. Normal operation diagnosis: graphs of the FDA discriminant functions $g_i(x)$ for each of the normal and seven fault classes, along the 24 h of the 140th day and the detailed representation revealing the class of the observations and the time moment of the firm normal operation identification.

It is worth noticing that for the first part of the testing day (no. 140) when sensor faults are not yet fully developed, the discrimination between normal operation and faults is less evident.

In Figure 3, the bias fault type diagnosis is performed after 2.5 h, for a time period of 5.75 h. After that period, the automatic controlled WWTP operation is driven progressively (due to the integral component of the PI controller) to low aeration and to values of the process variables that are becoming similar to those corresponding to the maximum values of the DO sensor and, as a result, the FDA discrimination is revealing the complete failure maximum class for the last part of the day.

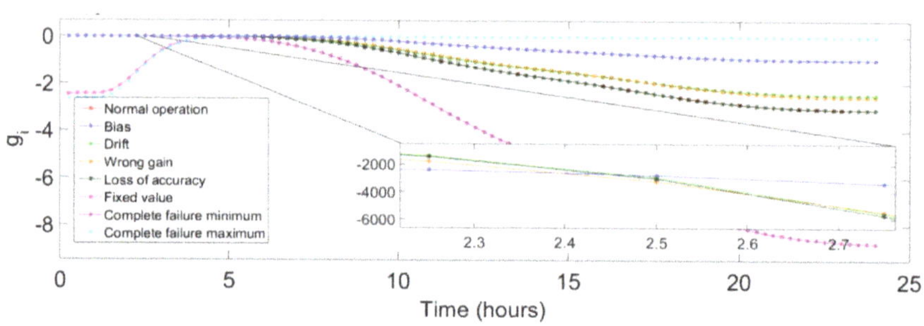

Figure 3. Bias diagnosis: graphs of the FDA discriminant functions $g_i(x)$ for each of the normal and seven fault classes, along the 24 h of the 140th day and the detailed representation revealing the class of the observations and the time moment of the firm fault type identification.

Drift diagnosis is firmly confirmed after 13.75 h, as it is shown in Figure 4. Drift fault effects are growing in time due to the intrinsic nature of this fault.

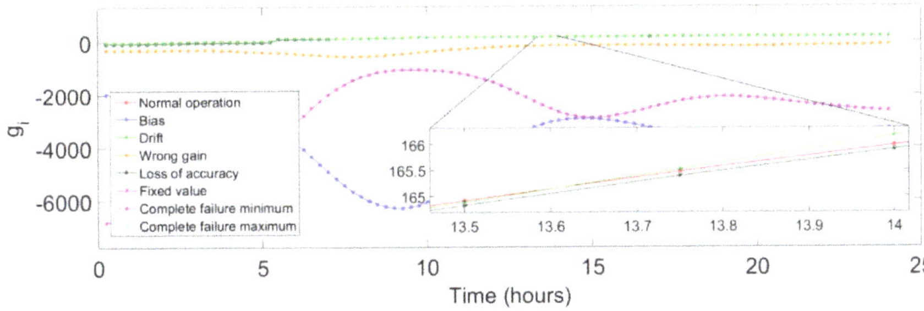

Figure 4. Drift diagnosis: graphs of the FDA discriminant functions $g_i(x)$ for each of the normal and seven fault classes, along the 24 h of the 140th day and the detailed representation revealing the class of the observations and the time moment of the firm fault type identification.

The diagnosis of wrong gain fault was confirmed after 14 h of the fault incidence moment. The graphical representation is shown in Figure 5. As the wrong gain fault was introduced by a time lag constant, the identification was relatively promptly achieved.

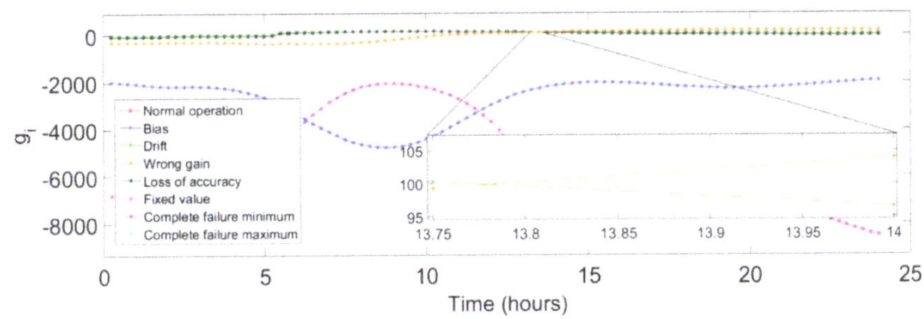

Figure 5. Wrong gain diagnosis: graphs of the FDA discriminant functions $g_i(x)$ for each of the normal and seven fault classes, along the 24 h of the 140th day and the detailed representation revealing the class of the observations and the time moment of the firm fault type identification.

The loss of accuracy fault was identified after 16.5 h of the fault start moment of action, as it is presented in Figure 6. Despite the irregular character of this fault, determined by the random component of the simulated faulty signal, the permanent discrimination type of fault was well achieved.

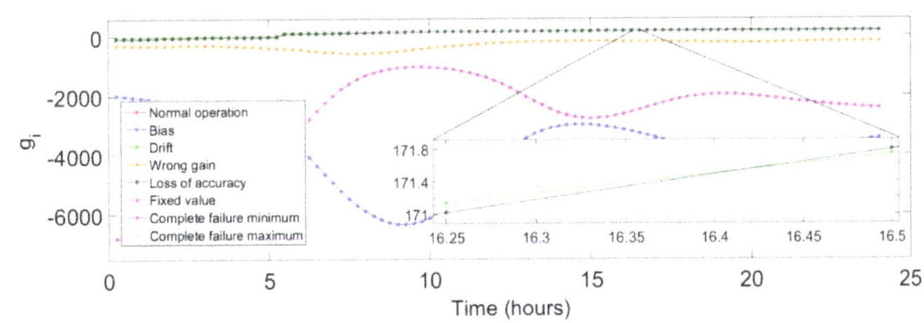

Figure 6. Loss of accuracy diagnosis: graphs of the FDA discriminant functions $g_i(x)$ for each of the normal and seven fault classes, along the 24 h of the 140th day and the detailed representation revealing the class of the observations and the time moment of the firm fault type identification.

Figure 7 shows the graphical representation of the fixed value fault, which was correctly and promptly identified after 6 h of the fault appearance.

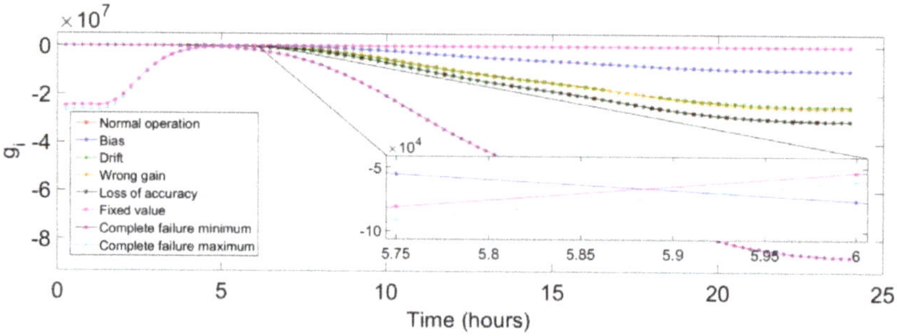

Figure 7. Fixed value diagnosis: graphs of the FDA discriminant functions $g_i(x)$ for each of the normal and seven fault classes, along the 24 h of the 140th day and the detailed representation revealing the class of the observations and the time moment of the firm fault type identification.

The complete failure minimum is correctly diagnosed after 6.75 h of its intervention, as it is revealed in Figure 8.

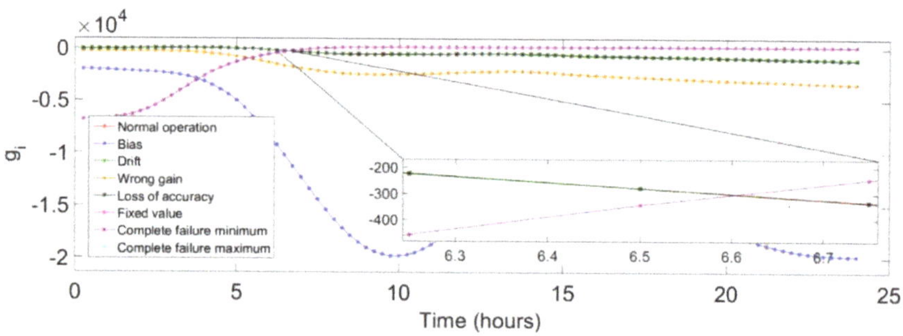

Figure 8. Complete failure minimum diagnosis: graphs of the FDA discriminant functions $g_i(x)$ for each of the normal and seven fault classes, along the 24 h of the 140th day and the detailed representation revealing the class of the observations and the time moment of the firm fault type identification.

The complete failure maximum diagnosis is diagnosed successfully after 9.5 h, as is shown in Figure 9.

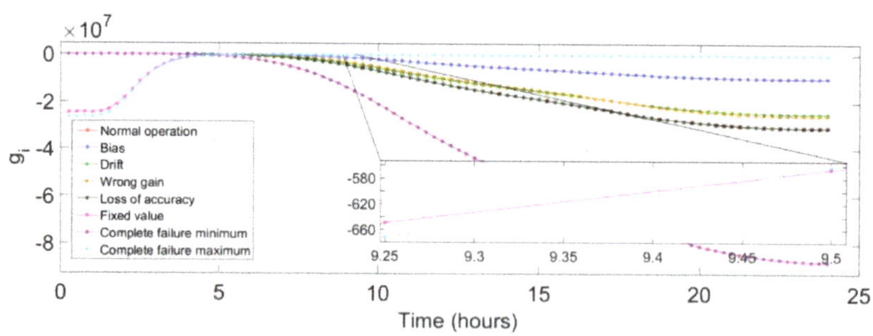

Figure 9. Complete failure maximum diagnosis: graphs of the FDA discriminant functions $g_i(x)$ for each of the normal and seven fault classes, along the 24 h of the 140th day and the detailed representation revealing the class of the observations and the time moment of the firm fault type identification.

The FDA-based identification methodology for the type of DO sensor fault proved to be successful for all of the investigated faults except the bias one (which has very similar effects with complete failure maximum failure). This means an accuracy of 87.5% correct identification of the eight considered cases. The time needed for obtaining the consolidated diagnosis decision varied from 2.5 h to 16.5 h.

From the rare previous studies that addressed fault diagnosis in WWTPs sensors, one showed good detection of bias, drift and precision degradation (loss of accuracy) of sensors used in a WWTP with classic and dynamic PCA-based methods but showed some limitations regarding the fault identification with variable reconstruction-based methods [42]. These limitations included the fact that a reconstruction-based method cannot identify the faulty sensor which causes process transition, so the approach is inappropriate for identifying the fault type connected to a control loop. Additionally, it did not study the fault identification time in detail. The present study investigated several fault types, and the identification time was determined for each of them. Another work proposed a complex-valued slow independent component analysis (CSICA) based method for fault detection and diagnosis with applications to wastewater treatment processes. Despite the fact that the method had good performance in detecting and diagnosing incipient faults, it was inadequate for scenarios of multi-fault and large-scale nonlinear systems [23]. The sub-period division strategies combined with multiway principal component analysis for only two faults diagnosis showed that the used methods could manifest false identification results during normal operation periods and the inability to detect the fault during some time intervals [35]. The present investigations considered six different fault types that implied increased difficulty in the promptitude and accuracy of the fault diagnosis.

The efficiency of the proposed FDA identification is substantiated by the fact that discrimination was achieved by the FDA discrimination model that was trained with faulty data from the five days subsequent to the first one, when the fault actually started to act, and testing was performed for this very first day.

3.3. Performance Indices

AE, PE and EQ were calculated for both normal and faulty operation cases. They were determined as a mean value over 28 days, i.e., for the period of the days ranging from day 140 to day 168. Table 2 shows the values of the AE, PE and EQ for the normal operation and for the six different faults.

Table 2. Values of the performance indices for the normal and faulty operation cases.

Operating Regime	AE (kWh/day)	PE (kWh/day)	Total Energy Demand (kWh/day)	EQ (kg PU/day)
Normal operation	16,992	1329	18,321	16,852
Bias fault	14,206	2415	16,621	21,461
Drift fault	15,569	1746	17,315	17,134
Wrong gain fault	15,866	1593	17,459	16,706
Loss of accuracy fault	9150	2403	11,553	219,189
Fixed value fault	1968	2415	4383	338,737
Complete failure minimum fault	23,537	1039	24,576	19,804
Complete failure maximum fault	1968	2415	4383	338,750

It can be observed that for the bias, loss of accuracy, fixed value and complete failure maximum types of faults, the PE and EQ indices values are much higher than the normal operation values, but AE is lower. The AE values can be explained by the fact that the four fault types are characterized by high values of the DO sensor signal, sent as a wrong feedback signal to the DO controller. As a result, the DO controller diminishes the aeration with the aim of reducing the value of the faulty DO signal, and the aeration energy drops.

This is followed by a decrease in nitrification, which induces a lower concentration of nitrates and nitrites in the aerated bioreactor and, subsequently, in the anoxic bioreactor. Consequently, the NO control loop reacts to this change by increasing the internal recycle flow rate, which determines higher pumping energy. Additionally, the quality of the effluent drastically deteriorated due to the inefficiency of the treatment process, affected by the air-lacking circumstances, and higher amounts of pollutants are discharged with the clean effluent. The high AE value in the case of the complete failure minimum type of fault is due to the increase in the DO controller manipulated airflow rate as a response to the low but faulty value of the DO sensor signal.

It is worth mentioning that in cases of the faulty operation of the DO sensor that is characterized by reduced values of the DO signal, compared to the true one, the DO controller will drive the aeration to increased airflow rates with straightforward consequences on the rise of the AE index values.

3.4. Energy Costs Assessment for Operation Affected by Faults

Operating a sewage treatment plant is highly energy intensive, with implicitly high economic costs. The price of energy is determined by a variety of supply and demand factors, such as the geopolitical situation, the national particular energy mix, distribution network expenses, environmental protection taxes, severe weather conditions or excise and taxation levels. The cost of energy also depends on the energy source. Energy prices vary from 4.8 to 12.10 eurocents depending on the energy source, as can be seen in Table 3. These values were extracted from an International Energy Agency (IEA) report on the mean costs of generating energy. This report provided data for a total of 243 plants in 24 countries from Europe, Africa and Asia [75].

Table 3. Costs of different energy sources.

Source/Technology	Eurocents/kWh
Lignite	9.12
Coal	8.80
Gas (CCGT [1])	6.81
Nuclear	6.62
Wind onshore	4.80
Wind offshore	8.45
Solar PV [2] commercial	5.38
Solar PV residential	12.10
Solar thermal (CSP [3])	11.62
Hydro reservoir	6.91
Hydro run of river	6.53
Geothermal	9.50
Biomass	11.33

[1] CCGT—combined cycle gas turbines, [2] PV—photovoltaic, [3] CSP—concentrating solar power.

The daily costs of normal operation and each type of DO sensor malfunction were determined for the sewage treatment plant considered in this study. The daily costs were individually evaluated for different potential energy sources or energy production technologies. The data in Table 4 were obtained by multiplying the costs of different energy sources [75] with the total energy demand values computed for the normal and faulty operation cases. Table 4 summarizes the findings.

Table 4. Source depending sum of AE and PE energy costs computed for normal and fault affected operation.

Source/Technology	Daily Operation Costs (EUR)							
	Normal Operation	Bias	Drift	Wrong Gain	Loss of Accuracy	Fixed Value	Complete Failure Minimum	Complete Failure Maximum
Lignite	1671	1516	1579	1592	1054	400	2241	400
Coal	1612	1463	1524	1536	1017	386	2163	386
CCGT	1249	1133	1180	1190	787	299	1675	299
Nuclear	1214	1101	1147	1156	765	290	1628	290
Onshore wind	879	798	831	838	555	210	1180	210
Offshore wind	1548	1404	1463	1475	976	370	2076	370
Solar PV commercial	985	894	931	938	621	236	1321	236
Solar PV residential	2216	2011	2094	2112	1397	530	2973	530
Solar thermal (CSP)	2128	1931	2011	2028	1342	509	2855	509
Hydro reservoir	1266	1149	1197	1207	799	303	1699	303
Hydro run of river	1196	1085	1130	1140	754	286	1604	286
Geothermal	1741	1580	1646	1659	1098	417	2336	417
Biomass	2075	1883	1961	1978	1309	497	2784	497

By analyzing the costs of electrical energy spent for the different cases of the faults, it may be observed that complete failure minimum fault of the DO sensor implies the largest energy costs, as the reduced faulty value of the DO sensor determines the control system to considerably increase the airflow rate. On the contrary, the fixed value and complete failure maximum faults imply the smallest and comparable energy costs due to the large values of the DO sensor signal that make the DO controller reduce the airflow rate. However, for these latter cases, the quality of the effluent dramatically deteriorates by a factor higher than 20 and ranks these faults as having the worst overall effect.

Values of the daily energy costs presented in Table 4 also reveal that onshore wind, solar PV commercial and hydro run of rivers are the most favorable sources of energy to be used from the economic costs point of view, while the solar PV residential, solar thermal (CSP) and biomass are the most expensive ones. The lignite (CSS) and coal (CSS) energy sources may be as well considered as having high costs.

Depending on the technical solution availability, switching to the alternative cheaper energy sources and implementing a customized program for the use of a mixed energy source in case of faulty sensor operation might reduce the implied energy costs before the replacement of the defective sensor.

3.5. Environmental Assessment of CO_2 and N_2O Emissions

The on-site and off-site emissions of CO_2 and N_2O, as main contributors to the Green House Gases released by the water line of the WWTP, were estimated for each type of faulty operation and also for the normal one. Their daily mean values are presented in Table 5.

The data in Table 5 revealed that on-site emissions are the most significant, accounting in the case of normal operation for about 75% of the total emissions, both for $P_{CO_2,total}$ and $P_{N_2O,total}$. Computed total CO_2 emissions, i.e., the sum of the on-site and off-site values, showed increased values for all cases of the DO sensor faults, with the exception of the complete failure minimum fault type. Very large total CO_2 emissions (more than one order of magnitude higher than those of the normal operation) were produced in the cases of the following faults: fixed value, complete failure maximum and loss of accuracy fault. Complete failure minimum fault type showed the least emission values, while drift and wrong gain had emissions that were slightly higher than those of the normal operation case.

Table 5. GHG emissions due to DO sensor defect.

Emissions Type	Source/Technology	Emitted Gas	Daily GHG Emissions							
			Normal	Bias	Drift	Wrong Gain	Loss of Accuracy	Fixed Value	Complete Failure Minimum	Complete Failure Maximum
Off-site emissions	Power generation	CO_2, $P_{CO_2,off-site}$, kg CO_2/day	3481	3158	3290	3317	2195	833	4669	833
	Biological degradation in the WWT effluent	N_2O, $P_{N_2O,off-site}$, kg N_2O/day	3.61	2.47	2.89	2.97	14.39	21.49	6.50	21.49
On-site emissions	Water-line aerobic biological processes	CO_2, $P_{CO_2,on-site}$, kg CO_2/day	13,689	30,459	17,851	16,178	461,439	921,028	10,604	921,168
		N_2O, $P_{N_2O,on-site}$, kg N_2O/day	10.35	10.81	10.64	10.05	6.07	3.27	9.20	3.27
Total emissions		CO_2, $P_{CO_2,total}$, kg CO_2/day	17,170	33,617	21,141	19,495	463,634	921,861	15,274	922,001
		N_2O, $P_{N_2O,\ total}$, kg N_2O/day	13.96	13.28	13.53	13.02	20.46	24.76	15.70	24.76
		CO_{2e}, $P_{CO_{2e},overall}$, kg CO_{2e}/day	21,330	37,574	25,173	23,375	469,731	929,239	19,953	929,379

Assessment of the total N_2O emissions for the fixed value, complete failure maximum and loss of accuracy faults also reveal increased values for the N_2O emissions. They are characterized by a factor ranging from 1.4 to 1.7 when compared to the normal operation case. It may be noticed that, contrary to the total CO_2 emissions observed trend, complete failure minimum fault type led to increased values of the N_2O emissions. N_2O emissions produced due to bias, drift and wrong gain faults were marginally smaller when compared to the normal reference case, especially due to the decreased values of the off-site N_2O emissions.

The overall CO_{2e} emission values were computed by cumulating the total CO_2 and total N_2O (as CO_2 equivalent) emissions. According to the results of the overall CO_{2e} emission, all faults show higher CO_2 emission levels than the normal operation case. The only exception is the complete failure minimum type of fault, but in this particular case, the effluent quality deteriorates by more than 15%, and the sum of the aeration and pumping energy has the highest values of all investigated cases. Assessment of the CO_2 and N_2O emissions, issued due to different DO sensor faults, provides valuable quantitative information on the extent and ranking of the most unfavorable sensor fault types that may affect the sustainable operation of the WWTP.

The results validate the logical assumption that the presence of all faults is detrimental to the WWTP operation, and the severity assessment of their consequences has to be considered in an integrated approach of energy, costs, water effluent quality and GHG issues. These evaluations constitute the foundation for the control and safety systems design aimed at achieving the plant sustainability objective.

4. Conclusions

The performance of the FDA-based sensor faults identification approach was proposed and assessed for the key DO sensor belonging to the automatically controlled A^2O WWTP with the typical dissolved oxygen and the nitrates and nitrites concentration control loops. The DO feedback control system performance is responsible for the WWTP nitrification process, directly determining the spent energy, effluent quality and GHG emissions of the entire plant.

The FDA fault identification was investigated for six different types of faults. They were: bias, drift, wrong gain, loss of accuracy, fixed value and complete failure minimum and complete failure maximum faults. The data sets of variables for both the normal and fault-affected operation of the DO sensor originated from simulations of the calibrated plant model in which the scenario of WWTP influent variations emerged from plant measurements.

The prompt and reliable FDA fault identification methodology was successful. The time needed for obtaining the consolidated diagnosis decision varied from 2.5 h to 16.5 h following the moment of the fault appearance. It is noteworthy that data used for building the training data matrix for all faulty classes, i.e., data from days 2 to 6, did not contain the data of the first day. The latter was only used for testing the promptitude and efficacy of the identification methodology. The complexity of the diagnosis increases when the number of fault types subject to investigation expands or the sensor faults produces comparable effects on the process variables.

The results showed that the WWTP effluent quality performance index depreciated during all of the DO sensor faults. From the EQ index perspective, the most detrimental cases were the complete failure maximum, fixed value and loss of accuracy faults, while the less affected case was the wrong gain fault type.

The WWTP normal and faulty operation modes were further investigated to assess the environmental effect of the on-site and off-site emissions of CO_2 and N_2O GHG and the economic impact of spent aeration and pumping energy. On-site emissions have the most significant GHG contribution, being responsible for about three-quarters of the total emissions, both for CO_2 and N_2O. The complete failure maximum, fixed value and loss of accuracy were the DO fault-affected operation modes that had the most undesired impact on the amount of GHG released emissions. The identification of simultaneous acting fault types needs specific investigations, and this is the subject of future research work.

The comprehensive evaluation of effluent quality, energy costs and GHG emissions is a useful quantitative assessment basis for the control and safety systems design aimed at satisfying multiple objective targets and the overall plant sustainability goal. Instrumentation maintenance tasks, metrological calibration or verification services and designers of intelligent DO sensors may benefit from the presented results for promoting the safe, efficient and environmentally friendly operation of the WWTP.

Author Contributions: Conceptualization, A.-V.L. and V.-M.C.; Methodology, A.-V.L. and V.-M.C.; Software, A.-V.L., M.S.-V. and N.-B.M.; Formal analysis, A.-V.L. and V.-M.C., Investigation, A.-V.L. and V.-M.C.; Resources, M.S.-V. and N.-B.M.; Data curation, A.-V.L., M.S.-V. and V.-M.C.; Writing—original draft preparation, A.-V.L.; Writing—Reviewing and Editing, V.-M.C.; Supervision, A.-V.L. and V.-M.C. All authors have read and agreed to the published version of the manuscript.

Funding: This research received no external funding.

Institutional Review Board Statement: Not applicable.

Informed Consent Statement: Not applicable.

Data Availability Statement: No data availability.

Acknowledgments: The present work has received support from the project: Entrepreneurship for innovation through doctoral and postdoctoral research, POCU/380/6/13/123886, co-financed by the European Social Fund through the Operational Program for Human Capital 2014–2020.

Conflicts of Interest: "The author Melinda Simon-Varhelyi is an employee of MDPI, however she does not work for the journal Applied Sciences at the time of submission and publication." The authors declare no conflict of interest.

References

1. Baklouti, I.; Mansouri, M.; Ben Hamida, A.; Nounou, H.; Nounou, M. Monitoring of wastewater treatment plants using improved univariate statistical technique. *Process. Saf. Environ. Prot.* **2018**, *116*, 287–300. [CrossRef]
2. Alex, J.; Benedetti, L.; Copp, J.; Gernaey, K.V.; Jeppsson, U.; Nopens, I.; Pons, M.N.; Rieger, L.; Rosen, C.; Steyer, J.P.; et al. *Benchmark Simulation Model No. 1 (BSM1)*; Lund University: Lund, Sweden, 2008; pp. 2–29.
3. Mamandipoor, B.; Majd, M.; Sheikhalishahi, S.; Modena, C.; Osmani, V. Monitoring and detecting faults in wastewater treatment plants using deep learning. *Environ. Monit. Assess.* **2020**, *192*, 148. [CrossRef] [PubMed]

4. Rieger, L.; Gillot, S.; Langergraber, G.; Ohtsuki, T.; Shaw, A.; Takacs, I.; Winkler, S. *Guidelines for Using Activated Sludge Models*; IWA Publishing: London, UK, 2012; pp. 11–147, ISBN 9781780401164.
5. Henze, M.; Gujer, W.; Mino, T.; van Loosedrecht, M. *Activated Sludge Models ASM1, ASM2, ASM2d and ASM3*; IWA Publishing: London, UK, 2002; pp. 1–120, ISBN 9781780402369.
6. Hauduc, H.; Rieger, L.; Ohtsuki, T.; Shaw, A.; Takacs, I.; Winkler, S.; Heduit, A.; Vanrolleghem, P.A.; Gillot, S. Activated sludge modelling: Development and potential use of a practical applications database. *Water Sci. Technol.* **2011**, *63*, 2164–2182. [CrossRef] [PubMed]
7. Jeppsson, U.; Rosen, C.; Alex, J.; Copp, J.; Gernaey, K.V.; Pons, M.N.; Vanrolleghem, P.A. Towards a benchmark simulation model for plant-wide control strategy performance evaluation of WWTPs. *Water Sci. Technol.* **2006**, *53*, 287–295. [CrossRef]
8. Jeppsson, U. The benchmark simulation modelling platform—Areas of recent development and extension. *Lect. Notes Civ. Eng.* **2017**, *4*, 81–91. [CrossRef]
9. Marais, H.L.; Nordlander, E.; Thorin, E.; Wallin, C.; Dahlquist, E.; Odlare, M. Outlining process monitoring and fault detection in a wastewater treatment and reuse system. In Proceedings of the 2020 European Control Conference (ECC), St. Petersburg, Russia, 12–15 May 2020; pp. 558–563. [CrossRef]
10. Tao, E.P.; Shen, W.H.; Liu, T.L.; Chen, X.Q. Fault diagnosis based on PCA for sensors of laboratorial wastewater treatment process. *Chemom. Intell. Lab. Syst.* **2013**, *128*, 49–55. [CrossRef]
11. Qin, S.J. Statistical process monitoring: Basics and beyond. *J. Chemom.* **2003**, *17*, 480. [CrossRef]
12. Gertler, J. Survey of model-based failure detection and isolation in complex plants. *IEEE Contr. Syst. Mag.* **1988**, *8*, 3–11. [CrossRef]
13. Benveniste, A.; Basseville, M.; Moustakides, G. The asymptotic local approach to change detection and model validation. *IEEE Trans. Automat. Contr.* **1987**, *32*, 583–592. [CrossRef]
14. Isermann, R. Process fault detection based on modeling and estimation methods—A survey. *Automatica* **1984**, *20*, 387–404. [CrossRef]
15. Frank, P. Fault diagnosis in dynamic systems using analytical and knowledge-based redundancy—A survey and some new results. *Automatica* **1990**, *26*, 459–474. [CrossRef]
16. Yoon, S.; MacGregor, J. Fault diagnosis with multivariate statistical models, part i: Using steady state fault signatures. *J. Process Contr.* **2001**, *11*, 387–400. [CrossRef]
17. Ching, P.M.L.; So, R.H.Y.; Morck, T. Advances in soft sensors for wastewater treatment plants: A systematic review. *J. Water Process Eng.* **2021**, *44*, 102367. [CrossRef]
18. Kano, M.; Nakagawa, Y. Data-based process monitoring, process control, and quality improvement: Recent developments and applications in steel industry. *Comput. Chem. Eng.* **2008**, *32*, 12–24. [CrossRef]
19. Sanchez-Fernandez, A.; Fuente, M.J.; Sainz-Palermo, G.I. Fault detection in wastewater treatment plants using distributed PCA methods. In Proceedings of the 20th Conference on Emerging Technologies & Factory Automation (ETFA), Luxembourg, 8–11 September 2015; pp. 1–7. [CrossRef]
20. Andersson, S.; Hallgren, F. Sensor fault detection methods applied on dissolved oxygen sensors at a full scale WWTP. In Proceedings of the 9th IWA Symposium on Systems Analysis and Integrated Assessment (Watermatex 2015), Gold Coast, Australia, 14–17 June 2015.
21. Tarcsay, B.L.; Bárkányi, Á.; Chován, T.; Németh, S. A Dynamic Principal Component Analysis and Fréchet-Distance-Based Algorithm for Fault Detection and Isolation in Industrial Processes. *Processes* **2022**, *10*, 2409. [CrossRef]
22. Villegas, T.; Fuente, M.J.; Sainz-Palmero, G.I. Fault diagnosis in a wastewater treatment plant using dynamic independent component analysis. In Proceedings of the 18th Mediterranean Conference on Control and Automation (MED'10), Marrakech, Morocco, 23–25 June 2010; pp. 874–879. [CrossRef]
23. Xu, C.; Huang, D.; Cai, B.; Chen, H.; Liu, H. A complex-valued slow independent component analysis based incipient fault detection and diagnosis method with applications to wastewater treatment processes. *ISA Trans.* **2022**, *128*, 1–20. [CrossRef] [PubMed]
24. Yang, C.; Zhang, Y.; Huang, M.; Liu, H. Adaptive dynamic prediction of effluent quality in wastewater treatment processes using partial least squares embedded with relevance vector machine. *J. Clean. Prod.* **2021**, *314*, 128076. [CrossRef]
25. Chen, A.; Zhou, H.; An, Y.; Sun, W. PCA and PLS monitoring approaches for fault detection of wastewater treatment process. In Proceedings of the 25th International Symposium on Industrial Electronics (ISIE), Santa Clara, CA, USA, 8–10 June 2016; pp. 1022–1027. [CrossRef]
26. Liu, H.; Yang, J.; Zhang, Y.; Yang, C. Monitoring of wastewater treatment processes using dynamic concurrent kernel partial least squares. *Process Saf. Environ. Prot.* **2021**, *147*, 274–282. [CrossRef]
27. Kazemi, P.; Giralt, J.; Bengoa, C.; Steyer, J.P. Data-driven fault detection methods for detecting small-magnitude faults in anaerobic digestion process. *Water Sci. Technol.* **2020**, *81*, 1740–1748. [CrossRef]
28. Schraa, O.; Tole, B.; Copp, J.B. Fault detection for control of wastewater treatment plants. *Water Sci. Technol.* **2006**, *53*, 375–382. [CrossRef]
29. Yang, J.; Zhang, Y.; Zhou, L.; Zhang, F.; Jing, Y.; Huang, M.; Liu, H. Quality-related monitoring of papermaking wastewater treatment processes using dynamic multiblock partial least squares. *J. Bioresour. Bioprod.* **2022**, *7*, 73–82. [CrossRef]
30. Ma, X.; Zhang, Y.; Zhang, F.; Liu, H. Monitoring of papermaking wastewater treatment processes using t-distributed stochastic neighbor embedding. *J. Environ. Chem. Eng.* **2021**, *9*, 106559. [CrossRef]

31. Li, Z.; Yan, X. Ensemble model of wastewater treatment plant based on rich diversity of principal component determining by genetic algorithm for status monitoring. *Control Eng. Pract.* **2019**, *88*, 38–51. [CrossRef]
32. Dong, Y.; Qin, Y. A novel dynamic PCA algorithm for dynamic data modeling and process monitoring. *J. Process Control* **2018**, *67*, 1–11. [CrossRef]
33. Corominas, L.; Villez, K.; Aguado, D.; Rieger, L.; Rosén, C.; Vanrolleghem, P.A. Performance evaluation of fault detection methods for wastewater treatment processes. *Biotechnol. Bioeng.* **2010**, *108*, 333–344. [CrossRef] [PubMed]
34. Garcia-Alvarez, D. Fault detection using principal component analysis (PCA) in a wastewater treatment plant (WWTP). In Proceedings of the International Student's Scientific Conference, Online Conference, 15 January 2009; pp. 55–60.
35. Zhou, J.; Huang, F.; Shen, W.; Liu, Z.; Corriou, J.-P.; Seferlis, P. Sub-period division strategies combined with multiway principal component analysis for fault diagnosis on sequence batch reactor of wastewater treatment process in paper mill. *Process Saf. Environ. Prot.* **2021**, *146*, 9–19. [CrossRef]
36. Duda, R.O.; Hart, P.E.; Stork, D.G. *Pattern Classification*; Wiley: New York, NY, USA, 2000; ISBN 978-0-471-05669-0.
37. Fuente, M.; Garcia, G.; Sainz, G. Fault diagnosis in a plant using fisher discriminant analysis. In Proceedings of the 16th Mediterranean Conference on Control and Automation, Ajaccio, France, 25–27 June 2008; pp. 53–58. [CrossRef]
38. He, Q.; Qin, S.; Wang, J. A new fault diagnosis method using fault directions in fisher discriminant analysis. *AIChE J.* **2005**, *51*, 555–571. [CrossRef]
39. Chiang, L.H.; Russell, E.L.; Braatz, R.D. Fault diagnosis in chemical processes using Fisher discriminant analysis, discriminant partial least squares, and principal component analysis. *Chemom. Intell. Lab. Syst.* **2000**, *50*, 243–252. [CrossRef]
40. Chiang, L.H.; Russell, E.L.; Braatz, R.D. *Fault Detection and Diagnosis in Industrial Systems*; Springer Science & Business Media: London, UK, 2001; ISBN 1-85233-327-8.
41. Du, Z.; Jin, X. Multiple faults diagnosis for sensors in air handling unit using Fisher discriminant analysis. *Energy Convers. Manag.* **2008**, *49*, 3654–3665. [CrossRef]
42. Lee, C.; Choi, S.W.; Lee, I.B. Sensor fault diagnosis in a wastewater treatment process. *Water Sci. Technol.* **2006**, *53*, 251–257. [CrossRef]
43. Li, X.; Chai, W.; Liu, T.; Qiao, J. Fault detection of dissolved oxygen sensor in wastewater treatment plants. In Proceedings of the 46th Annual Conference of the IEEE Industrial Electronics Society, Singapore, 18–21 October 2020; pp. 225–230. [CrossRef]
44. Chistiakova, T.; Zambrano, J.; Samuelsson, O.; Carlsson, B. Binary classifiers applied to detect DO sensors faults during washing events. In Proceedings of the 2nd New Developments in IT & Water, Rotterdam, The Netherlands, 8–10 February 2015. [CrossRef]
45. Samuelsson, O.; Bjork, A.; Zambrano, J.; Carlsson, B. Fault signatures and bias progression in dissolved oxygen sensors. *Water Sci. Technol.* **2018**, *78*, 1034–1044. [CrossRef] [PubMed]
46. Mali, B.; Laskar, S.H. Incipient fault detection of sensors used in wastewater treatment plants based on deep dropout neural network. *SN Appl. Sci.* **2020**, *2*, 2121. [CrossRef]
47. Luca, A.-V.; Simon-Várhelyi, M.; Mihály, N.-B.; Cristea, V.-M. Data driven detection of different dissolved oxygen sensor faults for improving operation of the WWTP control system. *Processes* **2021**, *9*, 1633. [CrossRef]
48. Focht, D.D.; Chang, A.C. Nitrification and denitrification processes related to waste water treatment. *Adv. Appl. Microbiol.* **1975**, *19*, 153–186. [CrossRef] [PubMed]
49. Simon-Varhelyi, M.; Cristea, V.-M.; Luca, A.-V. Reducing energy costs of the wastewater treatment plant by improved scheduling of the periodic influent load. *J. Environ. Manag.* **2020**, *262*, 110294. [CrossRef] [PubMed]
50. Nair, A.; Cristea, V.-M.; Agachi, P.S.; Brehar, M. Model calibration and feed-forward control of the wastewater treatment plant—Case study for Cluj-Napoca WWTP. *Water Environ. J.* **2018**, *32*, 164–172. [CrossRef]
51. Otterpohl, R.; Freund, M. Dynamic models for clarifiers of activated sludge plants with dry and wet weather flows. *Water Sci. Technol.* **1992**, *26*, 1391–1400. [CrossRef]
52. Takacs, I.; Patry, G.G.; Nolasco, D. A dynamic model of the clarification-thickening process. *Water Res.* **1991**, *25*, 1263–1271. [CrossRef]
53. Varhelyi, M.; Cristea, V.-M.; Brehar, M.; Nemes, E.D.; Nair, A. WWTP model calibration based on different optimization approaches. *Environ. Eng. Manag. J.* **2019**, *18*, 1657–1670. [CrossRef]
54. Cristea, V.-M. Counteracting the accidental pollutant propagation in a section of the river Somes by automatic control. *J. Environ. Manag.* **2013**, *128*, 828–836. [CrossRef]
55. Viktoryova, N.; Szarka, A. Hrouzkova, Recent developments and emerging trends in paint industry wastewater treatment methods. *Appl. Sci.* **2022**, *12*, 10678. [CrossRef]
56. Revollar, S.; Vilanova, R.; Vega, P.; Francisco, M.; Meneses, M. Wastewater treatment plant operation: Simple control schemes with a holistic perspective. *Sustainability* **2020**, *12*, 768. [CrossRef]
57. Viveros, P.; Miqueles, L.; Mena, R.; Kristjanpoller, F. Opportunistic strategy for maintenance interventions planning: A case study in a wastewater treatment plant. *Appl. Sci.* **2021**, *11*, 10853. [CrossRef]
58. Ostace, G.S.; Cristea, V.-M.; Agachi, P.S. Extension of activated sludge model no. 1 with two-step nitrification and denitrification processes for operation improvement. *Environ. Eng. Manag. J.* **2011**, *10*, 1529–1544. [CrossRef]
59. Schneider, M.Y.; Carbajal, J.P.; Furrer, V.; Sterkele, B.; Maurer, M.; Villez, K. Beyond signal quality: The value of unmaintained pH, dissolved oxygen, and oxidation-reduction potential sensors for remote performance monitoring of on-site sequencing batch reactors. *Water Res.* **2019**, *161*, 639–651. [CrossRef]

60. Teh, H.Y.; Kempa-Liehr, A.W.; Wang, K.I. Sensor data quality: A systematic review. *J. Big Data* **2020**, *7*, 11. [CrossRef]
61. Rosen, C.; Rieger, L.; Jeppsson, U.; Vanrolleghem, P.A. Adding realism to simulated sensors and actuators. *Water Sci. Technol.* **2008**, *57*, 337–344. [CrossRef]
62. Lv, Z.; Shan, X.; Xiao, X.; Cai, R.; Zhang, Y.; Jiao, N. Excessive greenhouse gas emissions from wastewater treatment plants by using the chemical oxygen demand standard. *Sci. China Earth Sci.* **2022**, *65*, 87–95. [CrossRef]
63. Nguyen, T.K.L.; Ngo, H.H.; Guo, W.; Nguyen, T.L.H.; Chang, S.W.; Nguyen, D.D.; Varjani, S.; Lei, Z.; Deng, L. Environmental impacts and greenhouse gas emissions assessment for energy recovery and material recycle of the wastewater treatment plant *Sci. Total Environ.* **2021**, *784*, 147135. [CrossRef]
64. Mannina, G.; Ekama, G.; Caniani, D.; Cosenza, A.; Esposito, G.; Gori, R.; Garrido-Baserba, M.; Rosso, D.; Olsson, G. Greenhouse gases from wastewater treatment—A review of modelling tools. *Sci. Total Environ.* **2016**, *551–552*, 254–270. [CrossRef]
65. Szulc, P.; Kasprzak, J.; Dymaczewski, Z.; Kurczewski, P. Life cycle assessment of municipal wastewater treatment processes regarding energy production from the sludge line. *Energies* **2021**, *14*, 356. [CrossRef]
66. Raghuvanshi, S.; Bhakar, V.; Sowmya, C.; Sangwan, K.S. Waste water treatment plant life cycle assessment: Treatment process to reuse of water. *Procedia CIRP* **2017**, *61*, 761–766. [CrossRef]
67. Corominas, L.; Foley, J.; Guest, J.S.; Hospido, A.; Larsen, H.F.; Morera, S.; Shaw, S. Life cycle assessment applied to wastewater treatment: State of the art. *Water Res.* **2013**, *47*, 5480–5492. [CrossRef] [PubMed]
68. Vallero, D.A. Air pollution biogeochemistry. In *Air Pollution Calculations*; Elsevier: Amsterdam, The Netherlands, 2019; pp. 175–206. ISBN 978-0-12-814934-8. [CrossRef]
69. Listowski, A.; Ngo, H.H.; Guo, W.S.; Vigneswaran, S.; Shin, H.S.; Moon, H. Greenhouse gas (GHG) emissions from urban wastewater system: Future assessment framework and methodology. *J. Water Sustain.* **2011**, *1*, 113–125.
70. Prendez, M.; Lara-González, S. Application of strategies for sanitation management in wastewater treatment plants in order to control/reduce greenhouse gas emissions. *J. Environ. Manag.* **2008**, *88*, 658–664. [CrossRef] [PubMed]
71. Intergovernmental Panel on Climate Change (IPCC). Guidelines for National Greenhouse Gas Inventories. In *Intergovernmental Panel on Climate Change*; IPCC Guidelines; Table 6.11; Intergovernmental Panel on Climate Change (IPCC): Geneva, Switzerland, 2006; p. 5. Available online: http://www.ipcc-nggip.iges.or.jp/public/2006gl/index.html (accessed on 14 December 2022).
72. Gori, R.; Jiang, L.-M.; Sobhani, R.; Rosso, D. Effects of soluble and particulate substrate on the carbon and energy footprint of wastewater treatment processes. *Water Res.* **2011**, *45*, 5858–5872. [CrossRef]
73. Akbarjon, N.; Huang, F.; Shen, W. Calculation of N_2O emissions in the wastewater treatment process of paper mill. In Proceedings of the 2nd International Conference on Sustainable Energy, Environment and Information Engineering (SEEIE 2019), Beijing, China, 24–25 March 2019. [CrossRef]
74. Baresel, C.; Andersson, S.; Yang, J.; Andersen, M.H. Comparison of nitrous oxide (N_2O) emissions calculations at a swedish wastewater treatment plant based on water concentrations versus off-gas concentrations. *Adv. Clim. Chang. Res.* **2016**, *7*, 185–191. [CrossRef]
75. IEA (International Energy Agency). Projected Costs of Generating Electricity. Available online: https://www.iea.org/reports/projected-costs-of-generating-electricity-2020 (accessed on 14 December 2022).

applied sciences

MDPI

Article

Multitarget Search Algorithm Using Swarm Robots in an Unknown 3D Mountain Environment

You Zhou [1], Shaowu Zhou [2,*], Mao Wang [2] and Anhua Chen [3]

1 College of Electrical and Information Engineering, Hunan Institute of Engineering, Xiangtan 411104, China
2 College of Information and Electrical Engineering, Hunan University of Science and Technology, Xiangtan 411201, China
3 College of Mechanical and Electrical Engineering, Hunan University of Science and Technology, Xiangtan 411201, China
* Correspondence: swzhou@hnust.edu.cn; Tel.: +86-165-0732-0111

Abstract: A multitarget search algorithm for swarm robot in an unknown 3D mountain environment is proposed. Most existing 3D environment obstacle avoidance algorithms are potential field methods, which need to consider the location information of all obstacles around the robot, and they easily fall into local optima, and their calculation is complex. Furthermore, they cannot well meet the requirements of real-time obstacle avoidance characteristics of swarm robots in multiobject searches. This paper first focuses on solving the obstacle avoidance problem of swarm robots in mountain environments. A new 3D curved obstacle tracking algorithm (3D-COTA) is designed by discretizing the mountains within the detection range of robot obstacles. Then, a task assignment model and virtual force model in 2D space are extended to 3D, and a particle swarm search model with kinematic constraints is constructed, which considers the kinematic constraints and the limitations of the communication ability of the robots. Finally, a new multitarget search algorithm for swarm robot in an unknown 3D mountain environment is proposed by means of the designed 3D surface obstacle tracking algorithm. Numerical simulation results demonstrate the effectiveness of the proposed algorithm.

Keywords: swarm robot; unknown complex environment; multitarget cooperative search; simplified virtual force model; particle swarm optimization

Citation: Zhou, Y.; Zhou, S.; Wang, M.; Chen, A. Multitarget Search Algorithm Using Swarm Robots in an Unknown 3D Mountain Environment. *Appl. Sci.* **2023**, *13*, 1969. https://doi.org/10.3390/app13031969

Academic Editor: Dimitris Mourtzis

Received: 4 November 2022
Revised: 22 January 2023
Accepted: 28 January 2023
Published: 2 February 2023

1. Introduction

A large number of studies are devoted to swarm robot multitarget search, which is widely used in postdisaster search and rescue, natural resources exploration, enemy position detection, underwater fishing, and other application fields [1]. ZENG et al. mapped particle swarm optimization (PSO) well with the target search process [2]. ZHANG et al. deployed the cooperation and competition to solve the spatial conflicts of swarm robots [3]. LI et al. introduced a probability-constrained finite state machine to effectively resolve individual resource conflicts and improve the efficiency of target search [4]. Taking UAV as the carrier, HE proposed a 3D adaptive inertia weight extended particle swarm optimization (IAEPSO) to realize the search of air targets in a mountain environment [5]. In order to search for the lost object, PHUNG et al. transformed the problem of target search into a probability problem based on the location of the last lost object and the creation of a Bayesian map, and proposed motion-encoded particle swarm optimization (MPSO) [6]. Aiming at the target search of underwater vehicles, CAO et al. proposed a multi-AUV collaborative team integration algorithm, which has the advantages of fewer parameters and no speed jump [7]. In order to reduce the communication pressure of swarm robots, TANG et al. realized the information interaction among swarm robots through indirect communication [8]. Brown et al. assumed that the target was discovered when it was within the detection range of individual UAV, and then proposed an ergodic target search method; under the background of this method, Brown et al. also proposed an approach to increase or decrease the number of UAV individuals [9,10].

The above research shows that existing research on swarm robot multitarget search is mainly aimed at 2D plane environments or a 3D environment with constant height [11–14]. However, in the practical application of environmental detection and postdisaster rescue, swarm robots may face complex mountain environments. For multitarget searches in 3D environments, many studies have implemented UAVs. For example, Dario [15] proposed a task planning strategy of a UAV swarm in a 3D environment for landslide monitoring and postdisaster search for survivors. Inspired by the gray wolf tracking strategy, Xie Yuxin [16] proposed an adaptive formation tracking control method applied to a UAV swarm system, which improved the system stability and accuracy of formation tracking. Wang [17] customized a UAV interactive decision-making mechanism that could switch the interaction method according to the distance between aircraft during a search for a cooperative UAV swarm search task under the condition of limited communication distance and realized search path planning in a dynamic environment. In view of the realistic environment faced by swarm robots in a targeted search, the premise of their search is to move safely in the task environment, so it is particularly important to consider the obstacle avoidance problem. BinKai Qi [18] proposed UAV path planning based on an improved 1artificial potential field to efficiently solve the UAV obstacle avoidance problem. YuWenqiang [19], in view of the traditional artificial potential field method in a complex environment and the problem of low efficiency of obstacle avoidance, proposed a traditional artificial potential field method as an improved potential field function and improved the traditional spherical potential field for the ellipsoid potential field. The experimental simulation proves that the improved artificial potential field method provides efficient and safe UAV obstacle avoidance path planning in a complex 3D environment.

A UAV has the advantages of information sharing, strong system survivability, and cost performance, which can better meet the needs of a targeted search in 3D space. However, it has some problems for ground targeted search in complex mountain environments. At present, there are few research results on swarm robot targeted ground search in 3D mountain environments. In view of existing 3D environment potential field methods, obstacle avoidance algorithms, the need to consider the obstacle position information around the robot, the complex calculation and ease of falling into local optima, not satisfying swarm robots well in the process of multirobot targeted ground search, and the insufficient real-time obstacle avoidance requirements, this paper proposes a simple 3D curved obstacle tracking algorithm that does not easily fall into local optima.

First, a task assignment model, particle swarm optimization algorithm with kinematic constraints, and a simplified virtual force model in a 2D search environment were extended to 3D space to solve the multiobjective search problem in a 3D scene [20–22]. Then, obstacle tracking was considered in the process of swarm robot completing the task under the condition of different robots according to the kinematic constraint using a particle swarm optimization algorithm, and a virtual force model was simplified to calculate the expected speed after the swarm robot 3D curved obstacle tracking algorithm to realize multitarget search in an unknown complex 3D mountain environment. The simulation results show that the proposed method is an effective method for swarm robots to search for multiple targets in unknown 3D mountain environments.

2. Ground Target Search Modeling in an Unknown Mountain Environment

To better study swarm robot ground target search in an unknown mountain environment, the corresponding environment hypothesis is made, and the corresponding mathematical model is established.

The search task is located in an $a \times b \times c$ mountain environment, which has a horizontal area of $a \times b$ and a height of c. Among them, the mountain height difference is less than c, and the mountain slope changes continuously and is always less than α degrees.

The task object includes the robots, target, and mountain. Robots are represented as set $\boldsymbol{Rob} = \{R_i | i = 1,2,\ldots,n_u; 30 \leq n_u \leq 100\}$, where R_i represents a robot, and the target is represented as set $\boldsymbol{T} = \{tar_j | j = 1,2,\ldots,n_T; n_T \geq 1\}$. The mountain is discretized in both the horizontal x- and y-axes with Δl as the unit distance, and the discrete points obtained are called obstacles. Obstacles are represented as set $\boldsymbol{S} = \{obs_k | k = 1,2,\ldots,n_s; n_s \geq 1\}$. The time t and the locations of R_i, tar_j, and obs_k are represented as $\boldsymbol{R_i(t)}$, $\boldsymbol{tar_j}$, and $\boldsymbol{obs_k}$, respectively, and the speed of R_i is $\boldsymbol{V_i(t)}$.

The Euclidean distance between each individual is expressed as follows: the distance between robots $dr_{i_1,i_2|t} = \|\boldsymbol{R_{i_1}(t)} - \boldsymbol{R_{i_2}(t)}\|$, the distance between a robot and the target $drt_{i,j|t} = \|\boldsymbol{R_i(t)} - \boldsymbol{tar_j}\|$, and the distance between a robot and obstacle $drs_{i,k|t} = \|\boldsymbol{R_i(t)} - \boldsymbol{obs_k}\|$.

Within the task environment, the search task can be described as follows: considering that the target reaches the threshold value d_0, if there is a robot with a target distance $drt_{i,j|t} < d_0$, it indicates that the target is found. When all targets have been found, the target search task is complete.

The robots involved in the search have certain characteristics. Assuming that each robot is isomorphic and the robot velocity $\boldsymbol{V_i(t)}$ satisfies $0 \leq V_i(t) \leq Vm$, each robot can reach any position close to the ground in the task environment. Considering maximum communication distance d_{com}, maximum obstacle sensing distance d_{obs}, and maximum target detection distance d_{tar}, each robot has the following functions: when $drt_{i_1,i_2|t} \leq d_{com}$, it can communicate between robots; when $drt_{i,j|t} \leq d_{tar}$, it can detect the target signal; and when $drs_{i,k|t} \leq d_{obs}$, according to the slope sensor sense obstacles and a robot's relative slope, a robot can climb slopes less than or equal to β and can drive on slopes less than or equal to α without rollover, and $\beta < \alpha$.

The target being searched for is stationary on the mountain surface within the mission mountain environment. When searching for a target, each robot can continuously detect the target signal using a sensor. The target signal and $drt_{i,j|t}$ meet an environmental interference function and should describe the function of the target as a response function [23]. The function can be set as Equation (1):

$$I_{i,j}(t) = \begin{cases} \frac{mQ}{drt_{i,|t}^2} + \eta & drt_{i,j|t} \leq d_{tar} \\ 0 & drt_{i,j|t} > d_{tar} \end{cases} \tag{1}$$

where $I_{i,j}(t)$ represents the target signal detected by R_i at time t; η is the environmental disturbance satisfying the standard normal distribution; $drt_{i,j|t}$ is the objective existence, which is unknown to the robots; m is the attenuation coefficient of the environment $(0 < m < 1)$; and Q is the constant signal power of the target.

The mountain surface is separated into obstacle points with spacing Δl, which are static, and the position of each obstacle point can be specifically expressed as $\boldsymbol{obs_k} = [xs_k, ys_k, zs_k]$.

In a 3D search task environment, each robot can locate itself through its own sensor position and speed information, can communicate through a communication device within the scope of communication with other robots, and can sense obstacle slope information in its detection scope. The robot pose and location information followed within the search environment is as Equation (2):

$$\begin{cases} \boldsymbol{R_i}(t) = \left[xu_{i|t}, yu_{i|t}, zu_{i|t}\right] \\ \boldsymbol{V_i}(t) = \left[x\dot{u}_{i|t}, y\dot{u}_{i|t}, z\dot{u}_{i|t}\right] \\ x\dot{u}_{i|t} = \frac{d(xu_{i|t})}{d(t)} = \|\boldsymbol{V_i}(t)\| \cos\theta \sin\phi \\ y\dot{u}_{i|t} = \frac{d(yu_{i|t})}{d(t)} = \|\boldsymbol{V_i}(t)\| \cos\theta \cos\phi \\ z\dot{u}_{i|t} = \frac{d(zu_{i|t})}{d(t)} = \|\boldsymbol{V_i}(t)\| \sin\theta \end{cases} . \tag{2}$$

where $xu_{i|t}$, $yu_{i|t}$ and $zu_{i|t}$ are the coordinate positions of the robot at time t in the Cartesian global coordinate system $xoyz$. $\boldsymbol{V}_i(t)$ is the movement speed of the robot at time t, ϕ is the angle between the projection vector of $V_i(t)$ in the xoy plane and the forward direction of the x-axis, and θ is the angle between $\boldsymbol{V}_i(t)$ and the forward direction of the z-axis. If the time change Δt is small enough, the relationship between the robot's position and speed in Equation (2) can be expressed as Equation (3):

$$\boldsymbol{R}_i(t+\Delta t) = \boldsymbol{R}_i(t) + \boldsymbol{V}_i(t+\Delta t) \tag{3}$$

To facilitate the planning of the trajectory of the robot, this study takes Δt as unit time, that is, $\Delta t = 1$, and the iterative relationship between the position and velocity of the group robot satisfies the following as Equation (4):

$$\boldsymbol{R}_i(t+1) = \boldsymbol{R}_i(t) + \boldsymbol{V}_i(t+1) \tag{4}$$

3. Robot Search Task Assignment Mechanism

3.1. Three Robot States

To make the robot swarm target search more coordinated and efficient, robots are divided into three states as shown in Figure 1: roaming search state, cooperative search state, and task completion state.

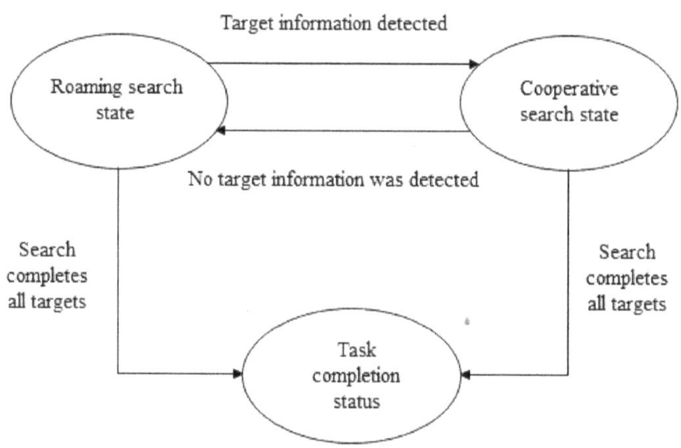

Figure 1. Three robot states relationship.

When the robots do not detect the target information, they are in a roaming search state; that is, the robots repel each other at the maximum speed to rapidly search the global environment [24,25]. When a robot detects the target signal, a multitarget task allocation model based on the response threshold is used to construct a suballiance. The robot members in the same suballiance search for the target corresponding to the suballiance, and the state of the robots forming the suballiance changes to the cooperative search state. When the robot and a target distance are less than the target reached threshold, the robot and the distance of a target $drt_{i,j|t} < d_0$, the target is regarded as a search success, and the robot changes to the task completed state. When all targets are successfully found, all robots change to the mission completed state.

3.2. Robot Task Assignment Model

In the process of task search, each robot participates in a task search process through self-organization and decides whether to choose task tar_1 or task tar_2 and whether to change the task between task selection and task completion. The process is as follows: First, the sensor detects the target response value of a robot in the detection range. If the robot detects multiple target signals in the detection range, the response probability of the robot to each target is calculated according to a response probability evaluation model, and then a roulette algorithm is used to make a decision regarding which target to search for [26]. The response probability assessment is expressed as Equation (5):

$$p(i,j) = \frac{I_{i,j}^2(t)}{\sum_{k=1}^{m} I_{i,k|t}^2}, \forall j,k = \{1,2,3,\cdots,m\} \tag{5}$$

where $I_{i,j}^2(t)$ is the target tar_j signal detected by robot R_i at time t. If the robot detects m target signals within its detection range, the probability of robot R_i responding to target tar_j excitation is $P(i,j)$, as Equation (6):

$$P(i,j) = \sum_{k=1}^{j} p_{ik}, j = 1,2,\cdots,m \tag{6}$$

When $P(i,j-1) < r_a < P(i,j)$, robot R_i selects target tar_j as the target of collaborative search, and $r_a \in (0,1)$.

Robots can obtain target information in two ways during driving. On the one hand, robots can directly detect target signals through their own sensors, which is called a class I robot. On the other hand, a robot fails to detect a target signal within its detection range but indirectly obtains the signal information of a target through communication with other robots. This kind of robot is called a class II robot [27]. If a target signal detected by two robots is the same target, they can participate in the target search process task together. When multiple robots participate in the same search task, these robots can form a suballiance to carry out a cooperative search for the target.

In the process of searching a 3D task environment, multiple robots will search for the same target, or only a few robots will search for the same target; that is, in the process of forming a suballiance, there will be an uneven distribution of robot resources. To avoid this situation, closed-loop regulation is introduced; that is, the resource allocation of each suballiance is re-evaluated after the first subtask assignment. When the number of members of a suballiance reaches an upper limit N_m, the suballiance preferentially selects N_m robots, and the remaining robots not selected will quit the suballiance and reselect other targets as their search tasks or switch to the roaming search state. When the number of members of a suballiance does not reach the upper limit N_m, suballiance members can be recruited from the surrounding robots to participate in the target task search. The priority principle of suballiance member selection is as follows: the priority of class I targets is greater than that of class II targets; if the priority is the same, the robot is evaluated according to the intensity of the target excitation signal; namely, the higher the intensity of the target signal is, the higher the dominant position. If the number of class I targets is less than N_m, a robot close to the class II communicating robot will be preferentially selected. If there is a robot with the same distance as the communicating robot, the robot with a strong signal will be preferentially selected. See Table 1 for details.

Table 1. Group drones rank the suballiance U_1 members at t = 36.

Serial Number	Robot	Target Type	Intensity of the Response	Nearest Communication Robot	Distance from Communication Robot	Priority Sorting
1	R_2	II	-	R_{14}	213.2349341	11
2	R_3	II	-	R_{19}	209.3224293	9
3	R_5	I	2.099988287	-	-	2
4	R_9	II	-	R_{14}	33.53008801	5
5	R_{11}	II	-	R_5	44.66655953	6
6	R_{14}	I	2.024188002	-	-	3
7	R_{17}	II	-	R_{19}	171.3542868	8
8	R_{18}	II	-	R_{14}	232.4477832	12
9	R_{19}	I	6.13611151	-	-	1
10	R_{21}	II	-	R_5	212.6859702	10
11	R_{22}	II	-	R_{14}	30.39406231	4
12	R_{23}	II	-	R_{19}	142.4618399	7

Swarm robots should not only avoid all obstacles but also complete all target searches in the process of movement. To complete all target search tasks quickly and effectively, the robots can form suballiances to search for the same target together according to the detected target signals and communicate with the surrounding robots. As presented in Table 1, the members of suballiances U_1 are sorted. Robots R_5, R_{14}, and R_{19} detect the signal of target tar_1 during their driving. Robots R_5, R_{14}, and R_{19} are class I robots. At this time, the number of class I robots is less than N_m, and a class I robot recruits the robots within its communication range as a communication robot. R_2, R_3, R_9, R_{11}, R_{17}, R_{18}, R_{21}, R_{22}, and R_{23} receive the recruitment information of class I robots and join one by one in the target tar_1 search task and form suballiance U_1 for this target. According to the principle of selecting members of suballiances, the priority of class I is higher than that of class II. Class I is sorted according to the corresponding intensity of the target.

The higher the target response intensity is, the higher the priority is. The class II robots are sorted according to the distance between them and communication robots. The closer the distance is, the higher the priority is. Therefore, suballiance U_1 is sorted by priority as R_{19}, R_5, R_{14}, R_{22}, R_9, R_{11}, R_{23}, R_{17}, R_3, R_{21}, R_2, and R_{18}. According to the priority order, R_{23}, R_{17}, R_3, R_{21}, R_2, and R_{18} quit the suballiance, and finally, R_{19}, R_5, R_{14}, R_{22}, R_9, and R_{11} form a suballiance and participate in the target tar_1 search task.

4. Multitarget Ground Search Algorithm for Swarm Robots in a 3D Mountain Environment

4.1. 3D Virtual Force Model Roaming Search

When no target signal is obtained, each robot conducts roaming search to quickly detect more areas. Here, a virtual force model is adopted [28]. When the distance between robots is less than $\min(d_{com}, d_{tar})$, the robots repel each other, making the robots spread quickly to quickly evaluate the search area. To simplify the calculation, a robot is repulsed only by the nearest two robots.

Assuming that the robot nearest to robot R_i is R_{i_1} as shown in Figure 2, the positions of the two robots are $\boldsymbol{R}_i(t) = \left[xu_{i|t}, yu_{i|t}, zu_{i|t}\right]$ and $\boldsymbol{R}_{i_1}(t) = \left[xu_{i_1|t}, yu_{i_1|t}, zu_{i_1|t}\right]$. In addition, $dr_{i,i_2|t} \le \min(d_{com}, d_{tar})$. The repulsive force of R_{i_1} on R_i is calculated using Equation (7), and the repulsive force direction is that the former points to the latter:

$$f_{i,i_1}(t) = \frac{c \cdot l_m^2}{{dv_{i,i_1|t}}^3}\left[\left(xu_{i|t} - xu_{i_1|t}\right), \left(yu_{i|t} - yu_{i_1|t}\right), \left(zu_{i|t} - zu_{i_1|t}\right)\right] \tag{7}$$

where $f_{i,i_1}(t)$ is the repulsion of R_{i_1} on R_i at time t. l_m strengthens the obstacle avoidance distance, and c optimizes the robot movement path.

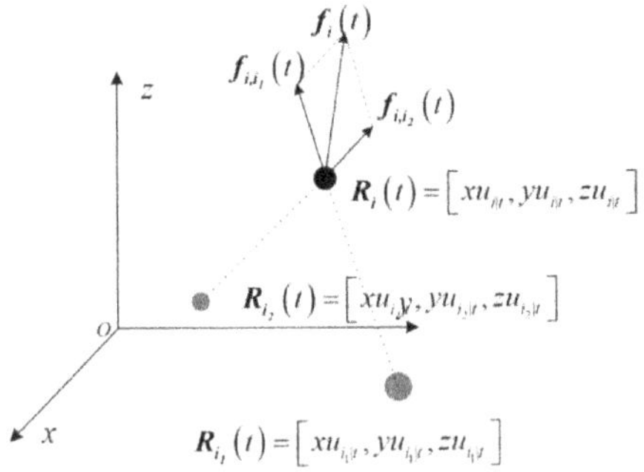

Figure 2. Virtual force model.

If the two robots closest to R_i satisfy $dv_{i,|t} \le min(d_{com}, 2d_{tar})$, then the virtual force on R_i is as shown in Figure 2. In the figure, $xu_{i_1|t} > xu_{i|t} > xu_{i_2|t}, yu_{i_1|t} > yu_{i|t} > yu_{i_2|t}, zu_{i|t} > zu_{i_2|t} > zu_{i_1|t}$, and the virtual forces $f_{i,i_1}(t)$ and $f_{i,i_2}(t)$ satisfy Equation (7). The virtual forces applied to the robot satisfy the vector sum $f_i(t) = f_{i,i_1}(t) + f_{i,i_2}(t)$. The speed of the robot in the roaming search state is the direction indicated by the virtual force; that is, the next expected speed of the roaming search is as Equation (8):

$$Ve_i(t+1) = Vm\frac{f_i(t)}{\|f_i(t)\|} \tag{8}$$

4.2. 3D Particle Swarm Cooperative Search Optimization with Motion Constraints

Swarm robot system is a typical distributed system. Comparing swarm robots with particle swarm optimization [29–31], a mapping relationship is found between the two. The particle swarm optimization algorithm can be applied to robot movement. Considering the movement constraints of a robot and the limitation of its communication ability, a particle swarm optimization model with kinematic constraints can be constructed to calculate the next expected velocity $Ve_i(t+1)$. The specific description is as Equation (9):

$$\begin{cases} Vp_i(t+1) = \omega V_i(t) + c_1r_1(R_i^*(t) - R_i(t)) + c_2r_2(g_i^*(t) - R_i(t)) \\ Ve_i(t+1) = V_i(t) + (Vp_i(t+1) - V_i(t))\cdot\lambda \end{cases} \tag{9}$$

where $R_i(t)$ and $V_i(t)$ represent the velocity and position vectors of robot R_i at time t, respectively; $Vp_i(t+1)$ is the velocity obtained by direct particle swarm iteration; $Ve_i(t+1)$ is the expected velocity vector of robot R_i at the next moment; the introduction of λ is to consider that the movement of the robot has a certain inertia; c_1 and c_2 are the individual cognitive coefficient and social cognitive coefficient of the robot, respectively; r_1 and r_2 are random variables in the interval (0,1); ω is the inertial weight; $R_i^*(t)$ represents the optimal position experienced by robot R_i thus far after joining the current suballiance; and $g_i^*(t)$ is the optimal position traversed by the suballiance cutoff time t.

4.3. 3D Curved Obstacle Tracking Algorithm (3D-COTA)

The search for ground targets in a 3D mountain environment is similar to that in a 2D environment. Curving a 2D search environment can obtain a mountain surface. When searching for a target, a robot needs to move along the mountain surface. The mountain surface is curved; therefore, the velocity direction of the robot at any time is the tangent direction of the surface corresponding to its position. Due to the limited climbing ability of the robot, it is necessary to avoid areas with high slopes. After the velocity of the robot in the roaming state or collaborative search is calculated according to the corresponding method, the velocity direction may not meet the speed requirements in the search environment. Therefore, further calculation is required after the expected velocity is obtained through the calculation of robots in different states. To ensure that the next velocity direction is the tangent direction of the curve, the mountain slope in the velocity direction must meet the requirements.

The robot discretizes the mountains within the detection range of obstacles and considers the discrete point obstacles. For example, the mountain detected by a robot shown in Figure 3a is discretized to obtain Figure 3b. The point set in the figure can be expressed as the obstacle set. The Euclidean distance between the robot and the obstacle can be expressed as Equation (10):

$$drs_{i,k|t} = \|\boldsymbol{R_i}(\boldsymbol{t}) - \boldsymbol{obs_k}\| = \sqrt{\left(xu_{i|t} - xs_k\right)^2 + \left(yu_{i|t} - ys_k\right)^2 + \left(zu_{i|t} - zs_k\right)^2} \quad (10)$$

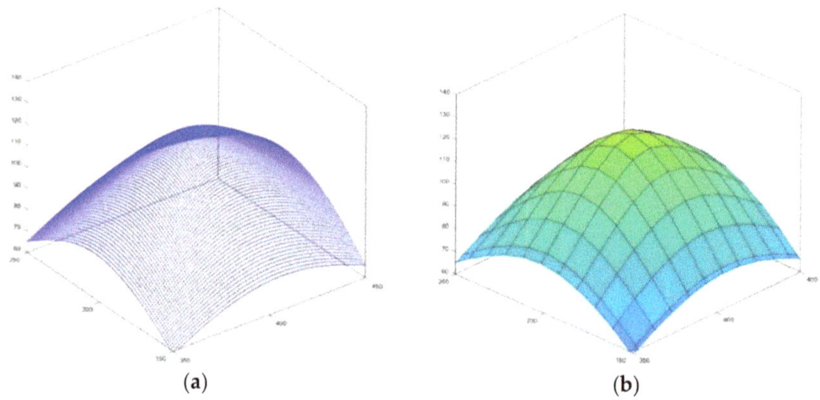

Figure 3. Discretization of a mountain. (**a**) Mountain, (**b**) mountain dispersion.

The slope of the obstacle relative to the robot can be expressed as Equation (11):

$$grd_{i,k|t} = \arctan\frac{zs_k - zu_{i|t}}{\sqrt{\left(xu_{i|t} - xs_k\right)^2 + \left(yu_{i|t} - ys_k\right)^2}} \quad (11)$$

4.3.1. Initial Obstacle Tracking

For robots in the roaming or collaborative search state, the expected velocity $\boldsymbol{Ve_i}(\boldsymbol{t+1}) = \left[\dot{xe}_{i|(t+1)}, \dot{ye}_{i|(t+1)}, \dot{ze}_{i|(t+1)}\right]$ is calculated according to the corresponding method. However, the expected velocity direction may not be tangent to the ground but may point to the air or the ground. Therefore, it is necessary to further calculate the velocity tangent to the ground.

Consider the nearest obstacle and two obstacle points not collinear to the nearest obstacle. In Equation (12), the nearest obstacle point to robot R_i is described as obs_{k_0}.

$$drs_{i,k_0|t} = \min_{obs_k \in S}\left(drs_{i,k|t}\right) \quad (12)$$

Based on obs_{k_0}, two other obstacle points, obs_{k_1} and obs_{k_2}, are found to satisfy the conditions described in Equation (13). According to Equation (13), points $\boldsymbol{obs_{k_0}}$, $\boldsymbol{obs_{k_1}}$, and $\boldsymbol{obs_{k_2}}$ are not collinear, so these three points can determine plane $fl_{i|t}$. For obstacle tracking to search for ground targets, the robot will tend to move parallel to plane $fl_{i|t}$.

$$\begin{cases} \boldsymbol{obs_{k_0}} = \left[xs_{k_0}, ys_{k_0}, zs_{k_0}\right] \\ \boldsymbol{obs_{k_1}} = \left[xs_{k_1}, ys_{k_1}, zs_{k_1}\right] \\ \boldsymbol{obs_{k_2}} = \left[xs_{k_2}, ys_{k_2}, zs_{k_2}\right] \\ xs_{k_1} = xs_{k_0} - \Delta l.\,\mathrm{sign}\left(xs_{k_0} - xu_{i|t}\right) \\ ys_{k_1} = ys_{k_0} \\ xs_{k_2} = xs_{k_0} \\ ys_{k_2} = ys_{k_0} - \Delta l.\,\mathrm{sign}\left(ys_{k_0} - yu_{i|t}\right) \end{cases} \tag{13}$$

At this time, plane $fl_{i|t}$ is shifted so that the resulting plane $fl_{i|t}{}'$ passes through $\boldsymbol{R_i(t)}$, and the coordinate system $\boldsymbol{R_{it}}xyz$ is established with $\boldsymbol{R_i(t)}$ as the origin. In this coordinate system, $\boldsymbol{R_i}(t) = [0,0,0]$, and $\boldsymbol{obs_{k_0}}$, $\boldsymbol{obs_{k_1}}$, and $\boldsymbol{obs_{k_2}}$ are expressed as Equation (14):

$$\begin{cases} \boldsymbol{obs'_{k_0}} = \left[\left(xs_{k_0} - xu_{i|t}\right), \left(ys_{k_0} - yu_{i|t}\right), \left(zs_{k_0} - zu_{i|t}\right)\right] \\ \boldsymbol{obs'_{k_1}} = \left[\left(xs_{k_1} - xu_{i|t}\right), \left(ys_{k_1} - yu_{i|t}\right), \left(zs_{k_1} - zu_{i|t}\right)\right] \\ \boldsymbol{obs'_{k_2}} = \left[\left(xs_{k_2} - xu_{i|t}\right), \left(ys_{k_2} - yu_{i|t}\right), \left(zs_{k_2} - zu_{i|t}\right)\right] \end{cases} \tag{14}$$

Under the $\boldsymbol{R_{it}}xyz$ coordinate system, plane $fl_{i|t}{}'$ is determined. Let Equation (15) of the plane be:

$$ax + by + cz = 0 \tag{15}$$

The vector normal to the plane for $\boldsymbol{nl_{i\,|\,t}} = [a,b,c]$ is a plane of two known vectors as Equation (16):

$$\begin{cases} \boldsymbol{p_{1,i\,|\,t}} = \left[\left(xs_{k_1} - xs_{k_0}\right), \left(ys_{k_1} - ys_{k_0}\right), \left(zs_{k_1} - zs_{k_0}\right)\right] \\ \boldsymbol{p_{2,i\,|\,t}} = \left[\left(xs_{k_2} - xs_{k_0}\right), \left(ys_{k_2} - ys_{k_0}\right), \left(zs_{k_2} - zs_{k_0}\right)\right] \end{cases} \tag{16}$$

The normal vector can be obtained as Equations (17) and (18):

$$\boldsymbol{nl_{i\,|\,t}} = [a,b,c] = \begin{bmatrix} [1,0,0] & [0,1,0] & [0,0,1] \\ \left(xs_{k_1} - xs_{k_0}\right) & \left(ys_{k_1} - ys_{k_0}\right) & \left(zs_{k_1} - zs_{k_0}\right) \\ \left(xs_{k_2} - xs_{k_0}\right) & \left(ys_{k_2} - ys_{k_0}\right) & \left(zs_{k_2} - zs_{k_0}\right) \end{bmatrix} \tag{17}$$

$$\begin{cases} a = \left(ys_{k_1} - ys_{k_0}\right)\left(zs_{k_2} - zs_{k_0}\right) - \left(ys_{k_2} - ys_{k_0}\right)\left(zs_{k_1} - zs_{k_0}\right) \\ b = \left(zs_{k_1} - zs_{k_0}\right)\left(xs_{k_2} - xs_{k_0}\right) - \left(zs_{k_2} - zs_{k_0}\right)\left(xs_{k_1} - xs_{k_0}\right) \\ c = \left(xs_{k_1} - xs_{k_0}\right)\left(ys_{k_2} - ys_{k_0}\right) - \left(xs_{k_2} - xs_{k_0}\right)\left(zs_{k_1} - zs_{k_0}\right) \end{cases} \tag{18}$$

After the normal vector $\boldsymbol{nl_{i\,|\,t}} = [a,b,c]$ of the plane is calculated, the robot begins to move in the direction parallel to plane $fl_{i|t}$, that is, motion tangential to the mountain ground at $\boldsymbol{R_i}(t)$. Considering the obstacle tracking velocity $\boldsymbol{Vo_i}(t+1) = \left[\dot{xo}_{i|(t+1)}, \dot{yo}_{i|(t+1)}, \dot{zo}_{i|(t+1)}\right]$ and expected velocity $\boldsymbol{Ve_i(t+1)} = \left[\dot{xe}_{i|(t+1)}, \dot{ye}_{i|(t+1)}, \dot{ze}_{i|(t+1)}\right]$ of the 3D curved obstacle tracking algorithm, the calculation process of the first obstacle tracking velocity $\boldsymbol{Vo_i}(t+1)$ of the 3D curved obstacle tracking algorithm is as Equation (19):

$$\begin{cases} \dot{x} = \dot{xe}_{i|(t+1)} \\ \dot{y} = \dot{ye}_{i|(t+1)} \\ \dot{z} = -\dfrac{a\dot{xe}_{\,i|(t+1)} + b\,\dot{ye}_{i|(t+1)}}{c} \end{cases} \tag{19}$$

(1) If the robot is in the roaming search state, it can be calculated as Equation (20):

$$\boldsymbol{Vo_i}(t+1) = \left[\dot{x}, \dot{y}, \dot{z}\right] \cdot \frac{V_m}{\sqrt{\dot{x}^2 + \dot{y}^2 + \dot{z}^2}} \tag{20}$$

(2) If the robot is in the cooperative search state, it can be calculated as Equation (21):

$$\boldsymbol{Vo_i}(t+1) = \begin{cases} [\dot{x}, \dot{y}, \dot{z}] \cdot \dfrac{Vm}{\sqrt{\dot{x}^2 + \dot{y}^2 + \dot{z}^2}}, & \sqrt{\dot{x}^2 + \dot{y}^2 + \dot{z}^2} > Vm \\ [\dot{x}, \dot{y}, \dot{z}], & \sqrt{\dot{x}^2 + \dot{y}^2 + \dot{z}^2} \leq Vm \end{cases} \tag{21}$$

4.3.2. Second-Obstacle Tracking

After $\boldsymbol{Vo_i}(t+1)$ is calculated using the corresponding method for the robot in the roaming or cooperative search state, the speed direction is adjusted according to the slope of the surrounding mountains, avoiding the movement direction of the mountain slope beyond the robot climbing ability range.

It is assumed that, according to the perception of the slope sensor, the distance near the robot is less than $\|\boldsymbol{Vo_i}(t+1)\|$, and in the direction of angle set $\boldsymbol{\Theta}$, the slope exceeds the climbing ability of the robot; that is, the slope is greater than β.

Within the set $\boldsymbol{\Phi} = \left\{ \frac{2\pi}{n_\phi} n \middle| n \in \left[-\frac{n_\phi}{2}, \frac{n_\phi}{2} \right] \cap \boldsymbol{Z} \right\}$, the sensor can identify mountain slopes in the n_ϕ angle directions, where $\left[-\frac{n_\phi}{2}, \frac{n_\phi}{2} \right]$ represents the set of numbers between $-\frac{n_\phi}{2}$ and $\frac{n_\phi}{2}$, and $\boldsymbol{Z}$ is the set of integers.

Suppose that the function $F(\varphi)$ has the following expression as Equation (22):

$$F(\varphi) = \varphi - 2\pi \cdot \text{sgn}(\varphi) \cdot \delta(|\varphi| - \pi) \quad -2\pi < \varphi < 2\pi \tag{22}$$

Among them:

$$\text{sgn}(\varphi) = \begin{cases} -1 & \varphi < 0 \\ 0 & \varphi = 0 \\ 1 & \varphi > 0 \end{cases} \tag{23}$$

$$\delta(\varphi) = \begin{cases} 0 & \varphi \leq 0 \\ 1 & \varphi > 0 \end{cases} \tag{24}$$

The second obstacle tracking velocity is denoted as $\boldsymbol{Vt_i(t+1)} = \left[\dot{xt}_{i|(t+1)}, \dot{yt}_{i|(t+1)}, \dot{zt}_{i|(t+1)} \right]$, and $\boldsymbol{Vo_i}(t+1) = \left[\dot{xo}_{i|(t+1)}, \dot{yo}_{i|(t+1)}, \dot{zo}_{i|(t+1)} \right]$. Subsequently, $\boldsymbol{Vt_i(t+1)}$ is calculated as:

(1) If $\arctan\left(\frac{\dot{yo}_{i|(t+1)}}{\dot{xo}_{i|(t+1)}} \right) + \delta\left(-\dot{xo}_{i|(t+1)} \right) \cdot \text{sgn}\left(\dot{yo}_{i|(t+1)} \right) \cdot \pi \notin \boldsymbol{\Theta}$

$$\boldsymbol{Vt_i(t+1)} = \boldsymbol{Vo_i(t+1)} \tag{25}$$

(2) If $\arctan\left(\frac{\dot{yo}_{i|(t+1)}}{\dot{xo}_{i|(t+1)}} \right) + \delta\left(-\dot{xo}_{i|(t+1)} \right) \cdot \text{sgn}\left(\dot{yo}_{i|(t+1)} \right) \cdot \pi \in \boldsymbol{\Theta}$

Calculation angle:

$$\varphi_0 = \min_{\varphi \in \boldsymbol{\Phi}, \varphi_n \in (\boldsymbol{\Phi} - \boldsymbol{\Theta})} |F(\varphi - \varphi_n)| \tag{26}$$

To calculate $\boldsymbol{Vt_i(t+1)}$:

$$\begin{cases} \dot{xt}_{i|(t+1)} = \left(\sqrt{\dot{xo}^2_{i|(t+1)} + \dot{xo}^2_{i|(t+1)}} \right) cos(\varphi_0) \\ \dot{yt}_{i|(t+1)} = \left(\sqrt{\dot{xo}^2_{i|(t+1)} + \dot{xo}^2_{i|(t+1)}} \right) sin(\varphi_0) \\ \dot{zt}_{i|(t+1)} = -\dfrac{a\, \dot{xt}_{i|(t+1)} + b\, \dot{yt}_{i|(t+1)}}{c} \end{cases} \tag{27}$$

a, b, and c are shown in Equation (18).

4.4. Robot Velocity and Position Iteration

When the robot moves on the mountain ground, the speed of the robot is along the tangent direction of the mountain surface at all times. When planning the path of swarm robots, there is a time interval between each iteration, and at the time between the two iterations, the velocity is also along the tangent direction of the mountain surface. Therefore, the position of the robot needs to be modified when updating its position. According to Equation (4), the velocity is corrected as the average velocity vector before the position is corrected.

It is assumed that the mapping relationship between coordinates $obs_k = (xs_k, ys_k, zs_k)$, zs_k, xs_k, and ys_k of the mountain surface is expressed as $zs_k = \text{Fs}(xs_k, ys_k)$. The robot calculates velocity $\boldsymbol{V}_i(\boldsymbol{t}+\boldsymbol{1})$ according to $\boldsymbol{Vt}_i(\boldsymbol{t}+\boldsymbol{1}) = \left[\dot{xt}_{i|(t+1)}, \dot{yt}_{i|(t+1)}, \dot{zt}_{i|(t+1)}\right]$. For robot R_i, the next velocity $\boldsymbol{V}_i(\boldsymbol{t}+\boldsymbol{1})$ is calculated as follows:

$$\boldsymbol{V}_i(\boldsymbol{t}+\boldsymbol{1}) = \left[\dot{xt}_{i|(t+1)}, \dot{yt}_{i|(t+1)}, \text{Fs}\left(\dot{xt}_{i|(t+1)} + xu_{i|t}, \dot{yt}_{i|(t+1)} + yu_{i|t}\right) - zu_{i|t}\right] \quad (28)$$

After calculating $\boldsymbol{V}_i(\boldsymbol{t}+\boldsymbol{1})$, the robot position is updated as follows:

$$\boldsymbol{R}_i(\boldsymbol{t}+\boldsymbol{1}) = \boldsymbol{R}_i(\boldsymbol{t}) + \boldsymbol{V}_i(\boldsymbol{t}+\boldsymbol{1}) \quad (29)$$

In summary, the multitarget ground search process in an unknown mountain environment is shown in Figure 4.

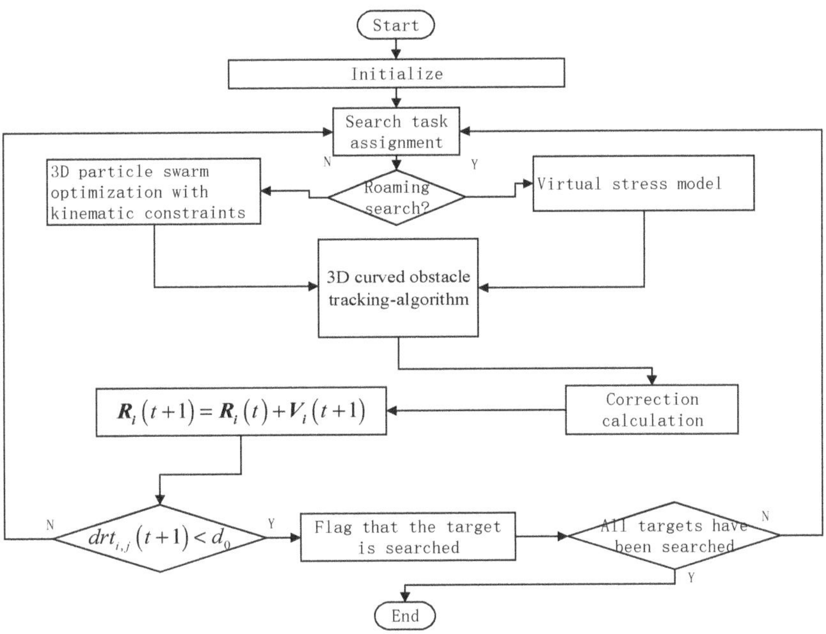

Figure 4. Target search process.

5. Simulation Experiment and Results

The parameters are set based on actual search requirements, as shown in Table 2.

Table 2. Parameter values.

Parameter	Value	Parameter	Value
α	40°	β	30°
n_u	30~60	N_m	6
n_T	10	m	0.1
n_ϕ	360	Δl	1
Vm	10	Q	10^5
r_{com}	300	c_1	1
r_{obs}	100	c_2	1.2
r_{tar}	100	ω	0.5
d_0	10	λ	0.1

As an example, when $n_u = 40$, assuming that slopes in the mountain environment are all less than or equal to β, a schematic of the target search process is shown in Figure 5. Figure 5a is a topographic map of the mountain area for target search. Figure 5b is the top view of the search area when $t = 1$, the robot is in a 100×100 area in the lower left corner, and the target is in an 800×800 area in the middle of the horizontal direction.

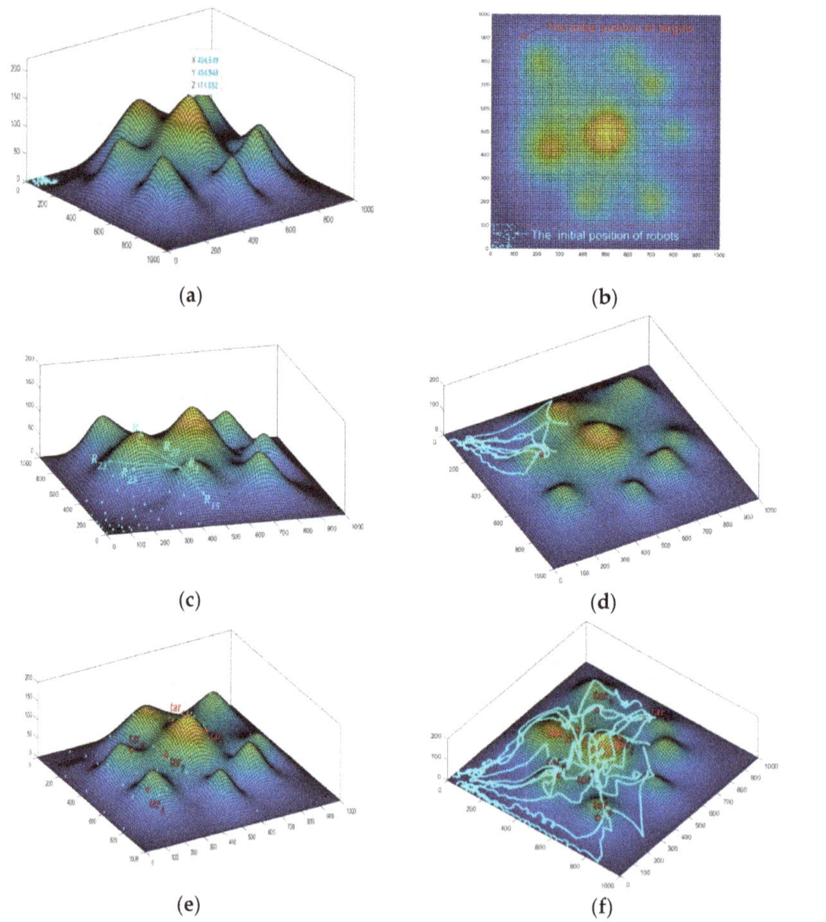

Figure 5. *Cont.*

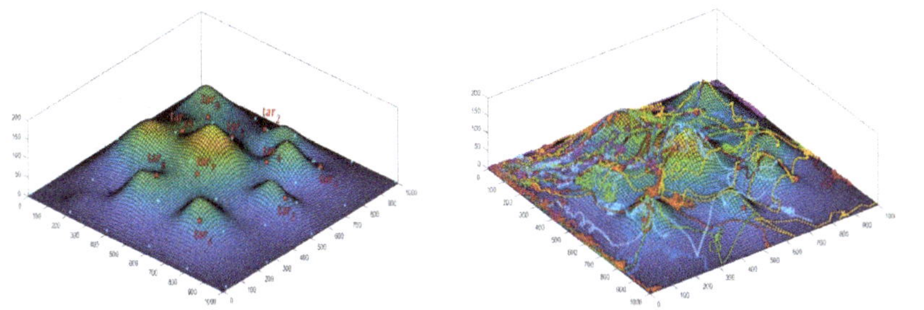

(g) (h)

Figure 5. 3D search simulation diagram. (**a**) Target search terrain, (**b**) When t = 1, the robot and the target are located, (**c**) When t = 37, a robot detects a target signal and forms a suballiance. (**d**) When t = 1~58, a robot searches the trajectory of target tar_8. (**e**) When t = 127, the target is found. (**f**) When t = 185, the trajectory of the robot. (**g**) When t = 255, all targets have been found. (**h**) When t = 1–255, the trajectories of all the robots.

In Figure 5c, R_2 detects the signal of target tar_8, and robots R_6, R_{15}, R_{22}, R_{23}, and R_{27} join the suballiance. In Figure 5d, from $t = 1 \sim 58$, R_2, R_6, R_{15}, R_{22}, R_{23}, R_{27}, R_8, R_{10}, R_{11}, and R_{17} in the suballiance participate in the search for tar_8, and R_{10} searches for R_{10} at t = 58.

In Figure 5e, when t = 127, the robots successfully complete the search for target tar_1. By this time, 5 targets (tar_1, tar_7, tar_6 and tar_8) have been found.

In Figure 5f, t = 1~185, 11 robots including, R_5, R_7, R_{11}, R_{17}, R_{18}, R_{19}, R_{20}, R_{22}, R_{24}, R_{28}, and R_{32}, participate in the collaborative search for target tar_2, and before this, multiple robots participated in the cooperative search for other targets.

In Figure 5g, t = 255, the robot swarm finally found all 10 targets. In Figure 5h, all robot movement tracks of swarm robots in the search for targets are shown, and the robots successfully found all ground targets.

Taking the number of robots $n_u = 30$, 40, 50, or 60 and the number of targets $n_T = 10$, the experiment was repeated 30 times, and the following data as shown in Table 3.

After verifying the effectiveness of the swarm robot target search in a mountain environment with a slope less than or equal to β, the existence of an environment with a slope greater than β in a mountain environment is verified. Assuming $n_u = 40$, the mountain slope is less than or equal to α. A diagram of the target search process is shown in Figure 6.

Figure 6a shows a mountain topographic map for the target search and the positions of the targets and robots when $t = 1$. Figure 6b is a top view of the search area. The areas marked in red indicate that each location within the region has a slope greater than β in one direction.

Figure 6c shows the process of searching for target tar_6. When $t = 48$, robot R_{37} detects the target signal of tar_6 and forms a suballiance with robots R_{27}, R_{30}, R_{40}, R_{23}, and R_{17}. The suballiance starts to search for target tar_6. At $t = 53$, R_{21} also detects the target signal of tar_6, joins the suballiance, and pushes R_{23} out of the suballiance. Finally, when $t = 62$, the target is found, and the suballiance is dissolved. It can be seen from the figure that when a robot is searching for a target, it can move in a direction with a smaller slope according to the 3D curved obstacle tracking algorithm and then smoothly search for a target in a region with a higher slope.

Figure 6d shows the trajectory of robot R_{21} during the period from $t = 1$ to the robot swarm finding all targets. As seen from the trajectory shown in the figure, when the upward slope of the robot's movement direction is too high for it to climb, the robot will adjust its movement direction to the climbing slope according to the 3D curved obstacle tracking algorithm and move as close to the original direction as possible.

Figure 6e shows the positions of the robots and targets at $t = 329$. All targets have been found by swarm robots at this point.

Figure 6f shows the trajectories of all robots in the swarm robot target search process. The figure shows that according to the proposed method, the robots can successfully find all targets in the task area. The robots will be more inclined to move in the region with a slower slope, but they can also move in the direction with a lower slope in a region with a higher slope.

Table 3. Target $n_T = 10$ and mountain slope less than or equal to β: the number of steps and energy consumption required for the robots to complete the task search.

n_u	**Step**				**Energy Consumption ($\times 10^4$)**			
	30	**40**	**50**	**60**	**30**	**40**	**50**	**60**
Data from 30 experiments	481	250	226	217	11.494	8.191	9.109	10.761
	370	283	220	215	8.857	9.197	8.868	10.483
	332	343	271	211	8.277	10.653	10.668	10.429
	455	249	232	219	11.385	8.128	9.474	10.784
	354	247	234	230	8.729	7.799	9.483	11.259
	287	249	242	205	6.956	7.951	9.887	10.260
	356	267	253	216	8.492	8.515	10.152	10.685
	286	230	245	190	7.085	7.466	9.694	9.456
	282	237	260	217	7.008	7.487	10.235	10.699
	311	215	207	227	7.611	7.024	8.480	10.972
	367	297	235	209	9.088	9.579	9.434	10.453
	282	232	235	194	6.962	7.501	9.244	9.474
	316	260	222	230	7.767	8.350	8.954	11.185
	343	295	244	206	8.513	9.532	9.731	10.289
	272	277	225	200	6.736	8.732	8.992	9.941
	240	248	227	220	5.883	8.009	9.217	10.916
	360	262	207	227	9.011	8.231	8.460	11.091
	294	280	244	195	7.068	8.923	9.709	9.728
	379	299	230	216	9.444	9.568	9.310	10.643
	355	269	239	225	8.679	8.590	9.537	10.901
	336	218	236	196	8.364	6.932	9.666	9.756
	253	352	275	221	6.225	10.924	10.894	10.800
	358	255	203	175	8.912	8.152	8.373	8.802
	259	239	227	229	6.288	7.729	9.191	11.157
	336	244	219	217	8.295	7.646	8.863	10.754
	457	251	260	216	11.264	8.030	10.370	10.591
	328	259	243	190	8.016	8.419	9.658	9.565
	261	329	234	203	6.615	10.298	9.465	9.996
	305	263	218	222	7.376	8.497	8.877	11.051
	340	280	227	206	8.399	8.946	9.098	10.251
Mean	**331.833**	**265.967**	**234.667**	**211.467**	**8.160**	**8.500**	**9.436**	**10.438**

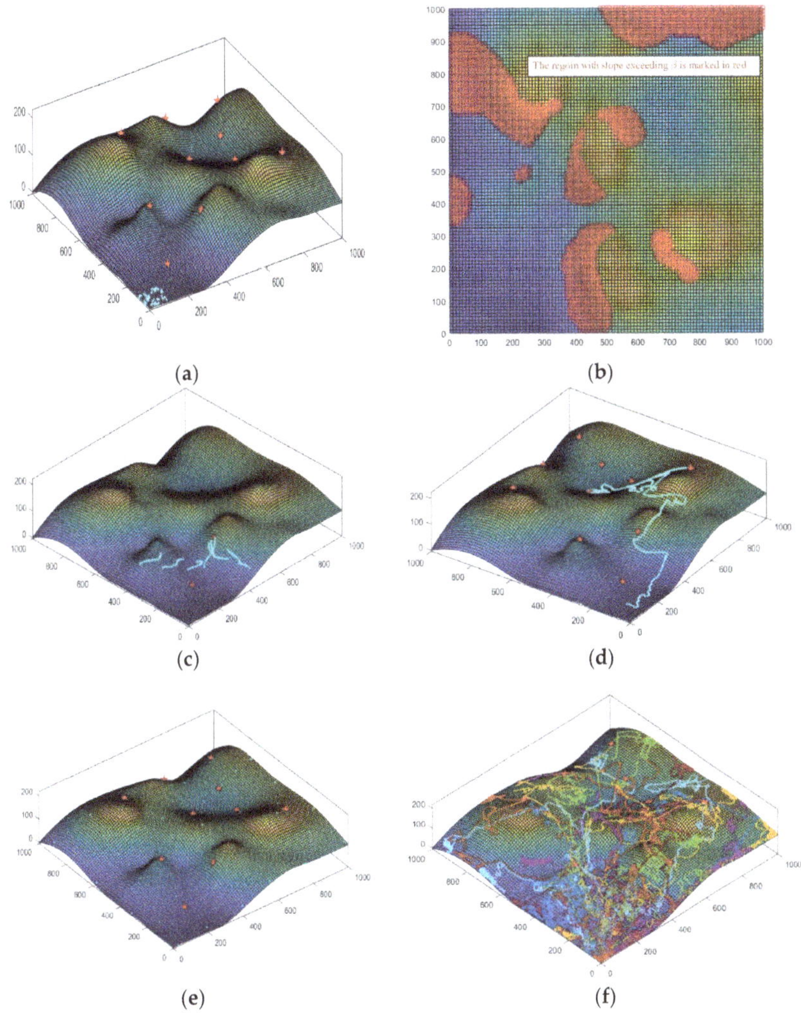

Figure 6. 3D simulation of a target search when the slope is greater than β. (**a**) Mountainous terrain, initial target positions. (**b**) Diagram of the slope over the β zone. (**c**) The robots search for tar_6. (**d**) Trajectory of robot R_{21} (**e**) When $t = 329$, the swarm robots have found all targets. (**f**) The trajectories of all the robots.

Taking the number of robots $n_u = 30$, 40, 50, or 60 and the number of targets $n_T = 10$, the experiment was repeated 30 times, and the following data as shown in Table 4.

Table 4. Target $n_T = 10$ and mountain slope greater than β: the number of steps and energy consumption required for the robots to complete the task search.

n_u	Step				Energy Consumption ($\times 10^4$)			
	30	40	50	60	30	40	50	60
Data from 30 experiments	598	328	452	289	16.480	12.548	21.803	16.916
	545	327	402	533	15.366	12.534	19.240	31.185
	464	401	397	370	12.544	15.245	19.237	21.520
	426	424	292	296	12.164	15.746	14.237	17.282
	496	311	461	544	13.248	11.843	22.182	31.659
	354	400	286	348	10.087	15.227	13.567	20.232
	430	334	456	336	12.090	12.779	21.769	19.447
	368	339	478	248	10.418	12.907	22.795	14.466
	425	508	385	311	11.934	19.194	18.640	18.117
	477	521	342	269	13.494	19.431	16.561	15.709
	436	376	366	310	12.268	14.428	17.494	18.032
	428	391	352	293	11.946	14.865	16.995	16.946
	377	515	650	307	10.343	19.732	30.874	17.752
	506	454	316	387	14.005	17.052	15.301	22.504
	528	452	310	313	14.284	17.315	15.036	18.232
	436	334	434	282	12.299	12.849	20.916	16.437
	502	366	306	393	13.941	13.977	14.680	22.798
	513	531	514	360	14.101	20.287	24.730	20.971
	425	353	269	291	11.521	13.482	12.969	16.984
	474	321	484	257	13.410	12.232	23.167	14.847
	520	586	403	379	14.135	21.730	19.310	21.910
	503	344	380	331	13.953	13.084	18.299	19.251
	346	323	421	410	9.752	12.457	20.163	23.883
	296	375	386	313	8.25	14.396	18.335	18.105
	638	376	342	399	17.944	14.457	16.606	23.201
	447	529	465	349	12.583	19.217	21.740	20.366
	428	321	369	383	11.826	12.105	17.919	22.415
	348	341	364	358	9.813	12.914	17.613	20.917
	509	531	308	538	14.502	20.242	14.828	30.778
	564	523	353	409	15.457	19.799	16.802	23.864
Mean	**460.233**	**407.833**	**391.433**	**353.533**	**12.805**	**15.469**	**18.794**	**20.558**

6. Conclusions

Aiming at the problem of robot swarm multitarget ground search in an unknown 3D mountain environment, this paper, based on unknown 2D environment robot swarm multiobject search research, extends the multiobjective task assignment model, particle swarm optimization algorithm, and virtual force model from a 2D environment to a 3D environment. A new multiobject ground search algorithm for swarm robots in a 3D mountain environment is proposed. Aiming at the problems of swarm robot's speed direction being tangent to the ground, each robot avoids a steep slope that cannot be climbed, and a 3D curved obstacle tracking algorithm that can effectively avoid conflict between the swarm robots and the mountain plans the speed based on the direction tangent to the 3D surface so that a robot can find the ground targets in a mountain environment more quickly and effectively. A 3D particle swarm optimization algorithm with kinematic constraints and a multiobjective task assignment model is used to complete multitarget search in the swarm robot system. A virtual force model is used to calculate the expected velocity during roaming search. During collaborative search, a 3D particle swarm optimization algorithm is used to calculate the expected velocity. After the expected velocity is calculated, the final planned velocity is calculated according to the 3D curved obstacle tracking algorithm. Simulation results show that the proposed method can not only find targets quickly but also avoid conflict effectively. The simulation results demonstrate the effectiveness of the proposed algorithm. However, the environment considered in this study is ideal, and problems such as environmental interference, communication delay, and energy consumption constraints in the swarm robot task execution are not considered. Therefore, in subsequent work, the above problems will be studied.

Author Contributions: Y.Z. and M.W.: conceptualization, methodology, writing—original draft preparation; S.Z.: conceptualization, methodology, writing—reviewing and editing; A.C.: conceptualization, software, data curation, writing—reviewing and editing. All authors have read and agreed to the published version of the manuscript.

Funding: This work was supported by the National Natural Science Foundation of China (62271199).

Institutional Review Board Statement: Not applicable.

Informed Consent Statement: Not applicable.

Data Availability Statement: Not applicable.

Conflicts of Interest: The authors declare no conflict of interest.

References

1. Senanayake, M.; Senthooran, I.; Barca, J.C.; Chung, H.; Kamruzzaman, J.; Murshed, M. Search and tracking algorithms for swarms of robots: A survey. *Robot. Auton. Syst.* **2016**, *75*, 422–434. [CrossRef]
2. Xue, S.; Zhang, J.; Zeng, J. Parallel asynchronous control strategy for target search with swarm robots. *Int. J. Bio-Inspired Comput.* **2009**, *1*, 151–163. [CrossRef]
3. Zhang, Y.; Xue, S.; Zeng, J. Cooperative and Competitive Coordination in Swarm Robotic Search for Multiple Targets. *Robot* **2015**, *37*, 142–151.
4. Li, J.; Tan, Y. A probabilistic finite state machine based strategy for multi-target search using swarm robotics. *Appl. Soft Comput.* **2019**, *77*, 467–483. [CrossRef]
5. He, X.; Zhou, S.; Zhang, H.; Wu, L.; Zhou, Y.; He, Y.; Wang, M. Multiobjective coordinated search algorithm for swarm of UAVs based on 3D-simplified virtual forced model. *Int. J. Syst. Sci.* **2020**, *51*, 2635–2652. [CrossRef]
6. Phung, M.D.; Ha, Q.P. Motion-encoded particle swarm optimization for moving target search using UAVs. *Appl. Soft Comput.* **2020**, *97*, 106705. [CrossRef]
7. Cao, X.; Sun, H.; Jan, G.E. Multi-AUV cooperative target search and tracking in unknown underwater environment. *Ocean Eng.* **2018**, *150*, 1–11. [CrossRef]
8. Tang, Q.; Yu, F.; Zhang, Y.; Zhang, J. A stigmergetic method based on vector pheromone for target search with swarm robots. *J. Exp. Theor. Artif. Intell.* **2020**, *32*, 533–555. [CrossRef]
9. Brown, D.; Sun, L. Exhaustive mobile target search and non-intrusive reconnaissance using cooperative unmanned aerial vehicles. In Proceedings of the 2017 International Conference on Unmanned Aircraft Systems (ICUAS), Miami, FL, USA, 13–16 June 2017; pp. 1425–1431.
10. Brown, D.; Sun, L. Dynamic exhaustive mobile target search using unmanned aerial vehicles. *IEEE Trans. Aerosp. Electron. Syst.* **2019**, *55*, 3413–3423. [CrossRef]
11. Pan, W.T. A new Fruit Fly Optimization Algorithm: Taking the financial distress model as an example. *Knowl. Based Syst.* **2012**, *26*, 69–74. [CrossRef]
12. Yang, X.-S. A new metaheuristic bat-inspired algorithm. In *Nature Inspired Cooperative Strategies for Optimization (NICSO 2010)*; González, J.R., Pelta, D.A., Cruz, C., Terrazas, G., Krasnogor, N., Eds.; Springer: Berlin/Heidelberg, Germany, 2010; pp. 65–74.
13. Mirjalili, S.; Mirjalili, S.M.; Lewis, A. Grey Wolf Optimizer. *Adv. Eng. Softw.* **2014**, *69*, 46–61. [CrossRef]
14. Dorigo, M.; Gambardella, L.M. Ant colony system: A cooperative learning approach to the traveling salesman problem. *IEEE Trans. Evol. Comput.* **1997**, *1*, 53–66. [CrossRef]
15. Guastella, D.C.; Cavallaro, N.D.; Melita, C.D.; Savasta, M.; Muscato, G. 3D path planning for UAV swarm missions. In *ICMSCE 2018: Proceedings of the 2018 2nd International Conference on Mechatronics Systems and Control Engineering*; Association for Computing Machinery: New York, NY, USA, 2018.
16. Xie, Y.; Han, L.; Dong, X.; Li, Q.; Ren, Z. Bio-inspired adaptive formation tracking control for swarm systems with application to UAV swarm systems. *Neurocomputing* **2021**, *453*, 272–285. [CrossRef]
17. Wan, N.; Li, Z.; Liang, X.L.; Wang, Y.B. Cooperative region search of UAV swarm with limited communication distance. *J. Systems Engineering and Electronics* **2022**, *44*, 1615–1625.
18. Qi, B.; Li, M.; Yang, Y.; Wang, X. Research on UAV path planning obstacle avoidance algorithm based on improved artificial potential field method. *J. Phys. Conf. Ser.* **2021**, *1948*, 012060. [CrossRef]
19. Yu, W.; Lu, Y. UAV 3D environment obstacle avoidance trajectory planning based on improved artificial potential field method. *J. Phys. Conf. Ser.* **2021**, *1885*, 022020. [CrossRef]
20. Zhou, Y.; Chen, A.; Zhang, H.; Zhang, X.; Zhou, S. Multitarget Search of Swarm Robots in Unknown Complex Environments. *Complexity* **2020**, *2020*, 8643120. [CrossRef]

21. Zhou, S.W.; Zhang, X.; Zhang, H.Q.; Zhou, Y.; Li, C.Y. Coordinated Control of Swarm Robots for Multi-target Search Based on a Simplified Virtual-Force Model. *Robots* **2016**, *11*, 641–650.
22. Zhou, Y.; Chen, A.; He, X.; Bian, X. Multi-Target Coordinated Search Algorithm for Swarm Robotics Considering Practical Constraints. *Front. Neurorobotics* **2021**, *15*, 753052. [CrossRef]
23. Pugh, J.; Martinoli, A. Inspiring and modeling multi-robot search with particle swarm optimization. In Proceedings of the Swarm Intelligence Symposium, 2007, Honolulu, HI, USA, 1–5 April 2007; pp. 332–339.
24. Viswanathan, G.M.; Buldyrev, S.V.; Havlin, S.; Da Luz, M.; Raposo, E.; Stanley, H.E. Optimizing the success of random searches. *Nature* **1999**, *401*, 911–914. [CrossRef]
25. Bénichou, O.; Loverdo, C.; Moreau, M.; Voituriez, R. Two-dimensional intermittent search processes: An alternative to Lévy flight strategies. *Phys. Rev. E Stat. Nonlinear Soft Matter Phys.* **2006**, *74*, 020102. [CrossRef] [PubMed]
26. Huang, T.Y.; Chen, X.B.; Xu, W.B.; Zhou, Z.W. A self-organizing cooperative hunting by swarm robotic systems based on loose-preference rule. *Acta Autom. Sin.* **2013**, *39*, 57–68. [CrossRef]
27. He, X.J.; Zhou, S.W.; Zhang, H.Q.; Zhou, Y. A 3D Parallel Multi-target Search Coordination Control Strategy for Swarm UAVS. *Inf. Control* **2020**, *49*, 605–614.
28. Zhang, H.Q. Research on Self-Organizing Cooperative Hunting by Swarm Robots Based on Simplified Virtual-Force Model. Ph.D. Thesis, Hunan University, Changsha, China, 2015.
29. Kennedy, J.; Eberhart, R. Particle swarm optimization. In Proceedings of the ICNN'95—International Conference on Neural Networks, Perth, WA, Australia, 27 November–1 December 1995; Volume 4, pp. 1942–1948.
30. Liang, J.J.; Suganthan, P.N. Dynamic multi-swarm particle swarm optimizer with local search. In Proceedings of the 2005 IEEE Congress on Evolutionary Computation, Edinburgh, UK, 2–5 September 2005; pp. 124–129.
31. Sun, W.; Lin, A.; Yu, H.; Liang, Q.; Wu, G. All-dimension neighborhood based particle swarm optimization with randomly selected neighbors. *Inf. Sci.* **2017**, *405*, 141–156. [CrossRef]

applied sciences

MDPI

Article

Dissipativity Analysis of Large-Scale Networked Systems

Yuanfei Sun [1], Jirong Wang [2,3], Huabo Liu [1,4,*]

1 School of Automation, Qingdao University, Qingdao 266071, China
2 School of Mechanical and Electrical Engineering, Qingdao University, Qingdao 266071, China
3 Weihai Innovation Insitute, Qingdao University, Weihai 264200, China
4 Shandong Key Laboratory of Industrial Control Technology, Qingdao 266071, China
* Correspondence: hbliu@qdu.edu.cn

Abstract: This paper investigates the dissipativity analysis of large-scale networked systems with linear time-invariant dynamics. The networked system is composed of a large number of subsystems whose connections are arbitrary, and each subsystem can have different dynamics. A sufficient and necessary condition for the strict dissipativity analysis of the networked system is derived, which takes advantage of the block-diagonal structure of the system parameter matrix and the sparseness characteristics of the subsystem interconnections. Then, a necessary condition and a sufficient condition that depend only on a single subsystem parameter are given separately. Numerical simulations illustrate that compared with the existing results, the conditions suggested in this paper have higher computational efficiency in the dissipative analysis of large-scale networked systems.

Keywords: dissipativity; large-scale system; linear matrix inequality; networked system; sparseness

Citation: Sun, Y.; Wang, J.; Liu, H. Dissipativity Analysis of Large-Scale Networked Systems. *Appl. Sci.* **2023**, 13, 1214. https://doi.org/10.3390/app13021214

Academic Editors: Xiongbo Wan, Sheng Du, Wei Wang and Hao Fu

Received: 18 December 2022
Revised: 13 January 2023
Accepted: 13 January 2023
Published: 16 January 2023

1. Introduction

In recent years, the research of large-scale networked systems has attracted great attentions [1–4]. The system can be considered as composed of a large number of subsystems with different spatial locations connected in a certain way [5]. Generally, subsystems exchange information with their neighbors straightforwardly and predictably, but the system often exhibits complicated dynamic behavior when seen as a whole. Such systems have extensive application background, including airplane formation flight [6], power network distributed system [7], automated highways [8], multi-agent formation systems [9], and so on. For such a complex system, the classic method of bringing all the subsystems together and analyzing it as a single large-scale system has very strict requirements on the calculation speed and accuracy of the computer, which will inevitably bring computational difficulties. Therefore, using the system structure to find more efficient computational conditions is of great engineering significance for the dissipativity analysis of large-scale networked systems.

There are already many results on the performance analysis of networked systems, but the research on dissipativity is not mature enough and needs to be further developed and improved. In 1972, the famous scholar Willems first put forward the concept of dissipativity [10,11]. Dissipativity describes the equilibrium relationship between the system's internal energy, which is a vital concept in theoretical research and practical application. Its essential meaning is that there is a non-negative energy function (called storage function) so that the energy supply rate of the system is always greater than the loss of energy inside the dynamic system. Based on Willems' work, many scholars have done in-depth studies on dissipativity and obtained rich results, which have played a significant role in the field of circuit, system, and control theory. Refs. [12,13] respectively studied the dissipative control problems of linear continuous-time and discrete systems based on linear matrix inequality (LMI) methods. A simplified mathematical model of the interconnected two-machine power system was established in [14], and its non-linear dynamic behavior such as dissipativity properties was analyzed. In [15], the analysis and improvement of the dissipativity performance of interconnected passive systems are studied. For networked control systems, Ref. [16] obtained some new sufficient conditions by utilizing Lyapunov stability theory and LMI technology to ensure that the closed-loop system is finite-time limited and dissipative. In [17], a distributed controller was created to ensure the dissipativity of a networked system made up of dynamically coupled subsystems. Its control synthesis is done locally at the subsystem level and doesn't involve the relationship among subsystems, hence it has certain drawbacks. The linear dynamic system with the interconnection structure specified by the directed graph is studied in [18]. Based on the dissipativity inequality, an LMI for calculating system performance is established and the concept of local dissipativity is defined. Using the knowledge of graph theory to analyze large-scale networked systems has certain constraints on the dynamic characteristics and connection modes of subsystems, which has certain limitations in practical application [19,20].

Considering that the system structure of large-scale networked systems usually has sparse characteristics or specific structural forms [21–23], in the large-scale connected systems discussed in [24], the concept of internal input and output is introduced to represent the connections and functions among subsystems, and the connection relationship among sub-units of the entire system is described by subsystem connection matrix. This description method takes into account the situation where the dynamic characteristics of the subsystems are different and the connection relationship of the subsystems is arbitrary. The previously mentioned UAV formation flight refers to the arrangement of multiple UAVs in a certain formation so that they maintain in formation or change their relative positions within a limited scope during the flight. To maintain a certain formation shape, information interaction is required among the UAVs. In a centralized strategy, each UAV has to know information about the whole formation, demanding substantial information interaction. It is computationally intensive and requires the high performance of the airborne computer. In fact, each UAV can interact with its position, speed, attitude, and motion target with only the UAVs connected to it in the formation. In this way, the amount of computation is greatly reduced, and the system is relatively simple to implement. It is this sparse property or specific structural form among subsystem connections that we exploit to give more computationally efficient dissipativity criteria for large-scale networked systems. Dissipativity explains some of the energy losses and control problems of control systems and is a more general performance indicator of system performance.

In this paper, our objective is to reduce the computational burden of dissipativity analysis for large-scale network systems with a large number of subsystems. In general, the large-scale networked systems studied in this paper have the following characteristics. The first is that the scale of the system is large, including many subsystems. The second and most important point is that the interaction among subsystems of large-scale networked systems is usually sparse or has a specific structural form. We introduce intermediate variables in networked systems to describe the relationship among subsystems, which is more general and explicitly characterizes the structural characteristics of large-scale systems. In this regard, this paper proposes several new LMI conditions, which effectively use the block diagonal structure of the system parameter matrix and the sparsity of the subsystem connection matrix, avoiding the inverse computation of high-dimensional matrices. The simulation results indicate that the conditions proposed in this paper are more efficient than the existing results.

The following is the structure of the paper. The model and the definition of dissipativity for the networked system and some preliminary results are given in Section 2. In Section 3, some conditions for dissipative analysis of networked systems are given, and the relationship between these conditions and existing conditions is discussed. Some numerical simulation results are presented in Section 4. The research results of this paper are summarized in Section 5, and the direction of further research is also proposed here.

Notation 1. *The symbol R is used to denote the set of real numbers, and the vector space produced by real numbers of appropriate dimensions is denoted as* $R^{\#}$. $col\{Z_i|_{i=1}^{L}\}$ *denotes the vector/matrix stacked by* $Z_i(i = 1, 2, \ldots, L)$*, and* $diag\{Z_i|_{i=1}^{L}\}$ *signifies a block diagonal matrix with* Z_i *as the i-th diagonal block.* $\left\{Z_{ij}|_{i=1,j=1}^{i=M,j=N}\right\}$ *stands a matrix with* $M \times N$ *blocks, and its i-th row j-th column block matrix being* Z_{ij}. 0_n *and* $0_{n\times m}$ *represent the n dimensional zero column vector and the* $n \times m$ *dimensional zero matrix respectively, the dimension subscript is omitted if there is no ambiguity, and the identity matrix I is the same. The superscript T represents the transpose of a matrix or vector, and* $(*)^T WZ$ *or* $ZW(*)^T$ *is shorthand for* $Z^T WZ$ *or* ZWZ^T.

2. System Description and Some Preliminaries

The networked system Γ is consisted of N linear time-invariant subsystems, and the i-th subsystem Γ_i is defined by the state-space model below,

$$\begin{bmatrix} \dot{x}(t,i) \\ z(t,i) \\ y(t,i) \end{bmatrix} = \begin{bmatrix} A_{\mathrm{TT}}(i) & A_{\mathrm{TS}}(i) & B_{\mathrm{T}}(i) \\ A_{\mathrm{ST}}(i) & A_{\mathrm{SS}}(i) & B_{\mathrm{S}}(i) \\ C_{\mathrm{T}}(i) & C_{\mathrm{S}}(i) & D_{\mathrm{T}}(i) \end{bmatrix} \begin{bmatrix} x(t,i) \\ v(t,i) \\ u(t,i) \end{bmatrix}, \quad (1)$$

in which t and i denote respectively for the temporal variable and the index number of a subsystem, $i = 1, 2, \ldots, N$. Moreover, $x(t,i)$ is the state vector of the i-th subsystem Γ_i at time t. $y(t,i)$ and $u(t,i)$ represent the external output vector and external input vector of the Γ_i, respectively. $z(t,i)$ and $v(t,i)$ are the output vector to other subsystems and input vector from others, which is also called internal output vector and input vector. The connection relationship among subsystems can be expressed as

$$v(t) = \Phi z(t), \quad (2)$$

here, $v(t) = col\left\{v(t,i)|_{i=1}^{N}\right\}$ and $z(t) = col\left\{z(t,i)|_{i=1}^{N}\right\}$. Φ is called the subsystem connectivity matrix. We assume that each row of the matrix Φ has only one non-zero element equal to one and there are no columns in which all of the items are equal to zero. That means the internal output channels of a subsystem can affect the internal inputs channels of other subsystems, and some subsystem internal input channels depend on the internal output of multiple subsystems. This assumption, as explained in [23], does not jeopardize the generality of the adopted system model. Approximate power-law degree distribution widely exists in engineering systems, such as protein interaction networks, gene regulatory networks, power systems, the Internet, etc. [23]. In these systems, the dimension of the subsystem connection matrix Φ is usually much smaller than the state dimension of the system, and the interactions among subsystems are sparse.

In this paper, we assume that the dimensions of vectors $x(t,i)$, $v(t,i)$, $z(t,i)$, $u(t,i)$ and $y(t,i)$ are m_{xi}, m_{vi}, m_{zi}, m_{ui} and m_{yi}, respectively. Based on the above assumptions and Equation (2), the dimension of the matrix Φ is $\sum_{i=1}^{N} m_{vi} \times \sum_{i=1}^{N} m_{zi}$. Then we can get $\Phi^T\Phi = \Sigma^2$,where $\Sigma^2 = \mathrm{diag}\left\{\Sigma_j^2\Big|_{j=1}^{N}\right\}$, $\Sigma_j^2 = \mathrm{diag}\left\{m(i)|_{i=M_{z,j-1}+1}^{M_{z,j}}\right\}$, $M_{z,i} = \sum_{k=1}^{i} m_{zk}$, $m(i)$ indicates the number of subsystems directly affected by the i-th element of the vector $z(t)$, $i = 1, \cdots, \sum_{k=1}^{N} m_{vk}$, $j = 1, \cdots, N$.

To simplify the mathematical derivation, we define the following matrix, $A_{*\#} = diag\left\{A_{*\#}(i)|_{i=1}^{N}\right\}$, $B_* = diag\left\{B_*(i)|_{i=1}^{N}\right\}$,$C_* = diag\left\{C_*(i)|_{i=1}^{N}\right\}$and $D_* = diag\left\{D_*(i)|_{i=1}^{N}\right\}$ in which $*, \# = \mathrm{T}, \mathrm{S}$. By exploiting the connection relationship among subsystems, the dynamic system Γ may be expressed equivalently in the following state-space form,

$$\begin{bmatrix} \dot{x}(t) \\ y(t) \end{bmatrix} = \begin{bmatrix} A & B \\ C & D \end{bmatrix} \begin{bmatrix} x(t) \\ u(t) \end{bmatrix}, \quad (3)$$

where

$$A = A_{\mathrm{TT}} + A_{\mathrm{TS}}\Phi(I - A_{\mathrm{SS}}\Phi)^{-1}A_{\mathrm{ST}},$$

$$B = B_T + A_{TS}\Phi(I - A_{SS}\Phi)^{-1}B_S,$$
$$C = C_T + C_S\Phi(I - A_{SS}\Phi)^{-1}A_{ST},$$
$$D = D_T + C_S\Phi(I - A_{SS}\Phi)^{-1}B_S.$$

Note: Well -posedness is very important in system design, and ill-posed systems are usually difficult to control or impossible to estimate [25–27]. Therefore, this paper assumes that System Γ is well-posed, which means that $(I - A_{SS}\Phi)^{-1}$ exists.

This paper intends to establish computationally effective conditions for the dissipativity analysis of large-scale networked systems Γ. The concept of dissipativity is very important in the system, whether from the perspective of theoretical research or the perspective of practical application. Roughly speaking, dissipative systems can be described as such properties. At any time, the energy that the system may provide cannot exceed the energy already supplied. We first describe the definition of dissipativity for System Γ.

The definition is related to the supply function. For the i-th subsystem Γ_i, its supply function is defined as

$$s_i(u(t,i), y(t,i)) = \begin{bmatrix} y(t,i) \\ u(t,i) \end{bmatrix}^T Q(i) \begin{bmatrix} y(t,i) \\ u(t,i) \end{bmatrix}, \tag{4}$$

where $Q(i)$ is a symmetric matrix of suitable dimensions.

Definition 1. *The large-scale networked system (1) and (2) with $x(0,i) = 0$ is said to be dissipative with supply function $s_i(u(t,i), y(t,i))$ if and only if there is a matrix $P(i) \geq 0$, such that,*

$$\int_{t_0}^{t_1} \sum_{i=1}^{N} s_i(u(t,i), y(t,i))dt \geq \sum_{i=1}^{N} x^T(t_1,i)P(i)x(t_1,i) - \sum_{i=1}^{N} x^T(t_0,i)P(i)x(t_0,i) \tag{5}$$

holds for all $t_0 \leq t_1$.

According to the definition, the supply function can be interpreted as the energy transferred to the system, which means that within a period of time $[t_0, t_1]$, as long as $\int_{t_0}^{t_1} \sum_{i=1}^{N} s_i(u(t,i), y(t,i))dt$ is positive, the system will work normally, otherwise, the system will not work. $\sum_{i=1}^{N} x^T(t_1,i)P(i)x(t_1,i) - \sum_{i=1}^{N} x^T(t_0,i)P(i)x(t_0,i)$ represents the actual energy consumption of the system after the time interval $t_1 - t_0$. Therefore, Equation (5) shows that in any time period $[t_0, t_1]$, the energy change inside the system will not exceed the energy supplied by the outside.

It can be seen from the following derivation that the definition of dissipativity for the networked system (1) and (2) are consistent with the one in [28] based on (3).

The supply function based on the large-scale networked system (1) and (2) is as follows,

$$s(u(t), y(t)) = \sum_{i=1}^{N} s_i(u(t,i), y(t,i)) = \begin{bmatrix} \begin{bmatrix} y(t,1) \\ u(t,1) \end{bmatrix} \\ \begin{bmatrix} y(t,2) \\ u(t,2) \end{bmatrix} \\ \vdots \\ \begin{bmatrix} y(t,N) \\ u(t,N) \end{bmatrix} \end{bmatrix}^T Q \begin{bmatrix} \begin{bmatrix} y(t,1) \\ u(t,1) \end{bmatrix} \\ \begin{bmatrix} y(t,2) \\ u(t,2) \end{bmatrix} \\ \vdots \\ \begin{bmatrix} y(t,N) \\ u(t,N) \end{bmatrix} \end{bmatrix}. \tag{6}$$

The supply function of System (3) is

$$
\begin{aligned}
s(u(t),y(t)) &= \begin{bmatrix} y(t) \\ u(t) \end{bmatrix}^T Q_1 \begin{bmatrix} y(t) \\ u(t) \end{bmatrix} \\
&= \begin{bmatrix} \begin{bmatrix} y(t,1) \\ y(t,2) \\ \vdots \\ y(t,N) \end{bmatrix} \\ \begin{bmatrix} u(t,1) \\ u(t,2) \\ \vdots \\ u(t,N) \end{bmatrix} \end{bmatrix}^T Q_1 \begin{bmatrix} \begin{bmatrix} y(t,1) \\ y(t,2) \\ \vdots \\ y(t,N) \end{bmatrix} \\ \begin{bmatrix} u(t,1) \\ u(t,2) \\ \vdots \\ u(t,N) \end{bmatrix} \end{bmatrix} = \begin{bmatrix} \begin{bmatrix} y(t,1) \\ u(t,1) \end{bmatrix} \\ \begin{bmatrix} y(t,2) \\ u(t,2) \end{bmatrix} \\ \vdots \\ \begin{bmatrix} y(t,N) \\ u(t,N) \end{bmatrix} \end{bmatrix}^T Q \begin{bmatrix} \begin{bmatrix} y(t,1) \\ u(t,1) \end{bmatrix} \\ \begin{bmatrix} y(t,2) \\ u(t,2) \end{bmatrix} \\ \vdots \\ \begin{bmatrix} y(t,N) \\ u(t,N) \end{bmatrix} \end{bmatrix},
\end{aligned}
\tag{7}
$$

in which $y(t) = col\{y(t,i)|_{i=1}^{N}\}$, $u(t) = col\{u(t,i)|_{i=1}^{N}\}$, $Q = diag\{Q(i)|_{i=1}^{N}\}$ and

$$
Q = \begin{bmatrix}
I & 0 & 0 & 0 & & 0 & 0 \\
0 & \vdots & I & \vdots & & \vdots & \vdots \\
\vdots & \vdots & 0 & \vdots & & \vdots & \vdots \\
\vdots & \vdots & \vdots & \vdots & & 0 & \vdots \\
0 & 0 & 0 & 0 & \cdots\cdots & I & 0 \\
0 & I & 0 & 0 & & 0 & 0 \\
\vdots & \vdots & \vdots & I & & \vdots & \vdots \\
\vdots & \vdots & \vdots & \vdots & & \vdots & \vdots \\
\vdots & \vdots & \vdots & \vdots & & \vdots & 0 \\
0 & 0 & 0 & 0 & & 0 & I
\end{bmatrix} Q_1(*)^T.
$$

For the convenience of the following discussion, we introduce the following preliminary results that need to be used.

Lemma 1 ([29]). *For matrices L and U with compatible dimensions, there is a scalar $\alpha > 0$ such that,*

$$
LU + U^T L^T \leq \alpha L L^T + \alpha^{-1} U^T U. \tag{8}
$$

Lemma 2 ([30]). *Given symmetric matrices F and G with appropriate dimensions, if $v^T F v > 0$ can be obtained for every non-zero vector v satisfying $v^T G v = 0$, then there must be a real number r such that $F+rG$ is positive definite, and vice versa.*

Lemma 3 ([29]). *For an LMI in the form of an $M \times M (M \geq 1)$ block matrix: $G(P) < 0$, except for the symmetric independent variable matrix P, other known coefficient matrices or constant matrices are all block diagonal matrices of appropriate dimensions, and all have $N (N > 1)$ diagonals. If it is divided into blocks, there is a full block feasible solution P for this LMI, and there must be a feasible solution for the diagonal division of the appropriate dimension.*

3. Dissipativity Analysis

In [28], the dissipativity criterion of System (3) is proposed.

Lemma 4. *Assume that the networked system Γ is controllable. Then, System (3) is strictly dissipative with the supply function $s(u(t),y(t))$ if and only if there exists a matrix $P > 0$ such that,*

$$
\begin{bmatrix} A^T P + PA & PB \\ B^T P & 0 \end{bmatrix} - \begin{bmatrix} C & D \\ 0 & I \end{bmatrix}^T Q \begin{bmatrix} C & D \\ 0 & I \end{bmatrix} < 0. \tag{9}
$$

Note that the matrices A, B, C and D in the condition of Lemma 4 all contain $(I - A_{SS}\Phi)^{-1}$ terms. Although the subsystem connection matrix Φ is sparse and the system parameter $A_{*\#}$, B_*, C_* and D_* with $*, \# = \mathrm{T}, \mathrm{S}$ are block diagonal, the matrix $(I - A_{\mathrm{SS}}\Phi)^{-1}$ is generally dense. When there are a large number of subsystems in large-scale networked systems, the calculation of matrices A, B, C, and D involves the inversion of high-dimensional matrices. Therefore, when the scale of the networked system increases, the computational complexity of Equation (9) will become very high.

Lemma 4 is a dissipative analysis condition based on lumped networked model. Due to the establishment of the lumped model, the connection relationship among subsystems is hidden inside the parameters, and its structural information is not effectively utilized. As a result, for large-scale networked systems, the use of this condition for dissipative testing will inevitably bring computational difficulties and even cannot be calculated.

Then, to reduce the computational difficulty caused by the increase of system scale, we establish a computationally efficient sufficient, and necessary condition for the strict dissipativity analysis of large-scale networked systems. This condition effectively utilizes the sparse structure of the subsystem connection matrix Φ in the networked system, that is, each subsystem is only connected to a limited number of other subsystems.

Theorem 1. *Assume that the networked system Γ is controllable. Then, System Γ is strictly dissipative with the supply function $s_i(u(t,i), y(t,i))$ if and only if there exists a symmetric positive definite matrix P and a positive scalar h such that,*

$$\begin{aligned} &(*)^T \begin{bmatrix} \begin{bmatrix} 0 & P \\ P & 0 \end{bmatrix} & \\ & -Q \end{bmatrix} \begin{bmatrix} 0 & I & 0 \\ A_{\mathrm{TS}} & A_{\mathrm{TT}} & B_{\mathrm{T}} \\ C_{\mathrm{S}} & C_{\mathrm{T}} & D_{\mathrm{T}} \\ 0 & 0 & I \end{bmatrix} \\ &-h \times (*)^T \begin{bmatrix} I & -\Phi \\ -\Phi^T & \Sigma^2 \end{bmatrix} \begin{bmatrix} I & \begin{bmatrix} 0 & 0 \end{bmatrix} \\ A_{\mathrm{SS}} & \begin{bmatrix} A_{\mathrm{ST}} & B_{\mathrm{S}} \end{bmatrix} \end{bmatrix} < 0. \end{aligned} \tag{10}$$

Proof of Theorem 1. Equation (9) can be expressed equivalently as follows,

$$(*)^T \begin{bmatrix} \begin{bmatrix} 0 & P \\ P & 0 \end{bmatrix} & \\ & -Q \end{bmatrix} \begin{bmatrix} \begin{bmatrix} I & 0 \\ A & B \end{bmatrix} \\ \begin{bmatrix} C & D \\ 0 & I \end{bmatrix} \end{bmatrix} < 0. \tag{11}$$

We express Equation (11) in the following equivalent form,

$$(*)^T \begin{bmatrix} \begin{bmatrix} 0 & P \\ P & 0 \end{bmatrix} & \\ & -Q \end{bmatrix} \begin{bmatrix} I & 0 & 0 & 0 \\ 0 & 0 & I & 0 \\ 0 & 0 & 0 & I \\ 0 & I & 0 & 0 \end{bmatrix} \begin{bmatrix} \begin{bmatrix} I & 0 \\ 0 & I \end{bmatrix} \\ \begin{bmatrix} A & B \\ C & D \end{bmatrix} \end{bmatrix} < 0. \tag{12}$$

Matrices A, B, C, and D can be written as follows,

$$\begin{bmatrix} A & B \\ C & D \end{bmatrix} = \begin{bmatrix} A_{\mathrm{TT}} & B_{\mathrm{T}} \\ C_{\mathrm{T}} & D_{\mathrm{T}} \end{bmatrix} + \begin{bmatrix} A_{\mathrm{TS}} \\ C_{\mathrm{S}} \end{bmatrix} \Phi (I - A_{\mathrm{SS}}\Phi)^{-1} \begin{bmatrix} A_{\mathrm{ST}} & B_{\mathrm{S}} \end{bmatrix}. \tag{13}$$

Substituting the above formula into Equation (12), we can get that,

$$\begin{aligned} &(*)^T \begin{bmatrix} \begin{bmatrix} 0 & P \\ P & 0 \end{bmatrix} & \\ & -Q \end{bmatrix} \begin{bmatrix} 0 & I & 0 \\ A_{\mathrm{TS}} & A_{\mathrm{TT}} & B_{\mathrm{T}} \\ C_{\mathrm{S}} & C_{\mathrm{T}} & D_{\mathrm{T}} \\ 0 & 0 & I \end{bmatrix} \\ &\times \begin{bmatrix} \Phi (I - A_{\mathrm{SS}}\Phi)^{-1} A_{\mathrm{ST}} & \Phi (I - A_{\mathrm{SS}}\Phi)^{-1} B_{\mathrm{S}} \\ I & 0 \\ 0 & I \end{bmatrix} < 0. \end{aligned} \tag{14}$$

Then we define matrices F, M, and K as follows,

$$F = (*)^T \begin{bmatrix} -\begin{bmatrix} 0 & P \\ P & 0 \end{bmatrix} & \\ & Q \end{bmatrix} \begin{bmatrix} 0 & I & 0 \\ A_{\mathrm{TS}} & A_{\mathrm{TT}} & B_{\mathrm{T}} \\ C_{\mathrm{S}} & C_{\mathrm{T}} & D_{\mathrm{T}} \\ 0 & 0 & I \end{bmatrix}, \tag{15}$$

$$M = \begin{bmatrix} \Phi(I - A_{\mathrm{SS}}\Phi)^{-1} \begin{bmatrix} A_{\mathrm{ST}} & B_{\mathrm{S}} \end{bmatrix} \\ \begin{bmatrix} I & 0 \\ 0 & I \end{bmatrix} \end{bmatrix}, \tag{16}$$

$$K = \begin{bmatrix} I & -\Phi \end{bmatrix} \begin{bmatrix} I & \begin{bmatrix} 0 & 0 \end{bmatrix} \\ A_{\mathrm{SS}} & \begin{bmatrix} A_{\mathrm{ST}} & B_{\mathrm{S}} \end{bmatrix} \end{bmatrix}. \tag{17}$$

Obviously, $M^T(-F)M < 0$ .When $v = M\zeta$, $\zeta \in R^{\#}$, for any $v \neq 0$, we can get that $Kv = 0$, which means $v^T F v > 0$. According to Lemma 2, there must be a real number h such that,

$$\begin{aligned} &(*)^T \begin{bmatrix} -\begin{bmatrix} 0 & P \\ P & 0 \end{bmatrix} & \\ & Q \end{bmatrix} \begin{bmatrix} 0 & I & 0 \\ A_{\mathrm{TS}} & A_{\mathrm{TT}} & B_{\mathrm{T}} \\ C_{\mathrm{S}} & C_{\mathrm{T}} & D_{\mathrm{T}} \\ 0 & 0 & I \end{bmatrix} \\ &+h \times (*)^T \begin{bmatrix} I & -\Phi \\ -\Phi^T & \Sigma^2 \end{bmatrix} \begin{bmatrix} I & \begin{bmatrix} 0 & 0 \end{bmatrix} \\ A_{\mathrm{SS}} & \begin{bmatrix} A_{\mathrm{ST}} & B_{\mathrm{S}} \end{bmatrix} \end{bmatrix} > 0. \end{aligned} \tag{18}$$

That is,

$$\begin{aligned} &(*)^T \begin{bmatrix} \begin{bmatrix} 0 & P \\ P & 0 \end{bmatrix} & \\ & -Q \end{bmatrix} \begin{bmatrix} 0 & I & 0 \\ A_{\mathrm{TS}} & A_{\mathrm{TT}} & B_{\mathrm{T}} \\ C_{\mathrm{S}} & C_{\mathrm{T}} & D_{\mathrm{T}} \\ 0 & 0 & I \end{bmatrix} \\ &-h \times (*)^T \begin{bmatrix} I & -\Phi \\ -\Phi^T & \Sigma^2 \end{bmatrix} \begin{bmatrix} I & \begin{bmatrix} 0 & 0 \end{bmatrix} \\ A_{\mathrm{SS}} & \begin{bmatrix} A_{\mathrm{ST}} & B_{\mathrm{S}} \end{bmatrix} \end{bmatrix} < 0. \end{aligned} \tag{19}$$

The characterization of the left term of Equation (19) shows that if the inequality has a solution, then there must be $h > 0$. So far, the necessity has been proved. Then, multiply the left and right sides of Equation (10) by the matrices M and M^T respectively, and direct algebraic operations can complete the sufficiency proof.

$$\begin{aligned} &(*)^T \begin{bmatrix} \begin{bmatrix} 0 & P \\ P & 0 \end{bmatrix} & \\ & -Q \end{bmatrix} \begin{bmatrix} 0 & I & 0 \\ A_{\mathrm{TS}} & A_{\mathrm{TT}} & B_{\mathrm{T}} \\ C_{\mathrm{S}} & C_{\mathrm{T}} & D_{\mathrm{T}} \\ 0 & 0 & I \end{bmatrix} \begin{bmatrix} \Phi(I - A_{\mathrm{SS}}\Phi)^{-1} \begin{bmatrix} A_{\mathrm{ST}} & B_{\mathrm{S}} \end{bmatrix} \\ \begin{bmatrix} I & 0 \\ 0 & I \end{bmatrix} \end{bmatrix} \\ &-h \times (*)^T \begin{bmatrix} I & -\Phi \\ -\Phi^T & \Sigma^2 \end{bmatrix} \begin{bmatrix} I & \begin{bmatrix} 0 & 0 \end{bmatrix} \\ A_{\mathrm{SS}} & \begin{bmatrix} A_{\mathrm{ST}} & B_{\mathrm{S}} \end{bmatrix} \end{bmatrix} \begin{bmatrix} \Phi(I - A_{\mathrm{SS}}\Phi)^{-1} \begin{bmatrix} A_{\mathrm{ST}} & B_{\mathrm{S}} \end{bmatrix} \\ \begin{bmatrix} I & 0 \\ 0 & I \end{bmatrix} \end{bmatrix} < 0. \end{aligned} \tag{20}$$

The proof is completed. □

It can be seen that the condition in Lemma 4 hides the connection relationship among subsystems inside the parameters, while the left side of Equation (10) in Theorem 1 linearly depends on the symmetric matrix P, and the structure of the system is specifically reflected in it, which can effectively make use of the sparse structure of the subsystem connection matrix. Furthermore, the matrices $A_{*\#}$, B_*, C_* and D_* with $*, \# = \mathrm{T}, \mathrm{S}$ are all block diagonal, and large-scale networked systems are sparse. Combined with the research on sparse semi-definite programming problems [31–33], when the system is relatively large, the computational complexity of solving the above sparse LMI is frequently lower than the condition in Lemma 4. This aspect can also be explained in subsequent numerical simulations. It is worth noting that the condition of Theorem 1 does not bring conservatism compared with Lemma 4, a dissipative criterion based on the lumped description.

When there are a huge number of subsystems, the strict dissipativity analysis using the condition in Theorem 1 may still encounter computational difficulties. To overcome this difficulty, we further explore the structural characteristics of the subsystem connection matrix Φ, and put forward the conditions for strictly dissipative analysis based on the parameters of each subsystem.

A simple derivation leads to the following relationship,

$$\begin{bmatrix} I & -\Phi \\ -\Phi^T & \Phi^T\Phi \end{bmatrix} \leq 2\left(\begin{bmatrix} I \\ 0 \end{bmatrix}\begin{bmatrix} I & 0 \end{bmatrix} + \begin{bmatrix} 0 \\ \Phi^T \end{bmatrix}\begin{bmatrix} 0 & \Phi \end{bmatrix}\right). \tag{21}$$

Combined with Lemma 4 and the properties of the subsystem connection matrix, on the basis of Equation (10), the necessary condition for the strict dissipativity analysis that only depends on the parameters of a single subsystem can be obtained.

Theorem 2. *Assume that the networked system Γ is controllable. A necessary condition for the strict dissipativity of System Γ with the supply function $s_i(u(t,i), y(t,i))$ is that each subsystem has a symmetric positive definite matrix $P(i)$ and a positive scalar h such that,*

$$\begin{aligned} &(*)^T\begin{bmatrix} \begin{bmatrix} 0 & P(i) \\ P(i) & 0 \end{bmatrix} & \\ & -Q(i) \end{bmatrix}\begin{bmatrix} 0 & I & 0 \\ A_{\mathrm{TS}}(i) & A_{\mathrm{TT}}(i) & B_{\mathrm{T}}(i) \\ C_{\mathrm{S}}(i) & C_{\mathrm{T}}(i) & D_{\mathrm{T}}(i) \\ 0 & 0 & I \end{bmatrix} \\ &-h \times (*)^T\begin{bmatrix} I & \\ & \Sigma_i^2 \end{bmatrix}\begin{bmatrix} I & \begin{bmatrix} 0 & 0 \end{bmatrix} \\ A_{\mathrm{SS}}(i) & \begin{bmatrix} A_{\mathrm{ST}}(i) & B_{\mathrm{S}}(i) \end{bmatrix} \end{bmatrix} < 0. \end{aligned} \tag{22}$$

For large-scale networked systems, sometimes the parameters of multiple subsystems are the same. In this case, using Theorem 2 is more efficient. A sufficient condition for strict dissipativity analysis which only depends on the parameters of a single subsystem is given below.

Theorem 3. *Assume that the networked system Γ is controllable. Then, System Γ is strictly dissipative with the supply function $s_i(u(t,i), y(t,i))$ if there exists a symmetric positive definite matrix $P(i)$ and real number $h_2 \geq h_1 \geq 0$ (or $h_1 \leq h_2 \leq 0$) for each subsystem such that,*

$$\begin{aligned} &(*)^T\begin{bmatrix} \begin{bmatrix} 0 & P(i) \\ P(i) & 0 \end{bmatrix} & \\ & -Q(i) \end{bmatrix}\begin{bmatrix} 0 & I & 0 \\ A_{\mathrm{TS}}(i) & A_{\mathrm{TT}}(i) & B_{\mathrm{T}}(i) \\ C_{\mathrm{S}}(i) & C_{\mathrm{T}}(i) & D_{\mathrm{T}}(i) \\ 0 & 0 & I \end{bmatrix} \\ &-(*)^T\begin{bmatrix} h_1 I & \\ & -h_2\Sigma_i^2 \end{bmatrix}\begin{bmatrix} I & \begin{bmatrix} 0 & 0 \end{bmatrix} \\ A_{\mathrm{SS}}(i) & \begin{bmatrix} A_{\mathrm{ST}}(i) & B_{\mathrm{S}}(i) \end{bmatrix} \end{bmatrix} < 0. \end{aligned} \tag{23}$$

Proof of Theorem 3. From Lemma 1, we can get

$$\begin{bmatrix} I & -\Phi \\ -\Phi^T & \Phi^T\Phi \end{bmatrix} \geq (1-\alpha)\begin{bmatrix} I \\ 0 \end{bmatrix}(*)^T + \left(1 - \frac{1}{\alpha}\right)\begin{bmatrix} 0 \\ \Phi^T \end{bmatrix}(*)^T. \tag{24}$$

Using the above formula and the conclusion in Theorem 1, one can obtain a sufficient condition for System Γ to be strictly dissipative is the existence of a symmetric positive definite matrix P and two real numbers $h > 0$, $\alpha > 0$, such that,

$$
\begin{aligned}
&(*)^T \begin{bmatrix} \begin{bmatrix} 0 & P \\ P & 0 \end{bmatrix} & \\ & -Q \end{bmatrix} \begin{bmatrix} 0 & I & 0 \\ A_{\mathrm{TS}} & A_{\mathrm{TT}} & B_{\mathrm{T}} \\ C_{\mathrm{S}} & C_{\mathrm{T}} & D_{\mathrm{T}} \\ 0 & 0 & I \end{bmatrix} \\
&-h \times (*)^T \left((1-\alpha) \begin{bmatrix} I \\ 0 \end{bmatrix} (*)^T + \left(1 - \tfrac{1}{\alpha}\right) \begin{bmatrix} 0 \\ \Phi^T \end{bmatrix} (*)^T \right) \begin{bmatrix} I & \begin{bmatrix} 0 & 0 \end{bmatrix} \\ A_{\mathrm{SS}} & \begin{bmatrix} A_{\mathrm{ST}} & B_{\mathrm{S}} \end{bmatrix} \end{bmatrix} < 0.
\end{aligned} \tag{25}
$$

Let $h_1 = (1-\alpha)h$, $h_2 = -(1-\alpha^{-1})h$, we can get $h_2 = \alpha^{-1}h_1$. Therefore, when $\alpha \leq 1$, $h_2 \geq h_1 \geq 0$; when $\alpha \geq 1$, $h_1 \leq h_2 \leq 0$. The proof can be completed by combining Lemma 3. □

Compared with Theorem 1, the left side of Equation (23) in Theorem 3 is linearly related to the matrix $P(i)$, and its dimension is entirely governed by the dimension of the subsystem Γ_i. When the state dimension of each subsystem is fixed, the computational complexity of Equation (23) only linearly depends on the number of subsystems N. Therefore, Theorem 3 has a substantially higher computing efficiency than Theorem 1 for large-scale networked systems. However, it should be noted that Theorems 2 and 3 are conservative.

4. Numerical Simulations

Several numerical simulations are employed in this section to demonstrate the efficacy of the strict dissipativity conditions presented in this paper. The simulation experiments are performed on a laptop computer with an Intel(R) Core(TM) i5-3230M CPU @ 2.60 GHz 2.60 GHz and 6 G RAM. In these simulations, we assume that $m_{ui} = m_{xi} = m_{vi} = m_{zi} = m_{yi} = 2$. Furthermore, all the parameters of the subsystem are independent of each other, and the parameters of each subsystem are randomly generated according to a continuous uniform distribution with an interval of $[-0.9, 0.9]$. The subsystem connection matrix is randomly generated, but there is only one non-zero element 1 in each row and column.

The conditions in Lemma 4, Theorem 1, and Theorem 3 are used to verify the strict dissipativity of the system. Among them, the conditions in Lemma 4 and Theorem 3 are calculated by the LMI toolbox provided by MATLAB, and the condition in Theorem 1 is calculated by the sparse solvers DSDP. For System Γ introduced in this paper, we generate 10 systems for calculation, and the average value and standard deviation of system dissipativity analysis calculation time are obtained. Tables 1 and 2 give some results when the number of subsystems is among 2 and 45.

Table 1. Average of calculation time.

Subsystem Number	Lemma 4 (s)	Theorem 1 (s)	Theorem 3 (s)
2	0.256809	0.113630	0.280457
10	0.506572	0.357334	0.332980
20	4.718364	1.825162	0.550354
30	49.843413	10.102070	1.132749
38	160.804551	25.194260	1.990903
40	201.484656	31.942283	2.235822

Table 2. Standard deviation of calculation time.

Subsystem Number	Lemma 4 (s)	Theorem 1 (s)	Theorem 3 (s)
2	0.005145	0.019887	0.003355
10	0.017458	0.038721	0.005182
20	0.035518	0.051297	0.005435
30	0.585151	0.160743	0.011004
38	1.305206	0.317135	0.017758
40	1.806196	0.389699	0.092795

The tables show that the calculation time of the above three methods all increases with the increase of the number of subsystems. When the number of subsystems is 10 or less, the computational efficiency based on Lemma 4 is comparable to that of Theorem 1 and Theorem 3. This is because the dimensionality of the matrix inequality in Theorem 1 is higher than that in Lemma 4, and Theorem 3 requires several inequalities to be verified. With the expansion of the number of subsystems, when the number of subsystems is 20, 30, 40, 45, the ratio of calculation time based on the conditions in Lemma 4 and Theorem 1 is 1.0957, 1.3414, 1.3911, 1.4581. The ratio of the average computation time becomes larger and larger, which means that the computational efficiency of Theorem 1 is improved to some extent. Because the condition in Lemma 4 requires operations such as inversion of high-dimensional matrices. Clearly, Theorem 3 is more computationally efficient than both Lemma 4 and Theorem 1. This is due to the fact that the conditions of Theorem 3 are tested based on individual subsystem parameters, and their computational complexity only increases linearly with the number of subsystems N. In addition, due to the limitation of computer memory, the conditions in Lemma 4 and Theorem 1 may not be calculated, but Theorem 3, which is tested independently for each subsystem, can still be calculated. Therefore, Theorem 3 has more computational advantages in the dissipativity analysis of large-scale networked systems. It should be noted that Theorem 3 is conservative compared to Lemma 4 and Theorem 1.

5. Conclusions

This paper investigates the strict dissipativity of networked systems composed of a large number of subsystems. At first, according to the model of large-scale networked systems, the definition of the dissipativity of networked systems is given in this paper. Then, we study the dissipative criteria of networked systems. For large-scale networked systems, when the number of subsystems is large, the performance analysis using the existing linear system theory will encounter computational difficulties. Some LMI-form conditions for dissipativity analysis of large-scale networked systems are derived. Among them, Theorem 1 is a necessary and sufficient condition, which effectively utilizes the block diagonal structure of the system parameter matrix and the sparsity of the subsystem connection matrix. Combined with the use of sparse semidefinite programming tools, it is more efficient than the lumped analysis method for medium-scale networked systems. In addition, the proposed sufficient condition, and necessary condition only depend on the parameters of a single subsystem, which are more suitable for the dissipative analysis of networked systems with a large number of subsystems, but they are conservative compared with other conditions.

Regarding large-scale networked systems, the design of distributed controllers to ensure the dissipativity of large-scale networked systems will be investigated in further research. For instance, in UAV formation flight, relying on a centralized controller to observe the entire formation and control all UAVs at once is both impractical and increases operational costs in engineering applications. A more reasonable option would be to decentralize the controller to each UAV platform and achieve the overall objective by interacting and sharing information between platforms. Therefore, if the structural information of the network topology is capable of being fully utilized and a distributed control strategy that relies on local information sharing is adopted, the amount of data transmission in the network will be greatly reduced and the computational efficiency will be raised. Furthermore, the presence of quantization errors, time delays, data packet loss, and other phenomena when communicating networked among subsystems or among subsystems and their local controllers will be explored. In practical engineering applications, network connections would possibly be non-idealized, and the arrival of information delivered is frequently unable to be achieved immediately.

Author Contributions: Conceptualization, Y.S. and H.L.; methodology, Y.S.; software, Y.S.; validation, H.L.; formal analysis, Y.S.; investigation, Y.S.; resources, J.W. and H.L.; data curation, J.W.; writing—original draft preparation,Y.S.; writing—review and editing, Y.S.; visuConceptualizationalization, J.W. and H.L.; supervision, J.W. and H.L.; project administration, J.W. and H.L.; funding acquisition, J.W. and H.L. All authors have read and agreed to the published version of the manuscript.

Funding: This work is supported by the National Natural Science Foundation of China under grant number 62273189; by the Shandong Provincial Natural Science Foundation under grant number ZR2019MF063 and ZR2020MF064.

Institutional Review Board Statement: Not applicable.

Informed Consent Statement: Not applicable.

Data Availability Statement: The data that support the findings of this study are available from the corresponding author upon reasonable request.

Conflicts of Interest: The authors declare no conflict of interest.

References

1. Liu, H.; Wang, D. Stability and stabilisation of a class of networked dynamic systems. *Int. J. Syst. Sci.* **2018**, *49*, 964–973. [CrossRef]
2. Montanari, A.N.; Duan, C.; Aguirre, L.A.; Motter, A.E. Functional observability and target state estimation in large-scale networks. *Proc. Natl. Acad. Sci. USA* **2022**, *119*, e2113750119. [CrossRef] [PubMed]
3. Yu, T.; Yu, L.; Xiong, J. Observer-based distributed control of large-scale systems under gossip communication protocol. *Asian J. Control* **2022**, *24*, 956–972. [CrossRef]
4. Rahman, M.H.; Al-Naeem, M.; Ali, M.N.; Sufian, A. TFBN: A cost effective high performance hierarchical interconnection network. *Appl. Sci.* **2020**, *10*, 8252. [CrossRef]
5. Bamieh, B.; Paganini, F.; Dahleh, M.A. Distributed control of spatially invariant systems. *IEEE Trans. Autom. Control* **2002**, *47*, 1091–1107. [CrossRef]
6. Zaeri Amirani, M.; Bigdeli, N.; Haeri, M. Distributed fault detection and isolation in time-varying formation tracking UAV multi-agent systems. *Asian J. Control.* **2022**, *25*, 604–622. [CrossRef]
7. Venkat, A.N.; Hiskens, I.A.; Rawlings, J.B.; Wright, S.J. Distributed output feedback MPC for power system control. In Proceedings of the 45th IEEE Conference on Decision and Control, San Diego, CA, USA, 13–15 December 2006; pp. 4038–4045.
8. Knies, C.; Diermeyer, F. Data-driven test scenario generation for cooperative maneuver planning on highways. *Appl. Sci.* **2020**, *10*, 8154. [CrossRef]
9. Lai, J.; Chen, S.; Lu, X.; Zhou, H. Formation tracking for nonlinear multi-agent systems with delays and noise disturbance. *Asian J. Control* **2015**, *17*, 879–891. [CrossRef]
10. Willems, J.C. Dissipative dynamical systems part I: General theory. *Arch. Ration. Mech. Anal.* **1972**, *45*, 321–351. [CrossRef]
11. Willems, J.C. Dissipative dynamical systems part II: Linear systems with quadratic supply rates. *Arch. Ration. Mech. Anal.* **1972**, *45*, 352–393. [CrossRef]
12. Xie, S.; Xie, L.; De Souza, C.E. Robust dissipative control for linear systems with dissipative uncertainty. *Int. J. Control* **1998**, *70*, 169–191. [CrossRef]
13. Tan, Z.; Soh, Y.C.; Xie, L. Dissipative control for linear discrete-time systems. *Automatica* **1999**, *35*, 1557–1564. [CrossRef]
14. Zhu, D.; Zhang, W.; Liu, C.; Duan, J. Fractional-order hyperbolic tangent sliding mode control for chaotic oscillation in power system. *Math. Probl. Eng.* **2021**, *2021*. [CrossRef]
15. Urata, K.; Inoue, M. Dissipativity reinforcement in feedback systems and its application to expanding power systems. *Int. J. Robust Nonlinear Control* **2018**, *28*, 1528–1546. [CrossRef]
16. Zhang, Q.; Yan, H.; Chen, S.; Zhan, X.; Jiang, X. Finite-time dissipative control for networked control systems with hybrid-triggered scheme. *Trans. Inst. Meas. Control* **2021**, *43*, 891–901. [CrossRef]
17. Agarwal, E.; Sivaranjani, S.; Gupta, V.; Antsaklis, P.J. Distributed synthesis of local controllers for networked systems with arbitrary interconnection topologies. *IEEE Trans. Autom. Control* **2020**, *66*, 683–698. [CrossRef]
18. Chen, B.D.; Lall, S. Dissipation inequalities for distributed systems on graphs. In Proceedings of the 42nd IEEE International Conference on Decision and Control (IEEE Cat. No. 03CH37475), Maui, HI, USA, 9–12 December 2003; Volume 3, pp. 3084–3090.
19. Ji, Z.; Lin, H.; Cao, S.; Qi, Q.; Ma, H. The complexity in complete graphic characterizations of multiagent controllability. *IEEE Trans. Cybern.* **2020**, *51*, 64–76. [CrossRef]
20. Ji, Z.; Yu, H. A new perspective to graphical characterization of multiagent controllability. *IEEE Trans. Cybern.* **2016**, *47*, 1471–1483. [CrossRef]
21. Yu-Geng, X. Large-scale systems control and complex networks-exploration and thinking. *Acta Autom. Sin.* **2013**, *39*, 1758–1768.
22. Liu, H.; Yu, H. Stability and stabilization of spatially interconnected systems. *Control Decis.* **2020**, *35*, 749–756.

23. Zhou, T.; Zhang, Y. On the stability and robust stability of networked dynamic systems. *IEEE Trans. Autom. Control* **2015**, *61*, 1595–1600. [CrossRef]
24. Liu, H.; Yu, H. Finite-time control of continuous-time networked dynamical systems. *IEEE Trans. Syst. Man, Cybern. Syst.* **2018**, *50*, 4623–4632. [CrossRef]
25. Zhou, T. Coordinated one-step optimal distributed state prediction for a networked dynamical system. *IEEE Trans. Autom. Control* **2013**, *58*, 2756–2771. [CrossRef]
26. van Schuppen, J.H.; Boutin, O.; Kempker, P.L.; Komenda, J.; Masopust, T.; Pambakian, N.; Ran, A.C. Control of distributed systems: Tutorial and overview. *Eur. J. Control* **2011**, *17*, 579–602. [CrossRef]
27. Khalil, I.; Doyle, J.; Glover, K. *Robust and Optimal Control*; Prentice Hall: Hoboken, NJ, USA, 1996.
28. Duan, G.R.; Yu, H.H. *LMIs in Control Systems: Analysis, Design and Applications*; CRC Press: Boca Raton, FL, USA, 2013.
29. Liu, H.; Yu, H. Decentralized state estimation for a large-scale spatially interconnected system. *ISA Trans.* **2018**, *74*, 67–76. [CrossRef]
30. Pólik, I.; Terlaky, T. A survey of the S-lemma. *SIAM Rev.* **2007**, *49*, 371–418. [CrossRef]
31. Andersen, M.S.; Pakazad, S.K.; Hansson, A.; Rantzer, A. Robust stability analysis of sparsely interconnected uncertain systems. *IEEE Trans. Autom. Control* **2014**, *59*, 2151–2156. [CrossRef]
32. Benson, S.J.; Ye, Y. *DSDP5 User Guide-Software for Semidefinite Programming*; Technical Repor; Argonne National Lab.(ANL): Argonne, IL, USA, 2006.
33. George, A.; Gilbert, J.R.; Liu, J.W. *Graph Theory and Sparse Matrix Computation*; Springer Science & Business Media: Berlin/Heidelberg, Germany, 2012; Volume 56,

applied sciences

MDPI

Article

Heat Load Forecasting of Marine Diesel Engine Based on Long Short-Term Memory Network

Rui Zhou, Jiyin Cao *, Gang Zhang, Xia Yang and Xinyu Wang

School of Mechanical & Electrical Engineering, Wuhan Institute of Technology, Wuhan 430205, China
* Correspondence: caojiyin@gmail.com; Tel.: +158-27595304

Abstract: High heat load on diesel engines is a main cause of ship failure, which can lead to ship downtime and pose a risk to personal safety and the environment. As such, predictive detection and maintenance measures are highly important. During the operation of marine diesel engines, operating data present strong dynamic, time lag, and nonlinear characteristics, and traditional models and prediction methods cause difficulties in accurately predicting the heat load. Therefore, the prediction of its heat load is a challenging and significant task. The continuously developing machine learning technology provides methods and ideas for intelligent detection and diagnosis maintenance. The prediction of diesel engine exhaust temperature using long short-term memory network (LSTM) is analyzed in this study to determine the diesel engine heat load and introduce an effective method. Spearman correlation coefficient method with the addition of artificial experience is utilized for feature selection to obtain the optimal input for the LSTM model. The model is applied to validate the ship data of the Shanghai Fuhai ship, and results show that the mean absolute percentage error (MAPE) of the model is lowest at 0.089. Compared with other models, the constructed prediction model presents higher accuracy and stability, as well as an optimal evaluation index. A new idea is thus provided for combining artificial knowledge experience with data-driven applications in engineering practice.

Keywords: diesel engine heat load; intelligent detection; long short-term memory network; prediction model; evaluation index

Citation: Zhou, R.; Cao, J.; Zhang, G.; Yang, X.; Wang, X. Heat Load Forecasting of Marine Diesel Engine Based on Long Short-Term Memory Network. *Appl. Sci.* **2023**, *13*, 1099. https://doi.org/10.3390/app13021099

Academic Editors: Sheng Du, Xiongbo Wan, Wei Wang and Hao Fu

Received: 19 November 2022
Revised: 6 January 2023
Accepted: 9 January 2023
Published: 13 January 2023

1. Introduction

In dealing with the increasing severity of fossil energy crisis and the strict emission requirements of internal combustion engines, the effective use of energy and environmental protection are also increasing in importance. If the diesel engine set has insufficient combustion, then the fuel-generated heat decreases, resources are wasted, black smoke and a large amount of CO and other harmful gases are discharged, and the environment is polluted, which will cause harm to human body through direct inhalation [1–3]. The diesel engine set is an important power source for ship navigation, and its normal working cycle is a major contributor to efficient transportation by sea, saving energy, and reducing emission [4].

Taking exhaust manifold as an example, the finite element method is used by Li et al. [5] to verify the effect of thermal load on its fatigue life. The high efficiency heat transfer model is used by Zhang et al. [6] to analyze the direct relationship between cylinder head fatigue life and average gas temperature. In addition, Chaboche model is established to analyze the local deformation and leakage of cylinder head under thermal cycle test [7]. The failure of the ship's exhaust valve was investigated and analyzed by EI-Bitar et al. [8], and it is determined that the high temperature environment would lead to the expansion of microcracks and easy fracture. According to the above research, main equipment of the diesel engine set will be damaged by high heat load, the ship will be stopped, which will greatly increase the navigation cost, and the safety of ship equipment and environment will be seriously affected [9,10]. Previously, a Belize foreign ship lost control of its main engine due to excessive heat load at the floating attachment of the Yangtze River No. 20. Fortunately it was timely assisted by the maritime department and did not cause a second accident At present, ship data are detected by sensors and transmitted to terminals. However when excessive heat load is detected by sensors, ship equipment and personal safety may have been damaged [11]. Therefore, the prediction of its heat load can achieve the preventive effect.

The heat load of diesel engine can be accurately characterized by exhaust temperature which can be estimated by predicting the exhaust temperature. However, factors affecting the exhaust temperature are typically influenced by uncertain dynamic environmental factors. As such, heat load prediction of marine diesel engine units is a challenging and meaningful task. It is usually used to analyze the heat load of marine diesel engine units through traditional methods, such as finite element analysis and linear regression modeling [12]. However, the accuracy of predicted parameters, results, and complex mapping relationships are difficult to model due to the complex process inside the combustion chamber. Complex and variable dynamic processes and nonlinear systems are modeled by neural network methods, and its continuous development led to various applications in marine diesel engines [13–16].

Artificial neural network (ANN) was used by Cay to replace traditional modeling to predict engine fuel consumption, effective power, and exhaust temperature. The mean error percentage (MEP) of the training test data was less than 2.7% [17]. Ignition timing, engine speed and air-fuel ratio were used as model inputs by Liu et al. [18] to analyze whether machine learning can be used to effectively predict engine exhaust temperature. Four different algorithm combinations were used to evaluate the applicability of ANN. ANN was used by Uslu et al. [19] to predict the emission and performance of an ether single-cylinder diesel engine. The maximum mean absolute error range of 5% was obtained, and the regression coefficient (R2) was in the range of 0.9640–0.9878. Despite the use of ANN has potential effectiveness on exhaust temperature, a large number of initial parameters are required in this study, and gradient explosion may lead to unsuccessful training, thereby requiring additional time in adjusting the hyperparameters. Moreover, heat load is usually characterized by nonlinear variations; collecting and obtaining these data are necessary under various conditions for analysis and prediction. However, only several influencing factors are analyzed, and the dependencies between the factors are ignored. Considering the shortage of ANN, long short-term memory network (LSTM) model is considered for prediction analysis.

LSTM network based on recurrent neural network (RNN) with three additional thresholds is a special form that can solve the problem of gradient explosion and disappearance in training [20]. Continuous development has led to the maturity of this neural network model. However, a large amount of raw data is not effective when processed by LSTM, so it is used together with other methods. The Spearman correlation coefficient method (SR) is utilized in neural networks for feature selection to effectively capture dependencies between variables by analyzing the correlation between two variables and removing redundant information. A LSTM network used to predict passenger flow at stations was proposed by Zhang et al. Spearman correlation features were used to select time and space factor data that significantly and effectively affect passenger flow, and the accuracy of the prediction model was improved [21]. Spearman correlation coefficient method was applied by Jiao et al. [22] to explore the temporal connection of nonresidential consumers under multiple time series. Spearman's correlation coefficient is a widely used feature selection method The correlation between multiple information sequences can be effectively analyzed by this method, and the best input of the network model can be provided. However, this purely data-driven method determines dependencies on the basis of only the correlation between feature variables, thereby leading to the exclusion of significant variables. Hence, artificial experience needs to be added when screening features in advance, and significance tests must be performed to ensure that accurate input is provided to the prediction model.

As such, a hybrid prediction model incorporating the artificial empirical Spearman correlation coefficient method (AESR) and long short-term memory network (LSTM) is proposed in this study to achieve accurate and stable predictions of exhaust temperature by using the AESR-LSTM model. Redundant information is eliminated through the Spearman correlation coefficient method, and the optimal input is derived by adding artificial empirical supplementary variables while retaining those with high correlation ratings. The hyperparameters are usually selected according to experience and then set in the combination. The combination of cross-validation and grid search methods is used to avoid the blindness of adjusting parameters. The hyperparameters combination of neural network is scientifically optimized and adjusted, and the robustness and accuracy of the prediction model are ensured. After the optimal parameter set is selected by grid search and cross-validation, the model is trained again using the optimal parameters. The trained LSTM model is utilized to predict the exhaust temperature and highlight the advantages of the AESR-LSTM model for data trend prediction compared with other models. The experimental results of the selected prediction model are consistent with the actual values. The prediction result of the model can be sent to the console as a feedback signal, and more convenience and information can be provided to the operator. The predicted results can be used to analyze the combustion conditions in the combustion chamber. Complex models do not need to be used to create analysis, and such signals are difficult to obtain by physical sensors. The predicted trend results can be adopted to analyze the working condition and emission substances of diesel engines, implement certain avoidance measures before failure occurs, reduce the risk of accidents, improve the safety of ship systems, and prevent serious personal injury and economic loss. The AESR-LSTM neural network modeling is simpler than conventional modeling analysis because the workload of heat load research is reduced, more comprehensive influencing factors are taken into account, complex changes in the combustion chamber are predicted by a small amount of experimental data, and more accurate prediction results are obtained. A new idea is provided in this study, which combines artificial experience with data driven application in engineering practice.

Accordingly, a method for predicting diesel engine exhaust temperature that integrates feature selection, parameter combination search, and comparative analysis of multiple model combinations is proposed in this study. The remainder of this paper is structured as follows. Methods used and the proposed hybrid prediction system model are briefly described in Section 2. Relevant data are collected and analyzed in Section 3, and the results of the proposed system used to predict the thermal load of the combustion chamber of the marine diesel engine set were displayed, and then the results were compared with those of other models. Finally, the conclusions of this study are drawn in Section 4.

2. Prediction Method

In this section, data preprocessing method, network model, and optimization method are introduced, and a method to predict the heat load of marine diesel engine combustion chamber is proposed. The AESR-LSTM method is developed, which mainly consists of the Spearman correlation coefficient method and the LSTM network, and is used to predict heat load.

2.1. Long Short-Term Memory Network

LSTM is a neural network proposed by Hochreiter and Schmidhuber in 1997 [23,24]. This model has been continuously developed to form a systematic and complete framework [25–27]. The LSTM is used in this study to compensate for the limitations of recurrent neural network (RNN) in dealing with the dependence problem at long distances and to solve the enlargement of and difficulty in updating partial derivatives W during training. The internal structure of the LSTM neural unit is shown in Figure 1. The LSTM adds three thresholds to the framework of the RNN as three logical control units, and the input and output information of the entire network is controlled and managed by the three thresholds. The three thresholds are described as follows:

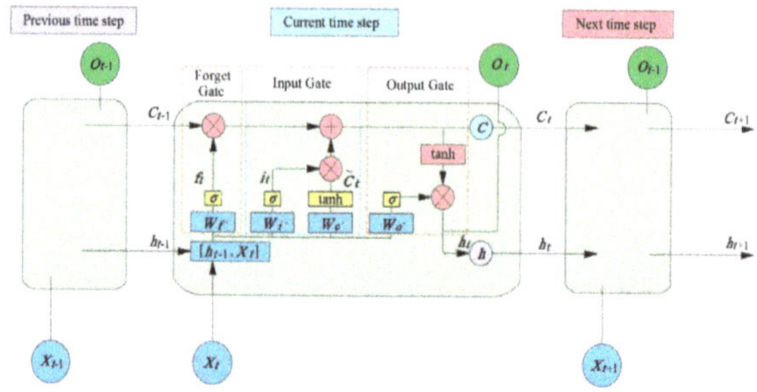

Figure 1. LSTM structure diagram.

Input Gate: Whether the information is stored in the storage unit is determined by the threshold and denotes it as i_t.

Forget Gate: Whether the information stored in the storage unit at the previous time is stored in the storage unit at the current time is determined by the threshold and denotes it as f_t.

Output Gate: Whether the information in the storage unit at the current moment enters the hidden state h_t is determined by the threshold and denotes it as o_t.

Historical information can be saved, read, updated, and reset by the unit; it is the core of the LSTM unit and is denoted as C_t.

The LSTM neural network at moment t is expressed as follows:

$$f_t = \sigma(W_f \cdot [h_{t-1}, X_t] + b_f), \tag{1}$$

$$i_t = \sigma(W_i \cdot [h_{t-1}, X_t] + b_i), \tag{2}$$

$$o_t = \sigma(W_o \cdot [h_{t-1}, X_t] + b_o), \tag{3}$$

$$\widetilde{C}_t = \tanh(W_C \cdot [h_{t-1}, X_t] + b_c), \tag{4}$$

$$C_t = f_t \times C_{t-1} + i_t \otimes \widetilde{C}_t, \tag{5}$$

$$h_t = o_t \times \tanh(C_t), \tag{6}$$

where f_t, i_t, o_t, and h_t are (1), (2), (3), and (6), respectively; W_f, W_i, W_o, and W_C denote the recursive connection weights of the corresponding thresholds; σ is the sigmoid function, which is the same as the tanh function for the activation in Equations (7) and (8).

$$\sigma(x) = \frac{1}{1 + e^{-x}}, \tag{7}$$

$$\tanh(x) = \frac{\sinh(x)}{\cosh(x)} = \frac{e^x - e^{-x}}{e^x + e^{-x}}, \tag{8}$$

The state at the previous point in time needs to be discarded, and the content saved to the memory unit is determined by the forgetting gate. The sigmoid function is used to decide whether C_{t-1} is cumulatively retained or not. Cumulative retention is achieved when the sigmoid function is equal to 1 but is absent when the function is equal to 0.

The input gate contains the output h_{t-1} from the previous moment and the input X_t at this time, and the sigmoid function is used to control how much to add to C_t. An alternative $\widetilde{C}_t$ is also created and then the tanh function can be used to control how much to add to $\widetilde{C}_t$. The two parts are then multiplied to determine the amount of influence C_t, and the influence of the forgetting gate is added to obtain the expression for C_t.

The output gate is a sigmoid function that can determine which parts of C_t need to be output to describe the o_t expression. C_t is placed into the tanh function to determine the final output C_t and then multiplied with o_t to obtain the final output h_t, which signals the end of the LSTM work for one moment. How many memory units are forgetten, retained, and outputted at each moment are determined and affected by the three thresholds, and they are finally transferred to the state of this moment at the end.

The prediction results of LSTM model are affected by the learning rate, weights, activation function, step size, and number of batches in the network. For example, convergence failure is caused by learning rate being set too high, while consuming a lot of training time to calculate the optimal value is caused by learning rate being set too low. Problems, such as gradient explosion and disappearance, can occur when the activation function is poorly chosen. Therefore, LSTM prediction model needs to be trained, and appropriate parameters are selected to improve the prediction accuracy.

2.2. Spearman Correlation Coefficient Method

As mentioned above, factors affecting the exhaust temperature are typically influenced by uncertain dynamic environmental factors. To find them, Spearman correlation analysis method was adopted by us. The change trend and correlation strength between the two variables were tested by Spearman's correlation coefficient method. This method is based on calculating the difference of each pair of equivalents of two columns of paired ranks as the basis. If the correlation coefficient between two variables is close to +1 and −1, then the surface correlation is strong. The Spearman correlation coefficient r_p can be expressed as follows:

$$r_p = 1 - \frac{6\sum d_i^2}{n(n^2-1)}, \quad (9)$$

where n is the sample size, d_i is the difference of bit values of the ith data pair. The values of r_p are within the range of [−1, 1]. If r_p = 1, then the correlation is perfectly positive; if r_p = −1, then the correlation is perfectly negative. The absolute value is used as the basis to judge the correlation. The strength of correlation between variables is divided into four categories, as shown in Table 1 [28].

Table 1. Correlation intensity.

Value of r	Strength of Relationship
−1.0 to −0.5 or 0.5 to 1.0	Strong
−0.5 to −0.3 or 0.3 to 0.5	Moderate
−0.3 to −0.1 or 0.1 to 0.3	Weak
−0.1 to 0.1	None or very weak

2.3. AESR-LSTM Hybrid Prediction Model

AESR-LSTM hybrid prediction model is proposed to combine Spearman correlation coefficient method with LSTM network, and artificial experience is added to conduct exhaust temperature prediction. First, sensor data is analyzed to eliminate overlapping features. Spearman correlation coefficient method is used to discard redundant information in the original data because exhaust temperature will be affected by various factors and there is correlation between various factors. Finally, the variables are supplemented by artificial experience, and the efficiency of the algorithm and the accuracy of prediction are improved. The cross-validation and grid search methods are used to optimize the hyperparameters of the neural network to obtain the optimal combination of parameters with maximum prediction accuracy. After the optimal parameter set is selected by grid search and cross-validation, the model is trained again using the optimal parameters. The overall framework and partial procedures of AESR-LSTM are shown in Figure 2 and Algorithm 1. The specific modeling steps are presented as follows.

Step 1: The influencing factors related to exhaust temperature are analyzed to collect relevant time series data X_t on the basis of engineering experience.

Step 2: The training and test sets are divided into pieces in a ratio of 7:3.

Step 3: Data is preprocessed, Spearman correlation coefficient is used for feature selection to process the original data, redundant information is eliminated, highly correlated variables are extracted, and variables are supplemented by mechanisms and human experience to obtain the best input X_t^*.

Step 4: The hyperparameters in the LSTM neural network model are adjusted through iterative optimization combined with cross-validation and grid search methods to select the optimal combination of parameters and improve its prediction accuracy.

Step 5: After the optimal parameter set is selected by grid search and cross-validation, the model is trained again using the optimal parameters.

Step 6: The test set samples are input into the prediction model to predict the combustion chamber exhaust temperature of marine diesel engine sets.

Step 7: The prediction performance of the proposed model is compared with those of other prediction models.

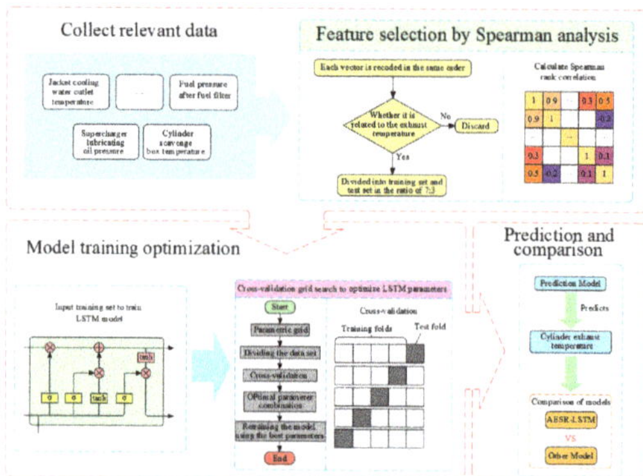

Figure 2. General Framework Structure.

Algorithm 1 Partial procedures

```
function coeff=mySpearman(X,Y)
if length(X)~=length(Y)
   error('Unequal dimensions');
   return;
end
N=length(X);
Xrank=zeros(1,N);
Yrank=zeros(1,N);
for i=1:N
   cont1=1;
   cont2=−1;
   for j=1:N
      if X(i)<X(j)
         cont1=cont1+1;
      elseif X(i)==X(j)
         cont2=cont2+1;
      end
   end
   Xrank(i) = cont1 + mean ([0:cont2]);
end
for i=1:N
   cont1=1;
   cont2=−1;
   for j=1:N
      if Y(i)<Y(j)
         cont1=cont1+1;
      elseif Y(i)==Y(j)
```

3. Case Study

3.1. Principle Analysis and Data Processing

In a ship, the power source is composed of the main engine and an auxiliary engine. The auxiliary power system is composed of machinery other than the diesel engine (main engine), including the fuel system, lubricating oil system, air system, cooling system, and other mechanical equipment. The main and auxiliary engines work together to propel the ship, and its composition structure is shown in Figure 3.

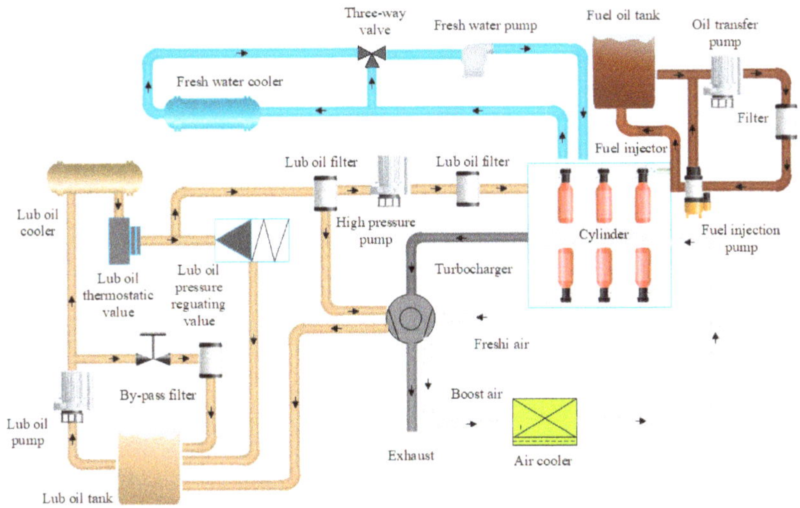

Figure 3. Sketch of the composition structure.

On the basis of the mechanism and data of the ship, the heat load of the marine diesel engine during operation is accurately reflected by the exhaust temperature. The amount, perfection, and timeliness of fuel combustion in the combustion chamber can be reflected by the exhaust temperature, as well as the high temperature heating time and brightness of combustion chamber components. Hence, exhaust temperature can be used to predict the heat load of diesel engine set.

The exhaust temperature of a single cylinder is predicted as an example in this study to analyze the trend of heat load variation and the operating performance of the combustion chamber. The high exhaust temperature of the cylinder is due to poor internal combustion, which is related to the amount of fresh air in the cylinder, cooler cooling effect, injector atomization quality, fuel viscosity, and cylinder compression pressure. The sensor is used to monitor its working condition and collect factors related to exhaust temperature, including high-temperature cooling, freshwater outlet temperature, cylinder liner cooling water inlet pressure, piston cooling oil outlet temperature, and fuel pressure after the fuel filter. Determining the correlation and dependence among these data is important to predict the exhaust temperature of marine diesel engine sets.

Sensor monitoring data of the Chinese vessel Shanghai Fuhai are used in this study, which are uploaded every 25 min. The sampled relevant data of initial variables are listed in Table 2. Field data for two months show that 28,160 pieces of ship data are measured via the ship's sensors and constitute the data set, which is randomly divided into training and test sets at a ratio of 7:3.

Table 2. Initial Variables.

Number	Variable	Description	Unit
1	T	Exhaust temperature	°C
2	T_{a1}	Cylinder scavenge box temperature	°C
3	T_{o1}	Diesel engine inlet oil temperature	°C
4	T_{o2}	lubricating oil outlet temperature	°C
5	P_{o1}	Diesel inlet oil pressure	Mpa
6	T_{o3}	Main engine inlet oil temperature	Mpa
7	T_{w1}	High temperature cooling fresh water outlet temperature	°C
8	P_{w1}	High temperature cooling fresh water inlet pressure	Mpa
9	T_{w2}	Cylinder liner cooling water outlet temperature	°C
10	T_{a2}	Pressurized air temperature after cooler	°C
11	T_{a3}	Exhaust temperature before supercharger	°C
12	T_{f1}	Fuel oil temperature at unit inlet	°C
13	T_{o4}	Outlet temperature of cylinder piston cooling oil	°C
14	P_{f1}	Fuel pressure after fuel filter	Mpa
15	P_{f2}	Fuel inlet pressure of main engine	Mpa
16	P_{w2}	Inlet pressure of cylinder liner cooling water	Mpa
17	P_{o2}	Pressurizer inlet oil pressure	Mpa
18	T_{a4}	Exhaust temperature after supercharger	°C
19	N_T	Turbocharger speed	rpm
20	P_{a1}	Exhaust valve air pressure	Mpa

The time series correlation data X_t' associated with the exhaust temperature are collected as follows.

$$X_t' = \left\{T_{a1}, T_{o1}, T_{o2}, P_{o1}, T_{o3}, T_{w1}, P_{w1}, T_{w2}, T_{a2}, T_{a3}, T_{f_1}, T_{o4}, P_{f_1}, P_{f_2}, P_{w2}, P_{o2}, T_{a4}, N_T, P_{a1}\right\}$$

The turbocharger is driven by the inertial impulse of the exhaust gas to drive the turbine, and then fresh air is pressurized into the cylinder. Thus, overlapping variables and supercharger front T_{a3} and rear T_{a4} exhaust temperatures must be eliminated to obtain time series data as follows.

$$X_t'' = \left\{T_{a1}, T_{o1}, T_{o2}, P_{o1}, T_{o3}, T_{w1}, P_{w1}, T_{w2}, T_{a2}, T_{f_1}, T_{o4}, P_{f_1}, P_{f_2}, P_{w2}, P_{o2}, N_T, P_{a1}\right\}$$

Spearman correlation coefficient method is used for feature selection of data, and the input of neural network is determined by the correlation between two factors, as shown in Figure 4.

Figure 4. Heat map of the correlation matrix.

According to the correlation matrix in Table 1 and the above figure, the correlation coefficient between the exhaust temperature and T_{o2} is 0.8997. Hence, the turbocharger lubricating oil outlet temperature is highly relevant to the exhaust temperature. This finding is consistent with the actual scenario. The viscosity of the lubricating oil will be affected by the temperature of the lubricating oil and increase the exhaust temperature. The correlation of the variable T_{w2} is 0.8639, and how much heat is taken away from the combustion chamber is determined by the outlet temperature of the cylinder liner cooling water, thereby indicating its sensitivity to changes in the exhaust temperature. The cylinder liner cooling water inlet pressure and the sweep box temperature are important factors affecting the exhaust temperature. Six variables with correlations higher than 0.5 are derived. The significance of their p-values is below 0.001.

If the temperature of the pressurized air after the cooler is excessively high, then the exhaust temperature rises because the fresh gas entering the diesel engine is cooled by the cooler after being pressurized by the turbocharger into the combustion chamber. With the increase in supercharger speed, the increase in exhaust energy is affected by the increase in exhaust temperature. The reason is that the high-temperature exhaust gas from the combustion chamber flows through the supercharger. Another factor to be considered is the fuel pressure after the diesel filter. This refers to whether the faulty filter is reflected by the fuel pressure. Fuel quality and exhaust temperature can be affected by damaged filters.

The three variables T_{a2}, N_T, and P_{f1} mentioned above are all important with a significance of less than 0.001, and the predictive variables will be affected, although their correlations are below 0.5, 0.1117, 0.1863, and 0.3574, respectively. Therefore, these factors are considered when deriving the final set of variables for the input model as follows.

$$X_t^* = \left\{T_{a1}, T_{o2}, T_{o3}, T_{w2}, T_{a2}, T_{o4}, P_{f_1}, P_{w2}, N_T\right\}$$

3.2. Analysis of Modeling and Prediction Results

On the basis of Spearman correlation analysis, the top nine positively correlated parameters are selected as model inputs in predicting the target output exhaust temperature T. The inputs are divided into training and test sets in a ratio of 7:3 given the impact of data volume on learning ability in the data drive. A combination of grid search and tenfold cross-validation methods is applied to improve the prediction performance of the model. The number of times to calculate the set of hyperparameters $X = \{X_1, X_2, \ldots, X_n\}$ is $\Pi_{i=1}^{i=n} | h_i |$, where ($i = 1, 2, \ldots$) and h_i is the number of hyperparameter values. Five parameters are selected in this study to set the hidden layers, hidden units, training rounds, learning rate, and batch size of the LSTM prediction network. The change trend of the loss function is affected by five super parameters, which are divided into two groups. See the change in loss function under the change in hyperparameters.

The influence of the number of units and learning rounds of the five-layer neural network on RMSE is shown in Figure 5. With the increase in the number of learning rounds, the RMSE decreases first and then increases, and the RMSE of 100 units is generally lower than that of other units. From Figure 6, we can see that the loss function is affected by different hidden layers. Usually, higher values are caused by the low learning rate of 0.001. Among the 0.01 learning rate and 0.005 learning rate, the number of hidden layers of five layers is better than other layers.

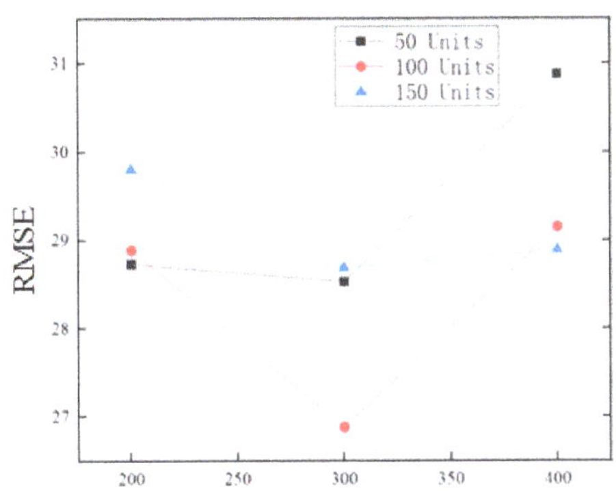

Figure 5. Influence of Unit Number and Learning Round Number on RMSE.

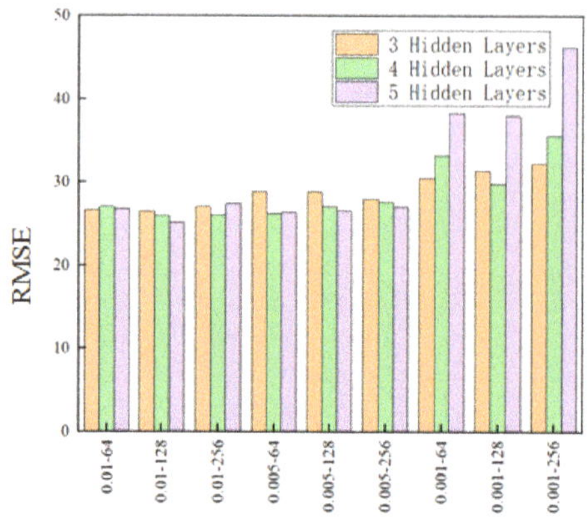

Figure 6. Influence of hidden layer on RMSE.

In the process of hyperparameters optimization, the combination with low RMSE value is selected as the best hyperparameters combination. Some adjustment results of cross-validation grid search optimization are shown in Table 3 below.

Table 3. Cross-Validation grid search optimization and tuning results.

Learning Rate	Hidden Layers	Hidden Units	Training Rounds	Batch Size	RMSE
0.01	4	150	200	128	27.82
0.005	5	50	400	64	27.92
0.001	4	150	200	256	34.25
0.005	3	150	300	128	26.44
0.001	3	50	300	64	33.35
0.005	4	100	200	64	26.35
0.01	5	100	300	128	25.21
0.001	3	100	300	64	30.48
0.01	5	50	200	256	29.15
0.001	4	50	400	128	32.19

After optimization, the best hyperparameter combination of RMSE is obtained. The hyperparameter candidate values and optimal values of the prediction model LSTM are shown in Table 4 below.

Table 4. Candidate and optimal sets of hyperparameters for the LSTM model.

Hyperparameter Name	Hyperparameter Values	Example of Optimal Hyperparameter Values
Learning rate	{0.01, 0.005, 0.001}	0.01
Hidden layers	{3, 4, 5}	5
Hidden units	{100, 150, 200}	100
Training rounds	{100, 200, 300}	300
Batch size	{64, 128, 256}	128

After the optimal parameter combination is selected, the training set is input into the LSTM model for training. At the same time, discard technology is introduced to prevent the model from over fitting. The training curve and training relative error scatter diagram are shown in Figures 7 and 8 below. From the figure, we can see that the predicted value basically coincides with the actual value in the training, and the error in the training finally approaches the zero line.

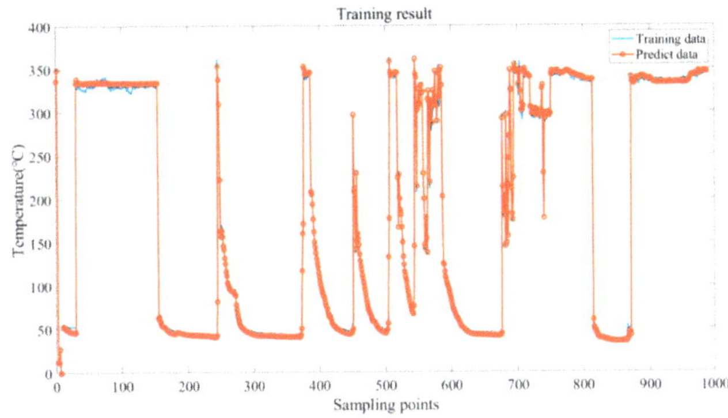

Figure 7. Model training result curve.

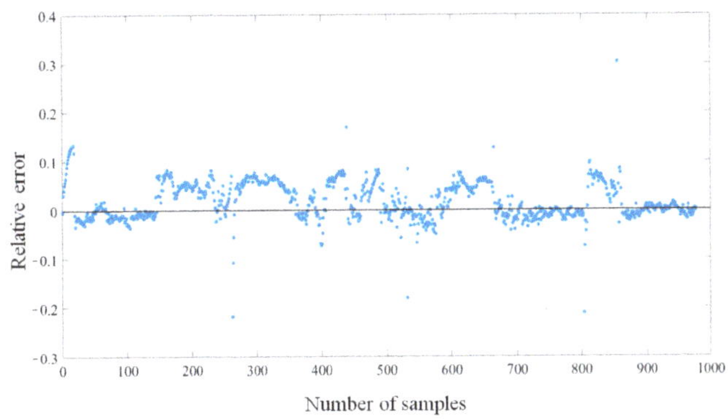

Figure 8. Relative error of model training.

The test set is fed into the trained model for exhaust temperature prediction. The prediction results are illustrated in Figure 9. The strong generalization ability of the prediction model is reflected by the consistency between the predicted and measured temperature values. The results of the selected forecasting model are subsequently analyzed by comparison with those of traditional forecasting methods, as described in detail below.

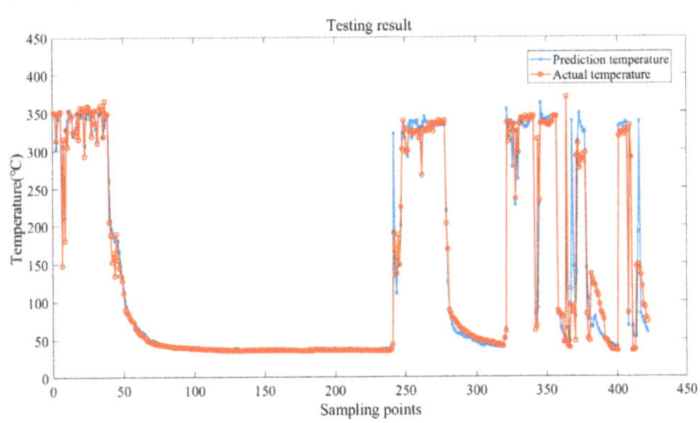

Figure 9. Model prediction outcomes.

3.3. Multimodel Comparative Analysis

In this study, Spearman correlation coefficient method and LSTM network are combined to predict a time series data. Other prediction models are input into the same data set, and the results of other prediction methods are compared with the results of the proposed methods for further analysis. The results of each prediction model are shown in Figures 10 and 11.

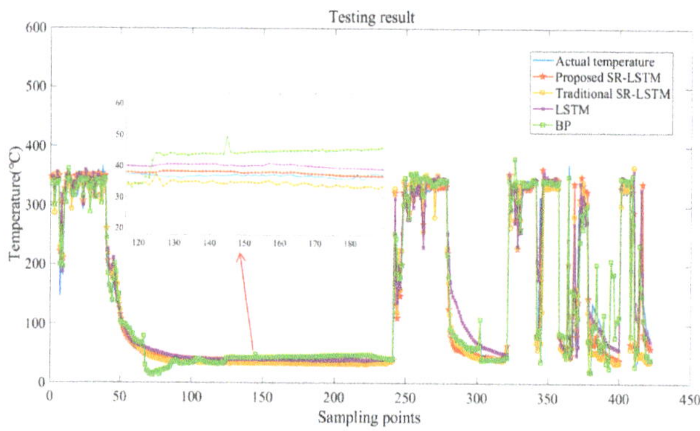

Figure 10. Comparison of prediction results.

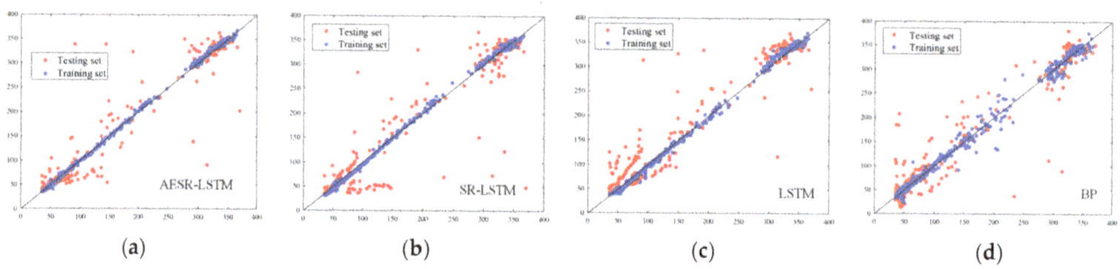

Figure 11. The forecasting and actual temperature for different models: (**a**) Training and test results of AESR-LSTM, (**b**) Training and test results of SR-LSTM, (**c**) Training and test results of LSTM, and (**d**) Training and test results of BP.

From Figure 10, the prediction curve (red line) of AESR-LSTM model with human experience is closer to the true value (blue line). As can be seen in Figure 11, except for a few predicted outliers, the system's scatter plot of forecasting and actual values is closest to the diagonal, which indicates that the difference between the forecasting value and actual value is the smallest.

At the same time, several commonly used evaluation indicators were cited to further verify the prediction performance of the AESR-LSTM model. The prediction performance of the four models is used for comparison, as shown in Table 5.

Table 5. Evaluation indicators.

Indicators	Formula
Mean absolute error (MAE)	$\frac{1}{N}\sum_{i=1}^{N}\left\|T_{ri}-T_{pi}\right\|$
Mean absolute percentage error (MAPE)	$\frac{100\%}{N}\sum_{i=1}^{N}\left\|\frac{T_{ri}-T_{pi}}{T_{ri}}\right\|$
Root-mean-square error (RMSE)	$\sqrt{\frac{1}{N}\sum_{i=1}^{N}(T_{ri}-T_{pi})^2}$

N is the number of predicted values, T_{ri} is the original data value, T_{pi} is the predicted value. The prediction performance of the prediction model is indicated by the value of MAPE, MAE and RMSE. The MAPE, MAE, and RMSE of the four models were calculated separately to reflect the goodness of the prediction model through the indexes. Figure 12 shows the values of the four prediction models the evaluation indexes. The error bars in the figure represent 95% confidence intervals. The mean absolute percentage, mean absolute, and root-mean-square errors of the proposed AESR-LSTM model are 0.089, 10.5403, and 27.5408, respectively, and the best indicators among several prediction models. The feature inputs selected by the improved AESR-LSTM model are better than those obtained by traditional methods for data trend prediction, so the method optimization is effective.

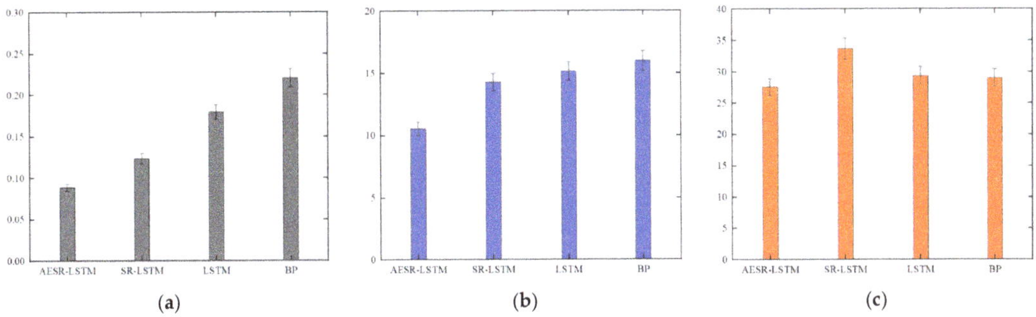

Figure 12. Comparative results histogram of model evaluation metrics: (**a**) MAPE value histogram of different models, (**b**) MAE value histogram of different models, and (**c**) RMSE value histogram of different models.

4. Conclusions

According to the data set collected in the marine cabin system, an AESR-LSTM data trend prediction model with artificial experience is constructed in this study. The model can be used for heat load prediction, fault detection, and diagnosis of marine diesel engines. Spearman correlation coefficient method is used to collect relevant raw data for feature selection, and the optimal input is selected by artificial empirical and significance check. The cross-validation and grid search methods are combined, and the hyperparameters are adjusted scientifically to avoid the randomness of the validation set. After the optimal parameter set is selected by grid search and cross-validation, the model is trained again with the optimal parameters, and the test set data is input into the training model to obtain the prediction results. The findings are subsequently compared and analyzed with those of other prediction models.

(1) The Spearman correlation coefficient method incorporating artificial experience was proposed to select features on the basis of operational monitoring data collected from the sensors. The correlation, redundancy, and significance of variable sets are analyzed separately, and the nine monitoring characteristic parameters with the maximum influence on the exhaust temperature are selected. Data-driven analysis and human experience are combined to provide optimal input features for the predictive models.

(2) The LSTM prediction model is trained with parameter tuning in combination with cross-validation grid search to obtain the prediction and evaluation metrics. The results and indicators of several models were compared. The results show that predicted value of AESR-LSTM are closest to the true value, and its evaluation indicators MAPE, MAE and RMSE are the best, which are 0.089, 10.5403, and 27.5408, respectively.

(3) The shortcomings of only using a single method can be overcome by the fusion of multiple methods, and the data can be scientifically and effectively screened to improve the effectiveness of the model in data prediction and fault diagnosis of marine diesel engines. Thus, the hybrid algorithm model is stable, and the error tolerance of the prediction results is reduced.

(4) The proposed method is based on the mechanism and data of the ship. All factors that may cause thermal load failure of the diesel engine are taken into account and can be used to analyze and refer to the working performance of the marine diesel engine. The prediction data can achieve effective fault detection and maintenance of ships for the implementation of preemptive corrective measures before ship failure, prevent ship downtime due to damaged components caused by excessive heat load, improve fuel economy and equipment reliability of ship diesel engines, and reduce economic losses.

A novel method combining artificial experience and data-driven is proposed. The selected optimal feature set is input into the model for prediction, and the better prediction results are obtained. As such, a feasible extended method of machine learning in marine diesel engine thermal load prediction and fault diagnosis is provided. Future research can focus on the optimization of methods, better operation parameter combination will be obtained through data mining techniques, and independent fault detection system will be developed to provide more convenience and information for ship operators.

Author Contributions: Conceptualization, R.Z.; methodology, R.Z.; software, R.Z.; validation, R.Z.; formal analysis, R.Z.; investigation, R.Z.; resources, R.Z.; data curation, R.Z. and J.C.; writing—original draft preparation, R.Z.; writing—review and editing, R.Z. and J.C.; visualization, R.Z. and X.W.; supervision, J.C. and G.Z.; project administration, J.C. and X.Y.; funding acquisition, J.C. All authors have read and agreed to the published version of the manuscript.

Funding: This research was supported by the Key Laboratory of Marine Power Engineering & Technology (Wuhan University of Technology), Ministry of Transport (No. KLMPET2018-10). This research was supported by the Scientific Research Program supported by Hubei Provincial Department of Education in 2021 (No. Q20211510).

Institutional Review Board Statement: Not applicable.

Informed Consent Statement: Not applicable.

Data Availability Statement: Not applicable.

Conflicts of Interest: The authors declare no conflict of interest.

References

1. Gabina, G.; Martin, L. Performance of marine diesel engine in propulsion mode with a waste oil-based alternative fuel. *Fuel* **2019**, *235*, 259–268. [CrossRef]
2. Wei, L.J.; Cheng, R.P. Combustion process and NOx emissions of a marine auxiliary diesel engine fuelled with waste cooking oil biodiesel blends. *Energy* **2018**, *144*, 73–80. [CrossRef]
3. Geng, P.; Tan, Q.M. Experimental investigation on NOx and green house gas emissions from a marine auxiliary diesel engine using ultralow sulfur light fuel. *Sci. Total Environ.* **2016**, *572*, 467–475. [CrossRef]
4. Gospic, I.; Glavan, I. Economic and Environmental Effects of the Marine Diesel Engine Trigeneration Energy Systems. *J. Mar. Sci. Eng.* **2021**, *9*, 773. [CrossRef]
5. Li, B.; Cui, Y. Marine diesel exhaust manifold failure and life prediction under high-temperature vibration. *Proc. Inst. Mech. Eng. Part C J. Mech. Eng. Sci.* **2022**, *236*, 6180–6191. [CrossRef]
6. Zhang, H.B.; Cui, Y. Fatigue life prediction analysis of high-intensity marine diesel engine cylinder head based on fast thermal fluid solid coupling method. *Braz. Soc. Mech. Sci. Eng.* **2021**, *43*, 327. [CrossRef]
7. Zhang, H.B.; Liang, G. Experimental and numerical study of inelastic behavior based on simulated cylinder head specimen under thermal cycling conditions. *Braz. Soc. Mech. Sci. Eng.* **2022**, *44*, 372. [CrossRef]
8. EI-Bitar, T.; EI-Meligy, M. Investigation of exhaust valve failure in a marine diesel engine. *Eng. Fail. Anal.* **2020**, *114*, 104574. [CrossRef]
9. Liu, Z.; Ning, X. Starved lubrication analysis for the top ring and cylinder liner of a two-stroke marine diesel engine considering the thermal effect of friction. *Int. J. Engine Res.* **2021**. [CrossRef]
10. He, T.; Lu, X.Q. Thermomechanical Fatigue Life Prediction for a Marine Diesel Engine Piston considering Ring Dynamics. *Adv. Mech. Eng.* **2014**, *6*, 429637. [CrossRef]
11. Safi, A.; Ahmad, Z. A Fault Tolerant Surveillance System for Fire Detection and Prevention Using LoRaWAN in Smart Buildings. *Sensors* **2022**, *22*, 8411. [CrossRef]
12. Dere, C.; Deniz, C. Effect analysis on energy efficiency enhancement of controlled cylinder liner temperatures in marine diesel engines with model based approach. *Energy Convers. Manag.* **2020**, *220*, 113015. [CrossRef]

13. Tosun, E.; Aydin, K. Comparison of linear regression and artificial neural network model of a diesel engine fueled with biodiesel-alcohol mixtures. *Alex. Eng. J.* **2016**, *55*, 3081–3089. [CrossRef]
14. Shin, S.; Lee, S. Deep learning procedure for knock, performance and emission prediction at steady-state condition of a gasoline engine. *Proc. Inst. Mech. Eng. Part D J. Automob. Eng.* **2020**, *234*, 3347–3361. [CrossRef]
15. Rezaei, J.; Shahbakhti, M. Performance prediction of HCCI engines with oxygenated fuels using artificial neural networks. *Appl. Energy* **2015**, *138*, 460–473. [CrossRef]
16. Huang, Q.; Liu, J.L. On the use of artificial neural networks to model the performance and emissions of a heavy-duty natural gas spark ignition engine. *Int. J. Engine Res.* **2022**, *23*, 1879–1898. [CrossRef]
17. Cay, Y. Prediction of a gasoline engine performance with artificial neural network. *Fuel* **2013**, *111*, 324–331. [CrossRef]
18. Liu, J.L.; Huang, Q. Machine learning assisted prediction of exhaust gas temperature of a heavy-duty natural gas spark ignition engine. *Appl. Energy* **2021**, *300*, 117413. [CrossRef]
19. Uslu, S.; Celik, M.B. Prediction of engine emissions and performance with artificial neural networks in a single cylinder diesel engine using diethyl ether. *Eng. Sci. Technol.* **2018**, *21*, 1194–1201. [CrossRef]
20. Pascanu, R.; Mikolov, T. On the difficulty of training recurrent neural networks. *arXiv* **2013**, arXiv:1211.5063.
21. Zhang, Z.; Wang, C. Passenger Flow Forecast of Rail Station Based on Multi-Source Data and Long Short Term Memory Network. *IEEE Access* **2020**, *8*, 28475–28483. [CrossRef]
22. Jiao, R.H.; Zhang, T.M. Short-Term Non-Residential Load Forecasting Based on Multiple Sequences LSTM Recurrent Neural Network. *IEEE Access* **2018**, *6*, 59438–59448. [CrossRef]
23. Hochreiter, S.; Schmidhuber, J. Long Short-Term Memory. *Neural Comput.* **1977**, *9*, 1735–1780. [CrossRef] [PubMed]
24. Hochreiter, S.; Schmidhuber, J. *Long Short-Term Memory Technical Report FKI-207-95*; Technische Universitat Munchen, Fakultat fur Informatik: Vienna, Austria, 1995.
25. Gers, F.A.; Schmidhuber, J. Learning to forget: Continual prediction with LSTM. *Neural Comput.* **2000**, *12*, 2451–2471. [CrossRef] [PubMed]
26. Graves, A.; Schmidhuber, J. Framewise phoneme classification with bidirectional LSTM and other neural network architectures. *Neural Netw.* **2005**, *18*, 602–610. [CrossRef] [PubMed]
27. Shobana, J.; Murali, M. An efficient sentiment analysis methodology based on long short-term memory networks. *Complex Intell. Syst.* **2021**, *7*, 2485–2501. [CrossRef]
28. Xiao, C.W.; Ye, J.Q. Using Spearman's correlation coefficients for exploratory data analysis on big dataset. *Concurr. Comput. Pract. Exp.* **2015**, *28*, 3866–3878. [CrossRef]

applied sciences

MDPI

Article

Multi-UAV Cooperative Path Planning with Monitoring Privacy Preservation

Yang Chen [1,2,*], Yifei Shu [1], Mian Hu [1] and Xingang Zhao [2]

[1] Engineering Research Center for Metallurgical Automation and Measurement Technology of Ministry of Education, Wuhan University of Science and Technology, Wuhan 430081, China

[2] State Key Laboratory of Robotics, Shenyang Institute of Automation, Chinese Academy of Sciences, Shenyang 110016, China

* Correspondence: chenyag@wust.edu.cn

Abstract: UAVs have shown great potential application in persistent monitoring, but still have problems such as difficulty in ensuring monitoring frequency and easy leakage of monitoring path information. Therefore, under the premise of covering all monitoring targets by UAVs, it is necessary to improve the monitoring frequency of the target and the privacy protection of the monitoring intention as much as possible. In response to the above problems, this research proposes monitoring overdue time to evaluate the monitoring frequency and monitoring period entropy in order to evaluate the ability to ensure monitoring privacy protection. It then establishes a multi-UAV cooperative persistent monitoring path planning model. In addition, the multi-group ant colony optimization algorithm, called overdue-aware multiple ant colony optimization (OMACO), is improved based on the monitoring overdue time. Finally, an optimal flight path for multi-UAV monitoring with high monitoring frequency and strong privacy preservation of monitoring intention is obtained. The simulation results show that the method proposed in this paper can effectively improve the monitoring frequency of each monitoring node and the privacy preservation of the UAV monitoring path and has great significance for enhancing security monitoring and preventing intrusion.

Keywords: persistent monitoring; privacy protection; path planning; monitoring frequency; overdue time

Citation: Chen, Y.; Shu, Y.; Hu, M.; Zhao, X. Multi-UAV Cooperative Path Planning with Monitoring Privacy Preservation. *Appl. Sci.* **2022**, *12*, 12111. https://doi.org/10.3390/app122312111

Academic Editors: Xiongbo Wan, Wei Wang, Hao Fu and Sheng Du

Received: 7 November 2022
Accepted: 23 November 2022
Published: 26 November 2022

1. Introduction

For the purposes of public safety, environmental protection, scientific research, etc., people need to observe, measure and collect information in certain areas over a long time, and then make decisions based on the results of these observations, measurements and collection. This is generally called a persistent monitoring problem [1–3]. Monitoring in person or by hand is usually constrained by weather, geography, working hours and labor costs, and intelligent equipment can greatly overcome the above deficiencies of human based monitoring. Unmanned aerial vehicles (UAV) are examples of one of these typical intelligent monitoring devices. Because they are free of human intervention and offer stable flight, a wide range of motion, and low cost, UAVs are often used to perform persistent monitoring tasks [4], target detection and tracking [5], and border patrols [6]. This research mainly studies the UAV path planning problem when they are used in persistent monitoring.

With the emergence of various complex environments and complex tasks, a single UAV will find it hard to meet the requirements of increasingly complex inspection operations. Consequently, there has been extensive research on multi-UAV cooperation. Compared with single-UAV operation, multi-UAV cooperation has demonstrated greater advantages. For example, multi-UAV cooperation [7,8] can obtain more comprehensive and wide information and can realize multi-angle monitoring of the target area. However, such problems as cooperation strategy, inconsistent monitoring frequency, unsynchronized monitoring information, and unsafe monitoring strategies still exists for multi-UAV cooperation. The task decisions of multi-UAV persistent monitoring have become popular issues in the application field of UAVs.

The multi-UAV persistent monitoring problem can be divided into two levels. One level is the monitoring frequency constraint, and the other is the persistent monitoring security, i.e., monitoring privacy preservation. The above two levels correspond to the two so-called modes of UAV persistent monitoring. One is the regular monitoring mode, that is, the route planned for the UAV to minimize the time delay between each adjacent visit of the task nodes and to improve their monitor frequency as much as possible. The other is the adversarial monitoring mode, which is to plan uncertain, unpredictable and non-periodic monitoring paths for UAVs in order to prevent any intelligent intruders from detecting the monitoring regularity [9]. If the monitoring frequency constraint is considered as the only criterion, the monitoring path is usually a certain periodic path. Once an intelligent intrusion appears in the monitoring environment, the privacy of the UAV monitoring intention cannot be protected, and the monitoring task is easily destroyed by intelligent intruders. On the other hand, when only the security of persistent monitoring is considered, it may be difficult to satisfy the monitoring frequency requirements of each node due to excessive consideration of path privacy security. Therefore, it is of great theoretical significance and practical value to study the joint optimization problem of monitoring frequency and privacy protection.

Portugal [10] reviewed the multi-robot cooperative patrol algorithms that has been studied in recent years and pointed out that a distributed, non-deterministic and cooperative strategy represents the future trend. Alamdari [11] studied the persistent monitoring problem of a single robot. The optimization goal is to minimize the revisit duration of the given monitoring tasks. Two approximate algorithms with complexity $O(\log \rho_G)$ and $O(\log n)$ were proposed, respectively. Elmaliach [12] studied the patrol problem in a closed area and proposed the patrol frequency optimization criterion for the first time, and each point in the area should be repeatedly visited by multiple robots. Smith [3,13] studied persistent monitoring problems in discrete and continuous environments, and established two optimization models, aiming to enhance the monitoring frequency. Wang [14] studied the persistent monitoring problem of multiple UAVs and established a mathematic model based on the optimization of the maximum environmental recognition accuracy, which was then solved by a heuristic algorithm. Kalyanam [15] studied a similar problem, i.e., UAV data collection, allowing UAVs to visit some targeted location with high priority more than once in a single cycle. An optimization by maximizing the average period reward was formulated, and the precise solution combining dynamic programming and mixed integer linear programming was achieved. Subsequently, considering the scalability of the algorithm and improving its efficiency, an approximate solution was proposed for the nodes with specific visiting times [16]. Von [17] also discussed the algorithm scalability where a genetic algorithm was used to obtain the approximate solution that showed better scalability than a precise method through experiments. Scherer [18] studied a multi-UAV cooperative path planning problem with monitoring data transport for the purpose of minimizing the time delay between data being captured by UAVs and the arrival of the data at the base station. Hari [19] considered the monitoring frequency constraint and set the fixed horizon to a given number, k, which assumes that the UAV can only access k nodes in each cycle. However, once there exists an intelligent intrusion in the monitoring environment, the monitoring privacy will have already been destroyed. The above persistent monitoring studies considered monitoring frequency constraints, but only focused on the monitoring performance or coverage rate of the given area [20] and did not consider monitoring security issues in an adversarial environment.

With regard to the concern for monitoring security, one also needs to consider how easy the monitoring strategy can be acquired by intelligent intruders. The privacy of the persistent monitoring process is of great concern, especially in some applications where intelligent adversaries or intruders might occur. At present, there are at least two ideas in the field of monitoring security. One is to improve the existing deterministic strategy for the path planning problem and use random algorithms instead, such as Markov chains, or random walk theory. The other is to establish a game model and a balance scheme between the competing players. Agmon [21] proposed a Markov strategy, which is a polynomial-time algorithm, and their research is motivated by reducing the probability of being invaded at a weak task position as much as possible. Entropy has also been introduced in path planning [22]. For example, George [23] and Duan [24] studied the entropy rate maximization problem based on Markov chains. Stackelberg game theory was used by Basilico [25] to formulate an optimal solution to the path planning problem for a single robot on a security patrol, while assuming only one intruder. Security game theory has been proposed for the study of the persistent monitoring path planning problem in ecological protection [26]. The main motivation for their study on patrol and monitoring strategies is to obtain an unpredictable trajectory, which was finally obtained through maximum entropy.

With the aforementioned observations, some studies on persistent monitoring path planning only concern the complete coverage rate, and some studies consider the monitoring frequency, but the final paths often fall in a fixed monitoring period which makes the monitoring regularity completely exposed to intrusions. The other study considers monitoring security, but they still do not consider monitoring frequency constraints. To bridge the gap between the monitoring frequency and monitoring security, this study will comprehensively consider both sides simultaneously, that is, improving monitoring path privacy while increasing monitoring frequency. The main contributions of this paper are as follows:

- Considering monitoring frequency and path privacy, this study shows how to formulate a multi-UAV cooperative persistent monitoring path planning problem with multiple constraints based on the monitoring of overdue time and of monitoring period entropy.
- A multi-group ant colony optimization algorithm, called overdue-aware multiple ant colony optimization (OMACO), is proposed to obtain an optimal flight path for UAV cooperation. The heuristic function and pheromone update method are improved based on the monitoring delay time and overdue time. In addition, a target exclusive mechanism and greedy strategy are proposed for ant node selection.
- Simulation experiments are carried out in complete and incomplete environments to verify the effectiveness and advantages of the designed algorithm. The simulation results show that the algorithm proposed in this paper can effectively improve both the monitoring frequency and the monitoring privacy protection.

2. Multi-UAV Cooperative Persistent Monitoring Path Planning Model

2.1. Problem Description

As the monitoring environment changes and the node quantity increases, computer resources onboard are often insufficient when performing persistent monitoring tasks in the stand-alone operation mode. As a result, the waiting time of nodes increase, causing some nodes to monitor overdue. Compared with a single drone, a drone group performing persistent monitoring tasks will face huge challenges. For example, each node will maintain a parameter that represents how long it has been waiting since its last monitoring. Once any drone visits a node position and completes that monitoring, the waiting-time parameter maintained by this node will be cleared— demonstrating a rigid nonlinearity. Other difficulties include collision avoidance between multiple drones, information synchronization, and collaborative work between drones.

This study focuses only on the multi-UAV cooperative path planning problem of persistent monitoring. A graph model is used to describe the distribution of the candidate nodes, i.e., $G = (V, E)$, where $V = \{1, 2, \ldots, N\}$ represents the nodes set, N represents the total number of nodes, and $E = \{e_{ij}, \forall i, j \in V\}$ represents the edges set of G. The UAV set is $M_{UAV} = \{1, 2, \ldots, M\}$, where M is the total number in the given UAV group, $M << N$. Here are some assumptions about the background of this study.

(1) For safety and efficiency purposes, the same nodes cannot exist for multiple drones at the same time. This means that different UAV are permitted to monitor the same node on different time.

(2) Without loss of generality, all UAVs fly with a constant speed, v.

(3) After a UAV accesses a node, the waiting time of the node is cleared, and all other UAVs need to be notified to ensure information synchronization.

This research tries to find the optimal flight path of a UAV group, so that the path meets the requirements of high monitoring frequency and strong monitoring path privacy.

2.2. Discretization of the Graph

Persistent monitoring needs to consider UAV movement synchronization. In order to solve the problem, a discrete approximation operation is introduced on the graph G. Several virtual nodes are inserted in an approximately uniform way to the edges of G leading to a discretized graph that includes many more edges of equal intervals, denoted by δ. This operation encourages good behavior in which any UAV will certainly move forward from its current node position to its neighbor node in G instead of staying between nodes at time step k. This is called UAV movement synchronization. Consequently, nodes can be divided into two categories, one is the **task node** set, V, which requires monitoring and the other is the **virtual node** set, U, which is generated during discrete approximation operation and does not to be monitored. The complete node set, called a **generalized node** set, is denoted as $V' = V \cup U = \{1, 2, \ldots, N + |U|\}$. It should be emphasized that all virtual nodes in U are not real monitoring tasks, so they do not need to record their monitoring delays. The final adjacency matrix of G is $A \in \mathbb{R}^{(N+|U|)\times(N+|U|)}$, where any element a_{ij} is binary. $a_{ij} = 1$ indicates that node i and j are adjacent to each other, otherwise $a_{ij} = 0$.

2.3. Multi-UAV Collaborative Monitoring Constraints

Let K denote the maximum length of the monitoring horizon. Let the binary variable matrix $Y^m \in \mathbb{R}^{K\times(N+|U|)}$ denote whether a node is monitored by UAV m, $m \in M_{UAV}$. For $\forall i \in V'$, the element $y^m_{k,i} = 1$ represents that the node i is monitored by UAV m at time k, and $y^m_{k,i} = 0$ represents that the node i is not monitored by UAV m at time k. Y^m represents the monitoring of all nodes by UAV m in the entire monitoring time horizon.

Let the binary variable matrix $X \in \mathbb{R}^{K\times(N+|U|)}$ represent whether a node is monitored by any UAV in the group, where the element $x_{k,i} = 1$ represents that there is at least one UAV monitoring node i at time k, and the element $x_{k,i} = 0$ represents that node i is not monitored by any UAV at time k. The matrix X stands for the monitored situation of all nodes in the monitoring time horizon, and can be obtained by combining all Y^m, $m = 1, 2, \ldots, M$. The relationship between X and Y^m is $X = Y^1 \cup Y^2 \ldots \cup Y^M$. The constraints are as follows:

$$x_{k,i} = \begin{cases} 0, \text{if } \sum\limits_{m=1}^{M} y^m_{k.i} = 0, \\ 1, \text{otherwise, i.e., } \sum\limits_{m=1}^{M} y^m_{k.i} = 1 \end{cases} \tag{1}$$

where $i \in V, k \in \{1, 2, \ldots, K\}$

$$\sum_{m=1}^{M} y^m_{k.i} \leq 1, i \in V', k \in \{1, 2, \ldots, K\} \tag{2}$$

$$\sum_{k=1}^{K} x_{k,i} \geq 1, i \in V \tag{3}$$

$$\sum_{i=1}^{N+|U|} x_{k,i} = M, k \in \{1, 2, \ldots, K\} \tag{4}$$

Equation (2) indicates that at any time k, a node is monitored by, at most, one UAV, that is, multiple UAVs cannot appear at the same location at the same time. Equation (3) indicates that within the monitoring horizon K, each node must be visited at least once. Equation (4) indicates that a UAV only has one position at a certain time k.

2.4. UAV Motion Constraints

Assuming that the initial moment k=1, all the UAVs need to start from the same given initial node $S_m \in V$. The following constraints are satisfied:

$$y^m_{1,S_m} = 1, m \in M_{UAV} \tag{5}$$

At the same time, the UAV m cannot visit the same node in two adjacent time steps:

$$y^m_{k,i} + y^m_{k+1,i} \leq 1, i \in V, k \in \{1, 2, \ldots, K-1\}, m \in M_{UAV} \tag{6}$$

2.5. The Waiting Time Constraint of the Task Node

Let $F \in \mathbb{R}^{(K-1)\times N}$ represent the whole task nodes' waiting time, in which the element is $f_{k,i} \geq 0$. In the interval between time step k-1 to k, all UAVs select a candidate node from their individual neighbor according to a certain movement strategy. After that, the waiting time of almost all nodes increases by one unit time except the arrived node i which is exactly a task node. That is, $i \in V$. The waiting time corresponding to the arrived node i will be cleared. Therefore,

$$f_{k,i} = \begin{cases} 0, & i \in V, k = 1 \\ (1 - x_{k,i})(f_{k-1,i} + c), & i \in V, k \in \{2, 3, \ldots, K\} \end{cases} \tag{7}$$

where c is a unit time constant, which represents the time consumed by the UAV when passing through each edge interval. This specific value is related to the accuracy of the discretization operation.

2.6. Min–Max Optimization for Multi-UAV Cooperative Monitoring

2.6.1. UAVs Monitoring Overdue Time Evaluation

Let the maximum monitoring interval of a task node i between two adjacent monitoring events be the expected period of the node, denoted by $T_i, i \in V$. Ideally, for any time k, the waiting time of node i should not exceed its expected period. That is

$$0 \leq f_{k,i} \leq T_i, i \in V, k \in \{1, 2, \ldots, K\} \tag{8}$$

However, in practical applications, since the number of UAVs is far less than the quantity of the task nodes, it is inevitable that some nodes' monitoring will be overdue. The overdue time can be expressed as $f_{k-1,i} + c - T_i$. Define the real monitoring period of the task node as $P \in \mathbb{R}^{K\times N}$:

$$p_{k,i} = \begin{cases} 0, & i \in V, k = 1 \\ x_{k,i}(f_{k-1,i} + c), & i \in V, k \in \{2, 3, \ldots, K\} \end{cases} \tag{9}$$

The above equation indicates that when the UAV arrives at node i at time step k, i.e., $x_{k,i} = 1$, the real monitoring period of this node is $f_{k-1,i} + c$. Otherwise, $p_{k,i}$ have no definition and it will be assigned to zero. Therefore, the maximum monitoring period of the task node i in the entire monitoring horizon is:

$$\max_{k \in \{1,2,\ldots,K\}} p_{k,i} \tag{10}$$

Then, the maximum overdue time of the task node i caused by exceeding its expected period T_i can be expressed as:

$$\max\left\{0, \max_{k\in\{1,2,\ldots,K\}}(p_{k,i} - T_i)\right\} \tag{11}$$

The following objective, J_1, is proposed for optimization by minimizing the normalized maximum overdue time of all task nodes.

$$\min_{Y,F} J_1 = \max_{i\in V}\left(\frac{1}{T_i}\max\left\{0, \max_{k\in\{1,2,\ldots,K\}}(p_{k,i} - T_i)\right\}\right) \tag{12}$$

2.6.2. UAVs Monitoring Path Privacy Criterion

As long as any UAV accesses a task node, its waiting time will be cleared. Therefore, it is necessary to evaluate the privacy of the monitoring path based on the actual visiting period of all task nodes. Since the uncertainty of the monitoring period indirectly reflects the monitoring privacy, this study proposes the concept of monitoring period entropy (MPE) which refers to the uncertainty when the UAV returns to the task node for monitoring again. The larger the MPE, the higher the randomness of the monitoring period. Define a vector $\widetilde{p}_i = \{p_{k,i} | p_{k,i} > 0, k = 1,2,\ldots,K\}$ to represent the vector composed of all the monitoring cycles of task node i in the entire monitoring horizon. The length of the vector, $\widetilde{p}_i$, is $l_{\widetilde{p}_i} = \sum_{k=1}^{K} x_{k,i}$. Define the monitoring period entropy of node i as:

$$H(\widetilde{p}_i) = -\sum_{j=1}^{l_{\widetilde{p}_i}} P(\widetilde{p}_i(j)) \log P(\widetilde{p}_i(j)) \tag{13}$$

where $P(\widetilde{p}_i(j))$ is the probability that the jth element in vector $\widetilde{p}_i$. One should note that $H(\widetilde{p}_i)$ is always positive. The minimum monitoring period entropy among all task nodes is:

$$\min_{i\in V} H(\widetilde{p}_i) \tag{14}$$

Therefore, in order to improve the randomness of the monitoring period, the optimization objective is designed to maximize the entropy of the smallest monitoring period among all task nodes, namely $\max_{Y,F}\left(\min_{i\in V} H(\widetilde{p}_i)\right)$. This criterion is also equivalent to the reciprocal of the minimum monitoring period entropy (because $H(\widetilde{p}_i)$ is a positive number), so the following optimization objectives can be designed:

$$\min_{Y,F} J_2 = \frac{1}{\min_{i\in V} H(\widetilde{p}_i)} \tag{15}$$

The dimension of the multi-UAV path solution $\boldsymbol{Y}$ is $K \times (N + |U|)$, and the algorithm time complexity of the calculation for the monitoring of overdue time and the evaluation of path privacy is $O(n^2)$.

2.6.3. Multi-UAV Persistent Monitoring Path Planning Model

The optimization problem of multi-UAV cooperative persistent monitoring path planning is expressed as follows:

$$\begin{aligned} \min_{Y,F} \quad & J = wJ_1 + (1-w)J_2 \\ \text{s.t.} \quad & (1) - (8) \end{aligned} \tag{16}$$

where $w \in (0,1)$ represents the weight coefficient, which will balance between the performance of overdue time and path privacy.

3. Improved Multi-Group Ant Colony Optimization Algorithms Based on Monitoring Overdue Time

From the perspective of reducing monitoring overdue time and improving path privacy, this section designs an improved ant colony optimization (ACO) algorithm based on the monitoring of overdue time, called an overdue-aware multiple ant colony optimization algorithm. Major improvements include the aspects:

- A greedy strategy for node selection is proposed, in which the ant colony heuristic function is modified using the expected period of the task nodes.
- Ant colony pheromone is updated based on monitoring overdue time and monitoring period entropy.
- A target exclusion mechanism is proposed to improve the utilization rate of multi-UAV in cooperative monitoring.

3.1. Heuristic Function Based on Monitoring Expectation Period

In order to increase the monitoring frequency and reduce the visiting delay of each task node, the improved heuristic function, η_{ij}, is as follows:

$$\eta_{ij} = \frac{1}{T_j d_{ij}} \tag{17}$$

where d_{ij} represents the distance between node i and j. Comparing with the traditional heuristic function in ACO, Equation (17) takes into account the expected period (T_j) of the neighbor task nodes, which is helpful in reducing its monitoring overdue time.

3.2. Target Exclusion Mechanism

When multiple UAVs perform tasks at the same time and do not consider the path privacy issue, multiple UAVs will be evenly distributed on the minimum Hamiltonian cycle of the graph [25]. The ants select generalized nodes (task nodes or virtual nodes are both possible) depending on stochastic probability. Therefore, there is a slim chance that the UAV follows its previous UAV when selecting its next node, which results in some nodes being monitored frequently while other task nodes are missed for a long time. Consequently, monitoring overdue events happen. In order to prevent UAVs from following synchronically, this research proposes a target exclusion mechanism, as shown in Figure 1.

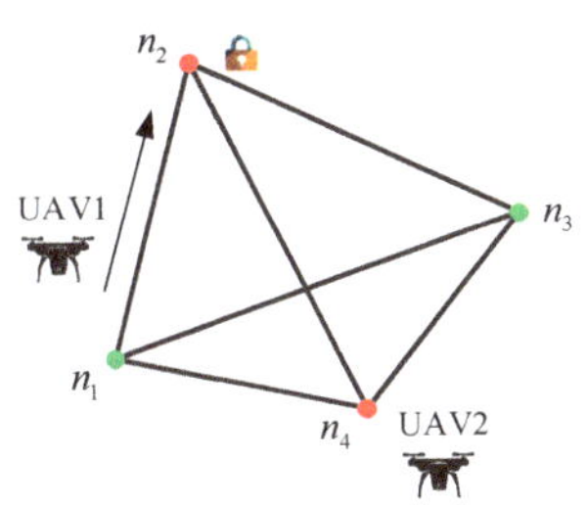

Figure 1. Target exclusion mechanism.

As an example, when UAV1 in Figure 1 selects node n_2 as the candidate task node, UAV1 exclusively occupies node n_2 and the node n_2 will be locked. However, UAV2, which is currently located at node n_4, cannot select the locked node as its candidate. Only one of n_1 and n_3 will be chosen as the UAV1's next waypoint. The target exclusive mechanism can fundamentally solve the UAV following problem.

3.3. Greedy Strategy for Node Selection

This section proposes a greedy strategy, which can help UAV select the optimal node among its neighbors. The strategy is motivated by the idea that the greater the overdue time of the ant's adjacent node j is, the greater the probability that node j will be selected by the ants in the next step. First calculate the overdue time of all adjacent nodes. Since some adjacent nodes may not be overdue, the calculated overdue time by $f_{k-1,j} + c - T_j$ is possibly negative and inconvenient to compute the transition probability. Therefore, this research constructs a pseudo-overdue time, $R_j(t)$, which is guaranteed to be positive.

$$R_j(t) = f_{k-1,j} + c - T_j + T_0, \forall k \in P \tag{18}$$

where j represents the adjacent node of the current node. T_0 represents the upper bound of the expected period of all monitoring nodes. Usually, it can be calculated by $T_0 = \max_{i\in V}\{T_i\}$ offline.

The transition probability is not only related to the overdue time of its neighbor node, but also related to the adjacency constraints, exclusive flags, and pheromone distribution of the ants' current adjacent nodes. The improved ant transition probability p_{ij}^z is as follows:

$$p_{ij}^z = \begin{cases} \frac{\tau_{ij}^{\alpha}(t)\eta_{ij}^{\beta}(t)R_j(t)a_{ij}(1-o_j)}{\sum\limits_{s\in allow_z} \tau_{is}^{\alpha}(t)\eta_{is}^{\beta}(t)R_s(t)a_{is}(1-o_s)}, & j \in allow_z \\ 0\,, & \text{other} \end{cases} \tag{19}$$

where i is the current node of the ant whose adjacent node is denoted by j. α and β stand for the importance factor of the pheromone and the heuristic function, respectively, $\tau_{ij}(t)$ represents the pheromone concentration on the edge e_{ij} after the optimization of each ant at the t-th iteration. a_{ij} stands for the adjacency relationship between node i and j, o_j represents the exclusive state of the node j, $z \in \{1, 2, 3, \ldots, Z\}$ represents the ant number, z is the ant quantity, and $allow_z$ represents the set of nodes that the ant z can visit next time. After the transition probability of the ants is calculated, the roulette method is used to select the next node according to the maximum probability.

3.4. Pheromone Update Based on Monitoring Overdue Time and Monitoring Period Entropy

The traditional ant colony algorithm updates the pheromone mainly based on the path length that ants travelled. In order to promote the evolution of the ant colony to the direction with the smallest cost function value, this study updates the pheromone according to the optimization objective (16).

$$\tau_{ij}(t+1) = (1-\rho)\tau_{ij}(t) + \sum_{z=1}^{Z} \Delta\tau_{ij}^z \tag{20}$$

$$\Delta\tau_{ij}^z = \begin{cases} \frac{Q}{J_z}, \text{ant } z \text{ from node } i \text{ to node } j \\ 0\,, \text{other} \end{cases} \tag{21}$$

where $\rho \in (0,1)$ represents the pheromone volatile factor. $\Delta\tau_{ij}^z$ represents the pheromone concentration released by the ant z on the edge between node i and j in the current iteration. Q is a constant, representing the total pheromone amount released by the ants at one time, and J_z represents the path cost of the ant z calculated according to (16).

To sum up, the scheme of the proposed OMACO algorithm is shown in Figure 2. The steps are as follows in Algorithm 1:

Algorithm 1: Overdue-aware multiple ant colony optimization (OMACO).

Step 1: Initialization (node quantity N, adjacency matrix A, ant quantity Z, maximum iterations N_c, pheromone importance factor α, heuristic function importance factor β, pheromone volatility factor ρ, pheromone quantity Q, and maximum monitoring horizon K, weight parameter w).
Step 2: Discretization of the graph.
Step 3: Calculate the target exclusion set O_0.
Step 4: Calculate the ant transition probability p_{ij}^{z} according to (19).
Step 5: Select the next node according to the roulette method, and update the node waiting time $f_{k,i}$.
Step 6: Update the ant's taboo table.
Step 7: Update the target exclusive flag o_i.
Step 8: Calculate the monitoring overdue time and monitoring period entropy according to (11) and (13).
Step 9: Update pheromone according to (20) and (21).
Step 10: Determine whether the iteration reaches the maximum iterations. If so, the procedure ends; otherwise, go to Step 3.

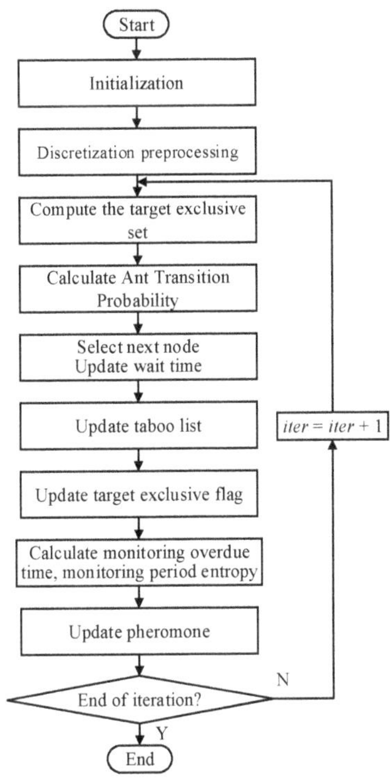

Figure 2. Flowchart of the OMACO algorithm.

4. Simulation Experiments and Discussions

In this section, simulation experiments are carried out for multi-UAV persistent monitoring tasks in complete and incomplete environments to evaluate the path planning model and solution algorithm proposed in this study.

4.1. Algorithm Feasibility Analysis

Assume that three UAVs perform tasks in a complete environment containing 10 task nodes with known locations to be monitored, which are labeled as numbers in Figure 3. Task nodes and virtual nodes are illustrated by red and green dots, respectively. The blue solid lines represent adjacency relationships within the graph. All the simulation parameters are listed in Table 1. The expected periods of the task nodes are shown in Table 2. All simulation examples in this paper are implemented on a computer with Matlab R2020a installed and the system configuration is Intel Core i7-9750H, 2.59 GHz, 16 GB RAM.

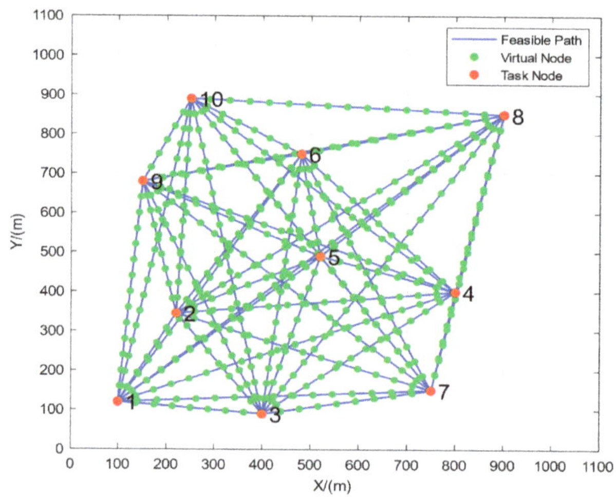

Figure 3. Discretization of a completely connected graph.

Table 1. Simulation parameters.

Parameters	Value	Notes
v	8 m/s	UAV speed
δ	40 m	interval for discretization
Z	15	ant quantity
c	5 s	constant
N_c	200	maximum iteration
α	1.2	pheromone importance factor
β	4	heuristic function importance factor
ρ	0.3	pheromone volatility factor
Q	10	pheromone quantity
K	500	monitoring Horizon
w	0.6	weight parameter

Table 2. Expected period of 10 task nodes.

Node	1	2	3	4	5	6	7	8	9	10
T_i (s)	370	380	350	375	365	390	380	380	375	360

Figure 4 shows the persistent monitoring flight path of the three UAVs obtained by the proposed method in this paper, where the x-axis represents the time, and the y-axis represents the node that the UAV arrived at the corresponding time step. The solid line represents the UAV flight path consisting of passing nodes.

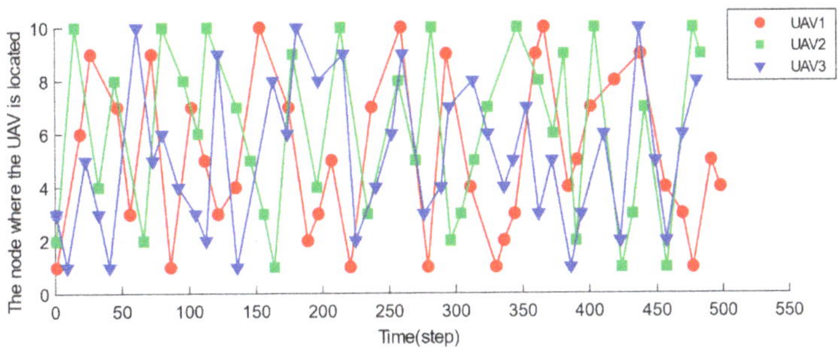

Figure 4. The persistent monitoring path obtained by the OMACO algorithm.

Figure 5 shows the expected period and the actual monitoring period of the task nodes. It can be seen that the actual monitoring period of all task nodes is less than the expected period, which indicates that the monitoring process of the UAV meets the monitoring frequency requirements of all nodes. Figure 5 also shows that each node has been visited multiple times in the monitoring horizon, obtaining multiple actual monitoring periods which are all lower than their expected periods, i.e., meeting the monitoring frequency requirements.

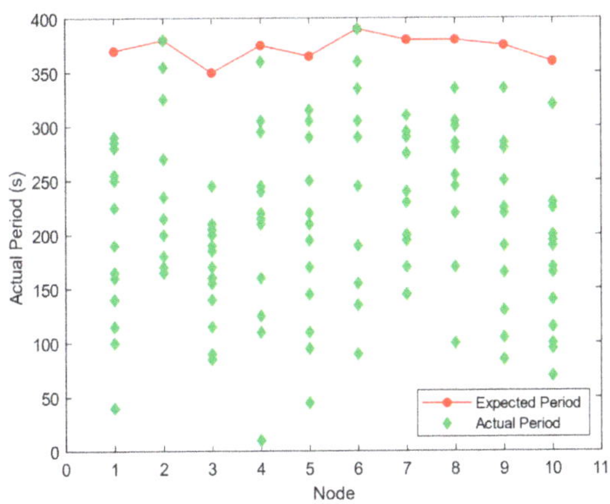

Figure 5. The actual monitoring period of task nodes obtained by the OMACO algorithm.

More importantly, the actual monitoring period of each node is different, that is, the waiting time when each node is monitored has a good random distribution. The simulation shows that the method proposed in this paper can cover all monitoring nodes, meet the monitoring frequency requirements, and also improve the privacy protection of the monitoring path.

4.2. Comparative Analysis with Traditional ACO

In order to evaluate the performance of the proposed OMACO algorithm, this section compares the optimization ability of OMACO and the traditional ACO. Figure 6 shows the monitoring path solved by the traditional ACO with the same parameters to Section 3.1. Different from Figure 4, the path sequences (node 6 $\rightarrow$ 7 $\rightarrow$ 9) repeat up to eight times in Figure 6, and the UAV3 trajectory (blue) between steps 450 and 500 can be seen following by UAV1 (red). This leads to the same monitoring period and is very harmful to the monitoring privacy protection. However, the UAV path in Figure 4 has no obvious repetitive path or circular trajectory, and there is no UAV following the other. Therefore, compared with the traditional ACO, the proposed OMACO algorithm can obtain better privacy protection performance.

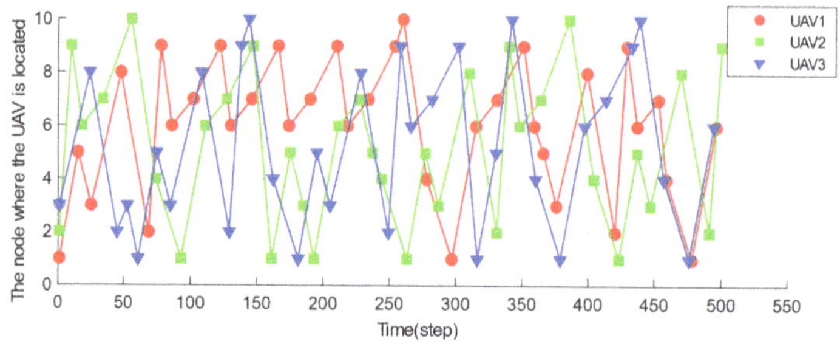

Figure 6. The persistent monitoring path obtained by ACO algorithm.

Figure 7 shows the actual monitoring period obtained by using the traditional ACO. There exist many nodes that have been monitored overdue many times, resulting in the waiting time of the task node frequently exceeding the expected period. Therefore, the proposed OMACO algorithm is superior to the traditional ACO in improving the monitoring frequency.

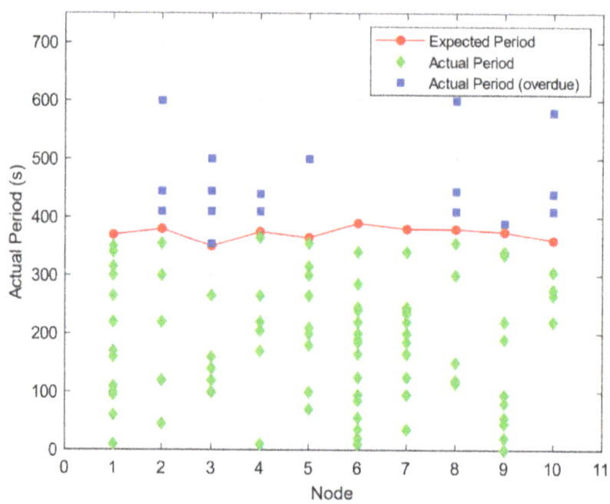

Figure 7. The actual monitoring period of each node obtained by ACO algorithm.

Table 3 shows a detailed comparison between OMACO and ACO on each task node monitoring data. Based on the proposed OMACO algorithm, most task nodes have been visited more times than that of ACO. Therefore, the average visit number is higher than the traditional ACO. Correspondingly, the average actual period will decrease and be less than ACO. Also, it is found that the ACO algorithm is not appropriate for our problem because the node No.10 exceeds its upper bound.

Figure 8 shows the iterative curves of the objective functions obtained by OMACO and ACO, and the related data are shown in Table 4. In the first iteration, the algorithm designed in this research has a lower value of objective function than ACO. This is because the waiting time of the task node has already been considered by OMACO when calculating the transition probability based on the greedy strategy. In fact, the node selection strategy has been optimized before the initial ant path. The traditional ACO only relies on the heuristic function and pheromone to decide the node transition probability. Consequently, the pheromone is equal on all path segments in the initial iteration which leads to a randomly path generated.

Table 3. Monitoring results comparison between OMACO and ACO.

Node	Number of Visits		Average of Actual Monitoring Period	
	OMACO	ACO	OMACO	ACO
1	12	12	198.33	198.75
2	9	7	253.33	350.00
3	14	8	167.14	278.75
4	11	8	232.50	286.25
5	12	9	204.17	242.22
6	9	15	260.00	165.00
7	10	12	220.00	188.33
8	9	7	266.11	335.00
9	11	15	219.09	166.33
10	14	6	170.00	365.00 *
Average	11.1	9.9	219.07	257.56

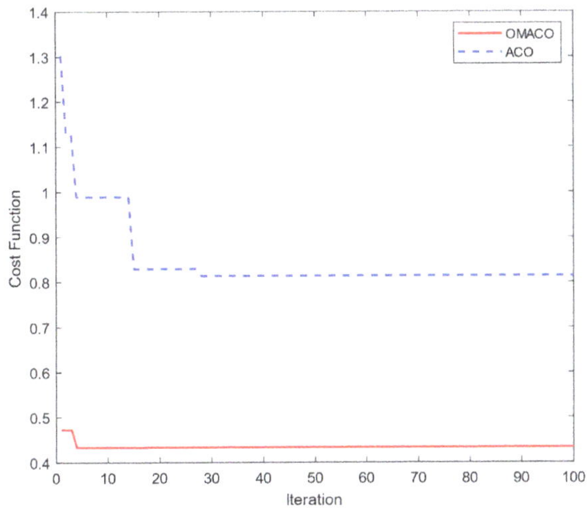

Figure 8. The objective function iteration curves of the two algorithms.

Table 4. Solution comparison between OMACO and ACO.

	OMACO	ACO
Iterations	4	28
Minimum Cost	0.433	0.814

The OMACO algorithm gets the optimal solution of 0.433 in the 4th iteration while the traditional ACO only obtains the optimal solution of 0.814 in the 28th iteration. Since the OMACO algorithm introduces the overdue time for optimization, it is significantly better than the ACO in terms of reducing the monitoring overdue time and improving the monitoring path privacy.

4.3. Algorithm Scalability Analysis

This section demonstrates the simulation experiments with three UAVs performing persistent monitoring in an incomplete environment which contains 15 task nodes. Other parameter settings are the same as in Section 3.1. Figure 9 shows the environment topology where 15 task nodes connected incompletely will be persistently monitored by the UAVs. The expected period of 15 nodes is shown in Table 5.

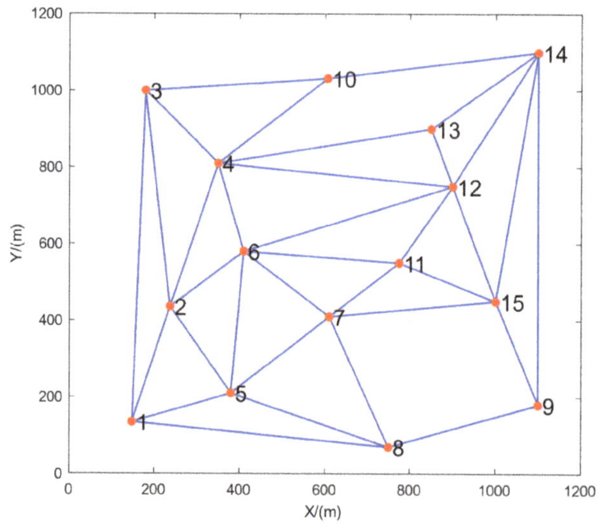

Figure 9. Incomplete environment including 15 task nodes.

Table 5. Expected period of 15 task nodes.

Node	**1**	**2**	**3**	**4**	**5**	**6**	**7**	**8**	**9**	**10**	**11**	**12**	**13**	**14**	**15**
T_i **(s)**	700	750	1050	950	950	850	950	850	700	850	850	750	700	700	750

In order to further evaluate the scalability of the OMACO algorithm, the algorithm is tested in the incomplete environment and the results are shown in Figures 10 and 11. It can be seen that the OMACO algorithm can obtain the optimal path of the UAV swarm in an incomplete environment, satisfying the objective that the actual monitoring period of each node be not higher than the expected period. It can be concluded that the OMACO algorithm can solve the problem of UAV flight paths in different monitoring environments, satisfying the requirements for monitoring overdue events and monitoring privacy.

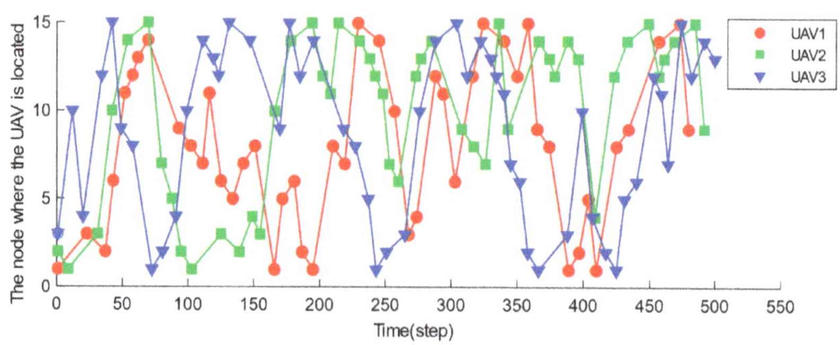

Figure 10. The monitoring path in incomplete environment obtained by OMACO algorithm.

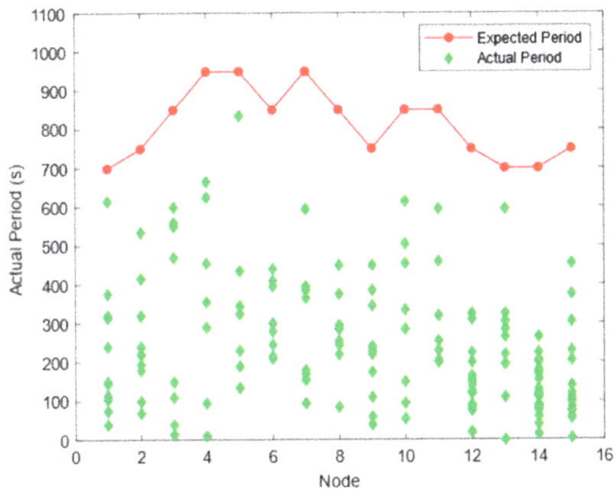

Figure 11. The monitoring period in an incomplete environment obtained by OMACO algorithm.

5. Conclusions

This research has studied the problem of multi-UAV persistent monitoring path planning from the perspective of monitoring privacy protection, reducing monitoring overdue events, and improving the privacy protection of the monitoring trajectory. A multi-UAV path planning mathematical model was established based on the monitoring overdue time and monitoring period entropy. Based on the overdue time, the heuristic function, transition probability and pheromone update, the strategy of the traditional ACO is improved. The simulation results show that the proposed OMACO algorithm can solve the optimal UAV flight path efficiently in both complete and incomplete monitoring environments and has better performance than ACO. This study is promising for the prevention of intelligent intrusions while meeting the requirements of regular monitoring.

However, as the complexity of the monitoring environment increases, there may be adversarial targets destroying monitoring tasks, and the privacy protection requirements may be more stringent. Subsequent consideration will be given to localize adversarial objects cooperatively while executing persistent monitoring assignments.

Author Contributions: Conceptualization, Y.C.; methodology, Y.S.; software, Y.C. and Y.S.; formal analysis, Y.C. and Y.S.; investigation, Y.C. and Y.S.; writing—original draft preparation, Y.S.; writing—review and editing, Y.C., M.H. and X.Z.; supervision, Y.C. All authors have read and agreed to the published version of the manuscript.

Funding: This research was funded by the Joint fund of Science & Technology Department of Liaoning Province and the State Key Laboratory of Robotics with Grant No. 2021-KF-22-14.

Institutional Review Board Statement: Not applicable.

Informed Consent Statement: Not applicable.

Data Availability Statement: Not applicable.

Conflicts of Interest: The authors declare no conflict of interest.

References

1. Cassandras, C.G.; Lin, X.; Ding, X. An optimal control approach to the multi-agent persistent monitoring problem. *IEEE Trans. Autom. Control* **2012**, *58*, 947–961. [CrossRef]
2. Wang, Y.W.; Wei, Y.W.; Liu, X.K.; Zhou, N.; Cassandras, C.G. Optimal persistent monitoring using second-order agents with physical constraints. *IEEE Trans. Autom. Control* **2018**, *64*, 3239–3252. [CrossRef]
3. Smith, S.L.; Rus, D. Multi-robot monitoring in dynamic environments with guaranteed currency of observations. In Proceedings of the 49th IEEE Conference on Decision and Control (CDC), Atlanta, GA, USA, 15–17 December 2010; IEEE: Manhattan, NY, USA, 2010; pp. 514–521.

4. Ostertag, M.; Nikolay, A.; Tajana, R. Trajectory planning and optimization for minimizing uncertainty in persistent monitoring applications. *J. Intell. Robot. Syst.* **2022**, *106*, 2. [CrossRef]
5. Zhu, S.; Wang, D. Adversarial ground target tracking using UAVs with input constraints. *J. Intell. Robot. Syst.* **2012**, *65*, 521–532. [CrossRef]
6. Girard, A.R.; Howell, A.S.; Hedrick, J.K. Border patrol and surveillance missions using multiple unmanned air vehicles. In Proceedings of the 43rd IEEE Conference on Decision and Control, Nassau, Bahamas, 14–17 December 2004; Volume 1, pp. 620–625.
7. Huang, L.; Qu, H.; Ji, P.; Liu, X.; Fan, Z. A novel coordinated path planning method using k-degree smoothing for multi-UAVs. *Appl. Soft Comput.* **2016**, *48*, 182–192. [CrossRef]
8. Yun, W.J.; Park, S.; Kim, J.; Shin, M.; Jung, S.; Mohaisen, D.A.; Kim, J.H. Cooperative multiagent deep reinforcement learning for reliable surveillance via autonomous multi-UAV control. *IEEE Trans. Ind. Inform.* **2022**, *18*, 7086–7096. [CrossRef]
9. Huang, L.; Zhou, M.; Hao, K.; Hou, E. A survey of multi-robot regular and adversarial patrolling. *IEEE/CAA J. Autom. Sin.* **2019**, *6*, 894–903. [CrossRef]
10. Portugal, D.; Rocha, R. A survey on multi-robot patrolling algorithms. In Proceedings of the Doctoral Conference on Computing, Electrical and Industrial Systems, Costa de Caparica, Portugal, 22–24 February 2011; Springer: Berlin/Heidelberg, Germany, 2011; pp. 139–146.
11. Alamdari, S.; Fata, E.; Smith, S.L. Persistent monitoring in discrete environments: Minimizing the maximum weighted latency between observations. *Int. J. Robot. Res.* **2014**, *33*, 138–154. [CrossRef]
12. Elmaliach, Y.; Agmon, N.; Kaminka, G.A. Multi-robot area patrol under frequency constraints. *Ann. Math. Artif. Intell.* **2009**, *57*, 293–320. [CrossRef]
13. Smith, S.L.; Schwager, M.; Rus, D. Persistent robotic tasks: Monitoring and sweeping in changing environments. *IEEE Trans. Robot.* **2011**, *28*, 410–426. [CrossRef]
14. Wang, T.; Huang, P.; Dong, G. Cooperative persistent surveillance on a road network by multi-UGVs with detection ability. *IEEE Trans. Ind. Electron.* **2022**, *69*, 11468–11478. [CrossRef]
15. Kalyanam, K.; Pachter, M.; Casbeer, D. Average reward dynamic programming applied to a persistent visitation and data delivery problem. In Proceedings of the Dynamic Systems and Control Conference, Tysons, VA, USA, 11–13 October 2017; American Society of Mechanical Engineers: New York, NY, USA, 2017; Volume 58295, p. V003T39A002.
16. Kalyanam, K.; Manyam, S.; Von Moll, A.; Casbeer, D.; Pachter, M. Scalable and exact MILP methods for UAV persistent visitation problem. In Proceedings of the 2018 IEEE Conference on Control Technology and Applications (CCTA), Copenhagen, Denmark, 21–24 August 2018; IEEE: Manhattan, NY, USA, 2018; pp. 337–342.
17. Von Moll, A.L.; Casbeer, D.W.; Kalyanam, K.; Manyam, S.G. Genetic algorithm approach for UAV persistent visitation problem. In Proceedings of the Dynamic Systems and Control Conference, Atlanta, GA, USA, 30 September–3 October 2018; American Society of Mechanical Engineers: New York, NY, USA, 2018; Volume 51913, p. V003T36A001.
18. Scherer, J.; Rinner, B. Multi-UAV surveillance with minimum information idleness and latency constraints. *IEEE Robot. Autom. Lett.* **2020**, *5*, 4812–4819. [CrossRef]
19. Hari, S.K.K.; Rathinam, S.; Darbha, S.; Kalyanam, K.; Manyam, S.G.; Casbeer, D. The generalized persistent monitoring problem. In Proceedings of the 2019 American Control Conference (ACC), Philadelphia, PA, USA, 10–12 July 2019; IEEE: Manhattan, NY, USA, 2019; pp. 2783–2788.
20. Lei, Z.; Chen, X.; Chen, X.; Chai, L. Radial coverage strength for optimization of monocular multicamera deployment. *IEEE/ASME Trans. Mechatron.* **2021**, *26*, 3221–3231. [CrossRef]
21. Agmon, N.; Kaminka, G.A.; Kraus, S. Multi-robot adversarial patrolling: Facing a full-knowledge opponent. *J. Artif. Intell. Res.* **2011**, *42*, 887–916.
22. Tao, X.; Lang, N.; Li, H.; Xu, D. Path planning in uncertain environment with moving obstacles using warm start cross entropy. *IEEE/ASME Trans. Mechatron.* **2021**, *27*, 800–810. [CrossRef]
23. George, M.; Jafarpour, S.; Bullo, F. Markov chains with maximum entropy for robotic surveillance. *IEEE Trans. Autom. Control* **2018**, *64*, 1566–1580. [CrossRef]
24. Duan, X.; George, M.; Bullo, F. Markov chains with maximum return time entropy for robotic surveillance. *IEEE Trans. Autom. Control* **2019**, *65*, 72–86. [CrossRef]
25. Basilico, N.; Gatti, N.; Amigoni, F. Patrolling security games: Definition and algorithms for solving large instances with single patroller and single intruder. *Artif. Intell.* **2012**, *184*, 78–123. [CrossRef]
26. Xu, H.; Ford, B.; Fang, F.; Dilkina, B.; Plumptre, A.; Tambe, M.; Driciru, M.; Wanyama, F.; Rwetsiba, A.; Nsubaga, M.; et al. Optimal patrol planning for green security games with black-box attackers. In Proceedings of the International Conference on Decision and Game Theory for Security, Vienna, Austria, 10–13 October 2017; Springer: Cham, Switzerland, 2017; pp. 458–477.

applied sciences

MDPI

Article

Altitude Control of Powered Parafoil Using Fractional Sliding-Mode Backstepping Control Combined with Extended State Observer

Erlin Zhu [1], Youwu Du [1,*], Wei Song [1] and Haitao Gao [2]

[1] School of Electrical & Information Engineering, Jiangsu University of Technology, Changzhou 213001, China
[2] College of Electrical and Electronic Engineering, Anhui Science and Technology University, Bengbu 233030, China
* Correspondence: duyouwu@jsut.edu.cn

Abstract: This paper presents a method of altitude control of the powered parafoil with uncertainties and disturbances based on sliding-mode backstepping control combined with a linear extended state observer (LESO). First, the dynamics of a powered parafoil is derived in the longitudinal plane using its inclination angle. The problem of altitude control is converted to the issue of angle control. Next, uncertainties and disturbances are considered as a total disturbance. An LESO is used to estimate the total disturbance and form an inner-loop compensation. Backstepping control is employed to regulate the inclination angle to follow the desired value. A fractional sliding surface is introduced to the backstepping control. This ensures the transient performance of altitude control of the powered parafoil. Then, stability analysis shows that the observation errors of the LESO are bounded and the control system is uniformly ultimately bounded. Simulation results of an 8 degree-of-freedom powered parafoil illustrate that the LESO can effectively estimate the states of the system and demonstrate the validity and the superiority of the presented method.

Keywords: powered parafoil; altitude control; sliding mode backstepping; fractional calculus; LESO

Citation: Zhu, E.; Du, Y.; Song, W.; Gao, H. Altitude Control of Powered Parafoil Using Fractional Sliding-Mode Backstepping Control Combined with Extended State Observer. *Appl. Sci.* **2022**, *12*, 12069. https://doi.org/10.3390/app122312069

Academic Editors: Sheng Du, Xiongbo Wan, Wei Wang and Hao Fu

Received: 10 November 2022
Accepted: 23 November 2022
Published: 25 November 2022

1. Introduction

The powered parafoil is a prevalent new type of air vehicle with a flex wing [1,2]. Because the engine is equipped with the payload, compared with traditional parafoils, the powered parafoil not only can glide but also cruise and climb, which traditional parafoils cannot perform [3–5]. The powered parafoil has been applied to paragliding sports, supply airdropping, pesticide spraying, and airport demisting, to name but a few, due to its excellent maneuverability [6]. In recent years, the control of powered parafoils has become a research hotspot. Compared with the horizontal control of traditional parafoils, the altitude control of powered parafoils has received extensive attention and has been a challenge because of its complex aerodynamics characteristics.

Many strategies have been proposed to control the altitude of powered parafoils. Yang et al. analyzed the flight performance of altitude control and derived the longitudinal model of a powered parafoil [7]. Aoustin and Martinenko designed a nonlinear control law for a powered parafoil based on partial feedback linearization to track a desired trajectory in the longitudinal plane [8]. Chen et al. presented a precise-gain method to handle the problem of longitudinal motion control of a powered parafoil [9]. The backstepping control was first employed to control the altitude of a powered parafoil using its lateral model [10]. Then, a fuzzy backstepping control was extended to improve the control performance of the altitude control based on a variable-gain scheme [11]. Tan et al. proposed an altitude-tracking control method for a powered parafoil using the coefficient adaptive control and the characteristic model of the parafoil [12]; however, the specification of the guidance was not provided. Zhu et al. presented a spatial path-tracking control of a powered parafoil using the guidance theory combined with the linear active disturbance control (LADRC) [13].

The above-mentioned methods mainly used a simplified model of powered parafoils to design a controller. However, this may impose limitations on control performance as there exist strong nonlinearities and complex model couplings in powered-parafoil systems. Although Tan and Zhu employed methods of the characteristic model and LADRC that do not rely on the precise model of a powered parafoil to remove the limitations, it increases the complexity of the system design and brings difficulty in analyzing the stability of control systems. The sliding mode control (SMC) is a commonly used method for the aircraft [14,15]; however, the control is susceptible to system uncertainties and disturbances. How to effectively suppress disturbances and improve the robustness of a control system is one of the key points in system design [16,17]. On the other hand, the fractional-order theory [18,19] is widely used in industrial control systems. A fractional calculus operator shows good robustness for systems with uncertainties and external disturbances due to its hereditary and memorability.

This paper presents an alternative way to address the problem of the altitude control of a powered parafoil using the fractional sliding-mode backstepping control combined with a linear extended state observer (LESO). First, the inclination angle model of a powered parafoil is derived according to the guidance law. The altitude control of the powered parafoil is converted into the issue of the inclination angle control. Then, an LESO is used to estimate system uncertainties and exogenous disturbances. A fractional dynamic sliding-mode surface is introduced into the backstepping control to improve the transient performance of the altitude tracking. A stability criterion is derived to guarantee that virtual control variables and the error of system states are bounded. Finally, the validity of the presented method is demonstrated by simulation results of an 8-degree-of-freedom powered parafoil.

The rest of the paper is organized as follows. Section 2 derives the inclination angle model of the powered parafoil according to the guidance law. Section 3 explains the design algorithm of the control system based on the fractional sliding-mode backstepping control (FSMBC) combined with an LESO. Section 4 analyzes the stability of the control system. Section 5 shows the validity of the method through simulation results. Section 6 gives some concluding remarks and points out future work.

2. Dynamics of Powered Parafoil and Problem Formulation

Figure 1 shows the structure of a powered parafoil and its three coordinate frames, that is, $O_dX_dY_dZ_d$ represents the earth reference frame, $O_sX_sY_sZ_s$ represents the parafoil reference frame, and $O_wX_wY_wZ_w$ represents the payload reference frame. The transformation between the earth reference frame and the parafoil reference frame is achieved through three Euler angles $\begin{bmatrix}\psi & \theta & \phi\end{bmatrix}^{\mathrm{T}}$; that is, ψ denotes the yaw angle, θ denotes the pitch angle, and ϕ denotes the roll angle.

The study in this paper focuses on the altitude control of a powered parafoil that involves the relative motion between a parafoil and a payload. The dynamics model is established according to the Kirchhoff motion equation. The specific modeling process and model parameters can be found in [20].

The inclination angle σ of the powered parafoil is defined as the angle between the velocity and the horizontal plane, shown as

$$\sigma = \arctan\left(\frac{-\dot{z}}{\sqrt{\dot{x}^2 + \dot{y}^2}}\right), \tag{1}$$

where $\begin{bmatrix}x & y & z\end{bmatrix}^{\mathrm{T}}$ is the position vector in the earth reference frame.

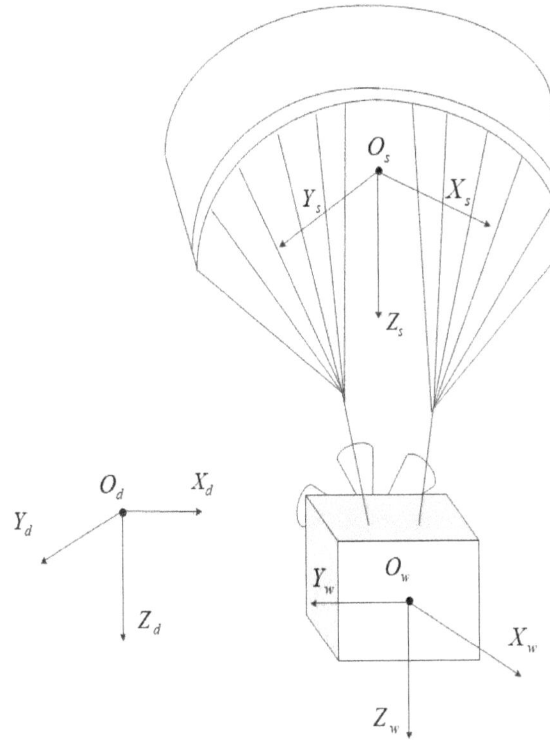

Figure 1. The structure of the powered parafoil and coordinate frames.

Lemma 1 ([13]). *The vertical error e_h between the powered parafoil and the desired point $\mathbf{p}_p$ is global uniformly asymptotically stable and local exponential stable, if the inclination angle of the powered parafoil changes with the following guidance law*

$$\sigma_d = \arctan\left(\frac{e_h}{k_h}\right), \tag{2}$$

where $e_h = H_d - H$ and k_h is an adjustable parameter. H_d and H are the desired altitude and actual altitude of the powered parafoil, respectively.

According to Lemma 1, the altitude control of the powered parafoil can be converted to the control of the inclination angle.

The velocity transformation from the parafoil frame to the earth frame can be described as

$$\begin{bmatrix} \dot{x} \\ \dot{y} \\ \dot{z} \end{bmatrix} = \boldsymbol{R}_{p-e} \begin{bmatrix} u_s \\ v_s \\ w_s \end{bmatrix}, \tag{3}$$

where $\begin{bmatrix} u_s & v_s & w_s \end{bmatrix}^{\mathrm{T}}$ is the velocity vector in the parafoil frame and the transformation matrix

$$\boldsymbol{R}_{p-e} = \begin{bmatrix} \cos\theta\cos\psi & \sin\phi\sin\theta\cos\psi - \cos\phi\sin\psi & \cos\phi\sin\theta\cos\psi + \sin\phi\sin\psi \\ \cos\theta\sin\psi & \sin\phi\sin\theta\sin\psi + \cos\phi\cos\psi & \cos\phi\sin\theta\sin\psi - \sin\phi\cos\psi \\ -\sin\theta & \sin\phi\cos\theta & \cos\phi\cos\theta \end{bmatrix}. \tag{4}$$

If the motion of the longitudinal plane is considered, then it is easy to obtain

$$v_s = 0, y = 0, \psi = 0, \phi = 0. \tag{5}$$

Substituting (4) and (5) into (3) yields

$$\dot{x} = u_s \cos\theta + w_s \sin\theta, \tag{6}$$

$$\dot{z} = -u_s \sin\theta + w_s \cos\theta. \tag{7}$$

Choose the inclination angle σ of the powered parafoil as the state variable of the system. According to (1), (6), and (7), the first derivative of σ is

$$\dot{\sigma} = \dot{\theta} + \frac{\dot{u}_s w_s - u_s \dot{w}_s}{u_s^2 + w_s^2}. \tag{8}$$

Let f_s be the derivative of $\frac{\dot{u}_s w_s - u_s \dot{w}_s}{u_s^2 + w_s^2}$. The second derivative of σ is obtained as

$$\ddot{\sigma} = \ddot{\theta} + f_s. \tag{9}$$

For the altitude control of the powered parafoil, the only control variable that affects the flight velocity is the thrust u provided by the power propulsion. According to the nonlinear dynamics model of the powered parafoil [20], the thrust control variable is coupled in f_s, which increases the difficulty in the design of the controller. To facilitate the design of the control system, the control variable should be separated out. Rewrite (9) as

$$\begin{aligned} \ddot{\sigma} &= \ddot{\theta} + f_s - bu + bu \\ &= f + bu, \end{aligned} \tag{10}$$

where $f = \ddot{\theta} + f_s - bu$ is viewed as a total disturbance.

The state space model of the dynamics of σ is obtained as

$$\begin{aligned} \dot{x}_1 &= x_2 \\ \dot{x}_2 &= f + bu \\ x_1 &= \sigma. \end{aligned} \tag{11}$$

The main objective of this study is to minimize the error of the inclination angle despite of the influence of system uncertainties of the dynamics model and external disturbances on the system. In the next section, a new control scheme is proposed to handle this issue. The LESO is used to estimate uncertainties and disturbance. A fractional sliding-mode backstepping control law is devised to improve control performance and the stability of the system.

3. Design of Control System

In this section, a method of FSMBC with LESO is proposed for the altitude control of a powered parafoil. The configuration of the control system of the powered parafoil is shown in Figure 2, which consists of guidance-based path following, FSMBC, LESO, and the powered parafoil. The outer-loop is a guidance loop, from where the desired inclination angle (2) is obtained according to altitude signals. The inner-loop is the control loop of the inclination angle. FSMBC is adopted, and LESO is used to observe and eliminate the total disturbance.

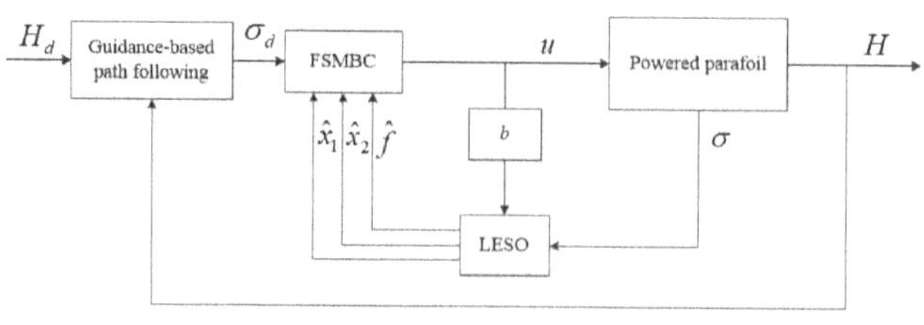

Figure 2. Configuration of the control system of the powered parafoil.

3.1. Design of LESO

Assume $\dot{f}_{total} = h$. Rewrite (11) as

$$\dot{X} = A_1 X + B_1 u + B_2 h, \tag{12}$$

where $A_1 = \begin{bmatrix} 0 & 1 & 0 \\ 0 & 0 & 1 \\ 0 & 0 & 0 \end{bmatrix}$, $B_1 = \begin{bmatrix} 0 & b & 0 \end{bmatrix}^{\mathrm{T}}$, $B_2 = \begin{bmatrix} 0 & 0 & 1 \end{bmatrix}^{\mathrm{T}}$, $X = \begin{bmatrix} x_1 & x_2 & f \end{bmatrix}^{\mathrm{T}}$. Construct an LESO as

$$\dot{\hat{X}} = A_1 \hat{X} + B_1 u + L(X - \hat{X}), \tag{13}$$

where $\hat{X} = \begin{bmatrix} \hat{x_1} & \hat{x_2} & \hat{f} \end{bmatrix}^{\mathrm{T}}$ is the estimated value of X and L is the gain matrix of LESO, which is given by

$$L = \begin{bmatrix} l_1 & 0 & 0 \\ l_2 & 0 & 0 \\ l_3 & 0 & 0 \end{bmatrix}. \tag{14}$$

Let $\tilde{e}$ represent the estimated error of the LESO. We have

$$\tilde{e} = X - \hat{X}. \tag{15}$$

According to (12), (13), and (15), the following differential equation with respect to $\tilde{e}$ can be obtained as

$$\dot{\tilde{e}} = A_2 \tilde{e} + B_2 h, \tag{16}$$

where $A_2 = A_1 - L$.

Assume that h is bounded, namely there exists a positive constant M_1 such that $|h| \leq M_1$. The estimated error of the LESO is always bounded, that is, there exists a positive constant M_2 such that $||\tilde{e}|| \leq M_2$ holds [21].

3.2. Design of Fractional Sliding Mode Backstepping Control

The estimated states $\hat{x}_1$ and $\hat{x}_2$ of the LESO are used in the design of fractional sliding-mode backstepping control. Define the tracking error of the inclination angle as

$$e_1 = \sigma_d - \hat{x}_1. \tag{17}$$

Calculating the derivative of (17) yields

$$\begin{aligned} \dot{e}_1 &= \dot{\sigma}_d - \dot{\hat{x}}_1 \\ &= \dot{\sigma}_d - \hat{x}_2. \end{aligned} \tag{18}$$

Define an auxiliary error e_2 as

$$e_2 = x_{2d} - \hat{x}_2. \tag{19}$$

where x_{2d} is a virtual control variable. It is easy to obtain

$$\dot{e}_1 = e_2 + \dot{\sigma}_d - x_{2d}. \tag{20}$$

According to the error system, let the virtual control variable satisfy

$$x_{2d} = \dot{\sigma}_d + k_1 e_1, \tag{21}$$

where $k_1 \in R^+$ is the feedback gain.

To avoid differential explosion, the algorithm of dynamic surface control is employed. Designing a low-pass filter and passing the virtual control variable x_{2d} through it yield

$$T\dot{\hat{x}}_{2d} + \hat{x}_{2d} = x_{2d},\ \hat{x}_{2d}(0) = x_{2d}(0), \tag{22}$$

where T is the time constant of the filter and $\hat{x}_{2d}$ is the filtered virtual control variable. Define the filter error as

$$\tilde{x}_{2d} = \hat{x}_{2d} - x_{2d}. \tag{23}$$

Calculating the derivative of (23) along (22) yields

$$\begin{aligned} \dot{\tilde{x}}_{2d} &= \dot{\hat{x}}_{2}d - \dot{x}_{2d} \\ &= -\frac{1}{T}\tilde{x}_{2d} - (\ddot{\sigma}_d + k_1\dot{e}_1) \\ &= -\frac{1}{T}\tilde{x}_{2d} + B(\ddot{\sigma}_d, \dot{e}_1), \end{aligned} \tag{24}$$

where $B(\ddot{\sigma}_d, \dot{e}_1)$ is a function about $\ddot{\sigma}_d$ and $\dot{e}_1$. There exists a positive constant B_M such that $|B| \leq B_M$ [22]. Substituting (21) into (20) yields

$$\dot{e}_1 = e_2 - k_1 e_1. \tag{25}$$

Choose a Lyapunov candidate to be

$$V_1 = \frac{1}{2}e_1^2. \tag{26}$$

It is easy to obtain

$$\dot{V}_1 = e_1\dot{e}_1 = e_1(e_2 - k_1e_1) = e_1e_2 - k_1e_1^2. \tag{27}$$

In this study, the methodology of the sliding surface and the backstepping control is used to ensure that e_1 and e_2 converge quickly. Moreover, the fractional calculus operator is employed to improve transient performance in the design of a sliding mode surface, which is denoted by

$$s = \lambda_1 e_1 +_a D_t^{\alpha} e_2. \tag{28}$$

where λ_1 is the sliding surface gain and ${}_aD_t^{\alpha}$ is the calculus operator

$${}_aD_t^{\alpha} = \begin{cases} \dfrac{d^{\alpha}}{dt^{\alpha}} & Re(\alpha) > 0 \\ 1 & Re(\alpha) = 0 \\ \displaystyle\int_a^t d(\tau)^{-\alpha} & Re(\alpha) < 0. \end{cases} \tag{29}$$

In (29), α is the order of the operator and can be used to adjust transient performance of the control system. For simplicity of implementation, we choose the Caputo fractional calculus [23–25]

$${}_aD_t^{\alpha} f(t) = \frac{1}{\Gamma(m-\alpha)} \int_a^t \frac{f^m(\tau)}{(t-\tau)^{1+\alpha-m}} d\tau, m-1 < \alpha < m, \tag{30}$$

where $\Gamma(\cdot)$ is Gamma Function, $\Gamma(\eta) = \int_0^{\infty} e^{-t} t^{\eta-1} dt$ and m is the least integer that is not less than α. The fractional differential operation is transformed to the particular form of integral operation; therefore, the fractional calculus has the heritability and the memorability. To simplify the notation, let D^{α} represent ${}_0D_t^{\alpha}$. This should not cause confusion.

In order to make the system state converge to the sliding-mode surface and further weaken the chattering of the system, the fractional reaching law is designed as

$$D^{\beta}s = -\epsilon sgn(s), \tag{31}$$

where ϵ is a positive constant. According to the properties of fractional calculus, we have

$$\dot{s} = D^{1-\beta}(-\epsilon sgn(s)). \tag{32}$$

Taking the derivative of the fractional sliding-mode surface yields

$$\dot{s} = \lambda_1 \dot{e}_1 + D^{\alpha} \dot{e}_2. \tag{33}$$

Choose a Lyapunov candidate to be

$$V_2 = V_1 + \frac{1}{2}s^2. \tag{34}$$

Calculating the derivative of V_2 gives

$$\begin{aligned} \dot{V}_2 &= \dot{V}_1 + s\dot{s} \\ &= e_1 e_2 - k_1 e_1^2 + s(\lambda_1 \dot{e}_1 + D^{\alpha} \dot{e}_2) \\ &= e_1 e_2 - k_1 e_1^2 + s(\lambda_1 \dot{e}_1 + D^{\alpha} (\dot{\hat{x}}_{2d} - f - bu)). \end{aligned} \tag{35}$$

The control law is designed to be

$$u = \frac{1}{b}\left(\dot{\hat{x}}_{2d} - \hat{f} + D^{-\alpha}\left(\lambda_1 \dot{e}_1 + ks + D^{1-\beta}\epsilon sgn(s)\right)\right), \tag{36}$$

where k is a positive adjustable parameter.

4. Stability Analysis

Theorem 1. *The errors of the system (11) with the control law (36) are uniformly ultimately bounded, if the parameter $k < 1$ and the time constant of the filter $T < 2$.*

Proof of Theorem 1. Assume there exists a positive definite matrix $\boldsymbol{P}$ such that $A_2^{\mathrm{T}}\boldsymbol{P} + \boldsymbol{P}A_2 = -\boldsymbol{I}$. Construct a Lyapunov function

$$V = \frac{1}{2}e_1^2 + \frac{1}{2}s^2 + \frac{1}{2}\tilde{x}_{2d}^2 + \tilde{\boldsymbol{e}}^{\mathrm{T}}\boldsymbol{P}\tilde{\boldsymbol{e}}. \tag{37}$$

Substituting (36) into (35) yields

$$\dot{V} = e_1 e_2 - k_1 e_1^2 + s\left(\hat{f} - f - ks - D^{1-\beta}\epsilon sgn(s)\right) + \tilde{x}_{2d}\dot{\tilde{x}}_{2d} + 2\tilde{\boldsymbol{e}}^{\mathrm{T}}\boldsymbol{P}\dot{\tilde{\boldsymbol{e}}}. \tag{38}$$

A proper selection of the gain $\boldsymbol{L}$ ensures that the error dynamics of the LESO is stable. Thus, it is reasonable to assume that $\hat{f} \approx f_{total}$. Then, rewrite (38) as

$$\begin{aligned} \dot{V} &= e_1 e_2 - k_1 e_1^2 - s\left(ks + D^{1-\beta}\epsilon sgn(s)\right) + \tilde{x}_{2d}\dot{\tilde{x}}_{2d} + 2\tilde{\boldsymbol{e}}^{\mathrm{T}}\boldsymbol{P}\dot{\tilde{\boldsymbol{e}}} \\ &= e_1 e_2 - k_1 e_1^2 - ks^2 - sD^{1-\beta}\epsilon sgn(s) + \tilde{x}_{2d}\dot{\tilde{x}}_{2d} + 2\tilde{\boldsymbol{e}}^{\mathrm{T}}\boldsymbol{P}\dot{\tilde{\boldsymbol{e}}} \\ &= e_1 e_2 - k_1 e_1^2 + (1-k)s^2 - s^2 + s\dot{s} + \tilde{x}_{2d}\dot{\tilde{x}}_{2d} + 2\tilde{\boldsymbol{e}}^{\mathrm{T}}\boldsymbol{P}\dot{\tilde{\boldsymbol{e}}}. \end{aligned} \tag{39}$$

Let $\boldsymbol{e}_{12} = \begin{bmatrix} e_1 & e_2 & D^{\alpha}e_2 \end{bmatrix}^{\mathrm{T}}$, and choose a symmetric matrix $\boldsymbol{Q}$

$$\boldsymbol{Q} = \begin{bmatrix} k_1 + (k-1)\lambda_1^2 & -0.5 & (k-1)\lambda_1 \\ -0.5 & 0 & 0 \\ (k-1)\lambda_1 & 0 & k-1 \end{bmatrix} \tag{40}$$

to make

$$\begin{aligned} \boldsymbol{e}_{12}^{\mathrm{T}}\boldsymbol{Q}\boldsymbol{e}_{12} &= -e_1 e_2 + k_1 e_1^2 - (1-k)(\lambda_1 e_1 + D^{\alpha}e_2)^2 \\ &= -e_1 e_2 + k_1 e_1^2 - (1-k)s^2 \end{aligned} \tag{41}$$

hold. Substituting (41) into (39) yields

$$\dot{V} = -\boldsymbol{e}_{12}^{\mathrm{T}}\boldsymbol{Q}\boldsymbol{e}_{12} - s^2 + s\dot{s} + \tilde{x}_{2d}\dot{\tilde{x}}_{2d} + 2\tilde{\boldsymbol{e}}^{\mathrm{T}}\boldsymbol{P}\dot{\tilde{\boldsymbol{e}}}. \tag{42}$$

Calculating the determinant of $\boldsymbol{Q}$ gives

$$|\boldsymbol{Q}| = -0.25(k-1). \tag{43}$$

It can be seen that k should satisfy $k < 1$ to guarantee that $|\boldsymbol{Q}| > 0$ holds. Thus $-\boldsymbol{e}_{12}^{\mathrm{T}}\boldsymbol{Q}\boldsymbol{e}_{12} < 0$ holds and there exists a positive constant μ such that the inequality

$$\dot{V} \leq -\mu e_1^2 - s^2 + s\dot{s} + \tilde{x}_{2d}\dot{\tilde{x}}_{2d} + 2\tilde{\boldsymbol{e}}^{\mathrm{T}}\boldsymbol{P}\dot{\tilde{\boldsymbol{e}}} \tag{44}$$

is satisfied.

Assume $\dot{s}$ is bounded and $|\dot{s}| \leq S_M$, where S_M is a positive constant. According to Yong's inequality, it is easy to obtain

$$\begin{aligned} \tilde{x}_{2d}\dot{\tilde{x}}_{2d} &= -\frac{1}{T}\tilde{x}_{2d}^2 + \tilde{x}_{2d}B \\ &\leq -\frac{1}{T}\tilde{x}_{2d}^2 + |\tilde{x}_{2d}B| \\ &\leq -\frac{1}{T}\tilde{x}_{2d}^2 + |\tilde{x}_{2d}|B_M \\ &\leq -\frac{1}{T}\tilde{x}_{2d}^2 + \frac{1}{2}\tilde{x}_{2d}^2 + \frac{1}{2}B_M^2 \end{aligned} \tag{45}$$

and

$$\begin{aligned} s\dot{s} &\leq \frac{1}{2}s^2 + \frac{1}{2}\dot{s}^2 \\ &\leq \frac{1}{2}s^2 + \frac{1}{2}S_M^2. \end{aligned} \tag{46}$$

According to (16), we have

$$\begin{aligned} 2\tilde{\boldsymbol{e}}^{\mathrm{T}}\boldsymbol{P}\dot{\tilde{\boldsymbol{e}}} &= 2\tilde{\boldsymbol{e}}^{\mathrm{T}}\boldsymbol{P}(\boldsymbol{A}_2\tilde{\boldsymbol{e}} + \boldsymbol{B}_2 h) \\ &= -\tilde{\boldsymbol{e}}^{\mathrm{T}}\tilde{\boldsymbol{e}} + 2\tilde{\boldsymbol{e}}^{\mathrm{T}}\boldsymbol{P}\boldsymbol{B}_2 h \\ &\leq -\tilde{\boldsymbol{e}}^{\mathrm{T}}\tilde{\boldsymbol{e}} + 2M_1M_2||\boldsymbol{P}\boldsymbol{B}_2||. \end{aligned} \tag{47}$$

Substituting (45)–(47) into (44) yields

$$\dot{V} \leq -\mu e_1^2 - \frac{1}{2}s^2 - \left(\frac{1}{T} - \frac{1}{2}\right)\tilde{x}_{2d}^2 - \tilde{\boldsymbol{e}}^{\mathrm{T}}\tilde{\boldsymbol{e}} + \frac{1}{2}S_M^2 + \frac{1}{2}B_M^2 + 2M_1M_2||\boldsymbol{P}\boldsymbol{B}_2||. \tag{48}$$

Assume $\frac{1}{T} - \frac{1}{2} > 0$. Then, let $\tau = min\left\{2\mu, 1, \frac{2}{T} - 1, \frac{1}{\lambda_{max}(\boldsymbol{P})}\right\}$, where $\lambda_{max}(\boldsymbol{P})$ denotes the maximum eigenvalue of $\boldsymbol{P}$. Rewrite (48) as

$$\dot{V} \leq -\tau V + v, \tag{49}$$

where $v = \frac{1}{2}S_M^2 + \frac{1}{2}B_M^2 + 2M_1M_2||\boldsymbol{P}\boldsymbol{B}_2||$.

Solving the differential inequality (49) yields

$$V \leq \frac{v}{\tau} + \left[V(0) + \frac{v}{\tau}\right]e^{-\tau t}. \tag{50}$$

This indicates that V is uniformly ultimately bounded.

It can be concluded that $|e_1| \le \sqrt{\frac{2v}{\tau}}$, $|s| \le \sqrt{\frac{2v}{\tau}}$, $|\tilde{x}_{2d}| \le \sqrt{\frac{2v}{\tau}}$, $||\tilde{e}|| \le \sqrt{\frac{v}{\lambda_{min}(\boldsymbol{P})\tau}}$, where $\lambda_{min}(\boldsymbol{P})$ denotes the minimum eigenvalue of $\boldsymbol{P}$. All signals in the powered-parafoil control system are uniformly ultimately bounded. This completes the proof. □

5. Simulation Verification

Simulation results of the altitude control of an 8-degree-of-freedom powered parafoil were used to verify the validity of the presented method. The main structure parameters of the powered parafoil are shown in Table 1. Simulation results are compared with LADRC and SMC in [13].

Table 1. Structure parameters of the powered parafoil.

Paremeter	Value/Unit
Span	10.5/m
Chord	3.1/m
Aspect ratio	3
Area of canopy	33/m^2
Length of lines	6.8/m
Rigging angle	10/deg
Mass of canopy	10/kg
Mass of payload	80/kg
Characteristic area of drag of payload	0.6/m^2

The bandwidth parameterization method was used to design the gain of the LESO.

$$l_1 = 90,\ l_2 = 2700,\ l_3 = 2700.$$

The time constant of the filter was selected to be $T = 0.025$ s.

The parameters of the designed fractional sliding mode backstepping controller were designed to be

$$\lambda_1 = 0.16,\ k_1 = 0.02,\ k = 0.015,\ \alpha = 0.82,\ \beta = 0.36,\ \epsilon = 0.01.$$

The saturation of the control input was considered. The maximum thrust provided by the engine was limited to 400 N.

The initial altitude of the powered parafoil was set to be 2000 m and the desired altitude H_d was set to be 1970 m. A gust disturbance (2 m/s) along the negative direction of the Z axis was added to the system during 100–115 s, which is shown in Figure 3.

Figure 4 shows the observed results of the LESO. The outputs of the observer can effectively observe each state of the system. For comparison, simulations for the LADRC and the SMC in [13] were also carried out. The design of the LADRC and the SMC was the same as that in [13]. Figure 5 shows the attitude angles of the powered parafoil in the longitudinal plane for the three methods. The pitch angles remained at 9° after the powered parafoil entered the steady-state. The pitch angles oscillated violently when the parafoil suffered from the gust disturbance. It can be seen that the transient performance is better for the parafoil with FSMBC than for that with other two methods.

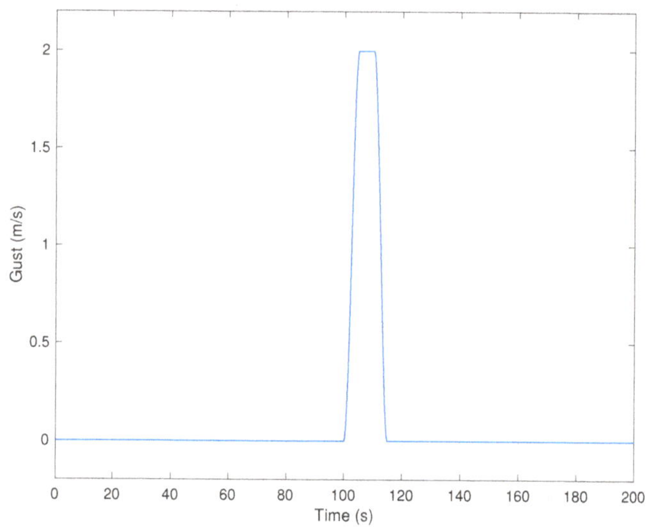

Figure 3. The gust disturbance.

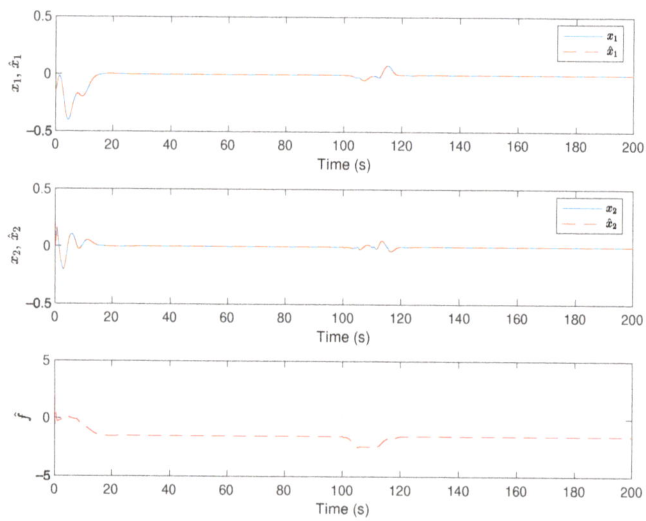

Figure 4. Observed results of LESO.

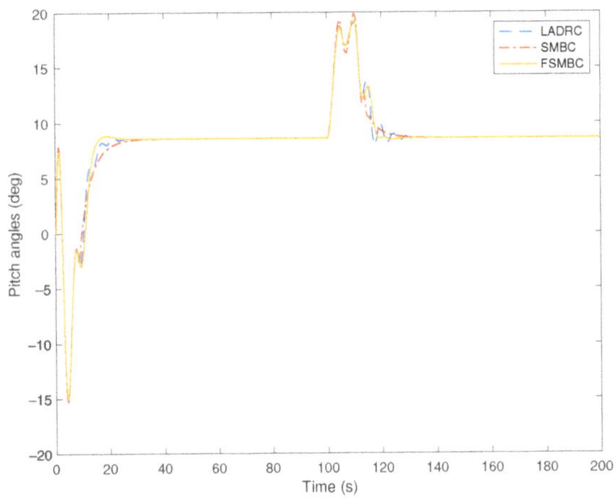

Figure 5. Simulation results of the pitch angle.

Figure 6 illustrates the simulation results of the altitude control. The altitude of the powered parafoil with the FSMBC stabilized at 1970 m after 13 s. The convergence time was less than that of the LADRC and SMC. Moreover, the recovery speed against the disturbance is faster for the FSMBC than for the LADRC and SMC. Compared with other two control methods, there exists about 1.4 m steady-state error for the SMC due to the lack of ESO. It can be seen from Figure 7 that the thrust input for the FSMBC stabilizes at 18 s, and the convergence speed is faster than that for the LADRC and SMC. It is obvious that the thrust input for the FSMBC is smoother than that for the LADRC and SMC. This is beneficial to energy saving and system stability. The SMC used a traditional integer order sliding surface, which caused the thrust input fluctuated. The specific performance indexes are shown in Table 2.

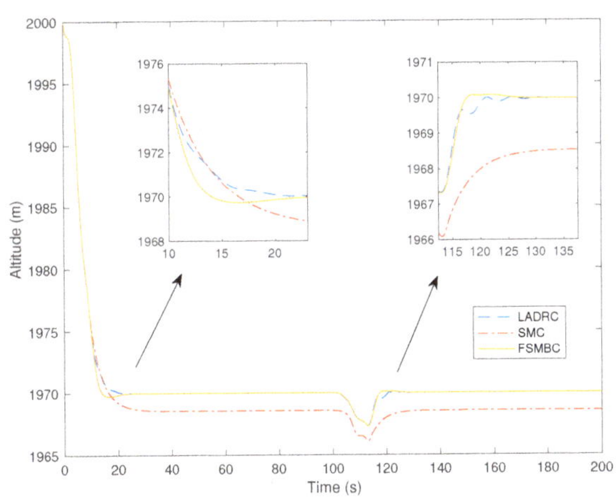

Figure 6. Simulation results of altitude control by LADRC, SMC, and FSMBC.

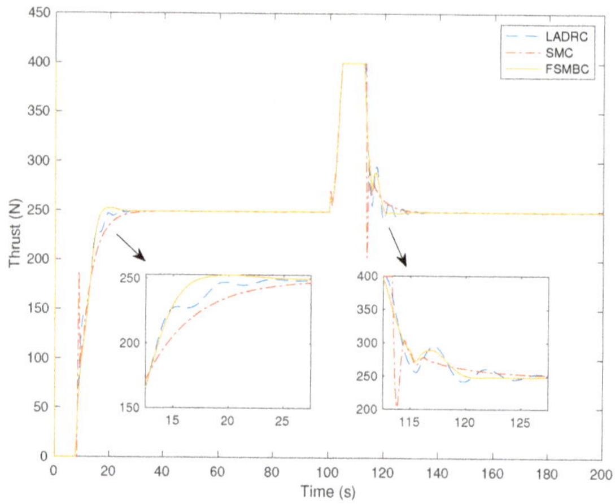

Figure 7. Thrust inputs for LADRC, SMC, and FSMBC.

Table 2. The performance indexes of the control system.

	FSMBC	LADRC	SMC
Transient time of the altitude	13 s	16 s	18 s
Steady-state error	0.00 m	0.00 m	1.44 m
Transient time of the thrust	18 s	29 s	35 s
Settling time under disturbance	16 s	20 s	28 s

The variable altitude control of the powered parafoil were also considered. The desired altitude changed from 1970 m to 1960 m at 50 s. Simulation results are shown in Figures 8–11.

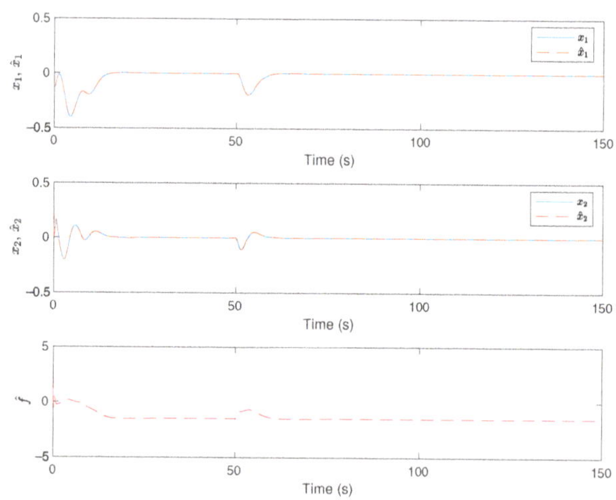

Figure 8. Observed results of LESO for variable altitude control.

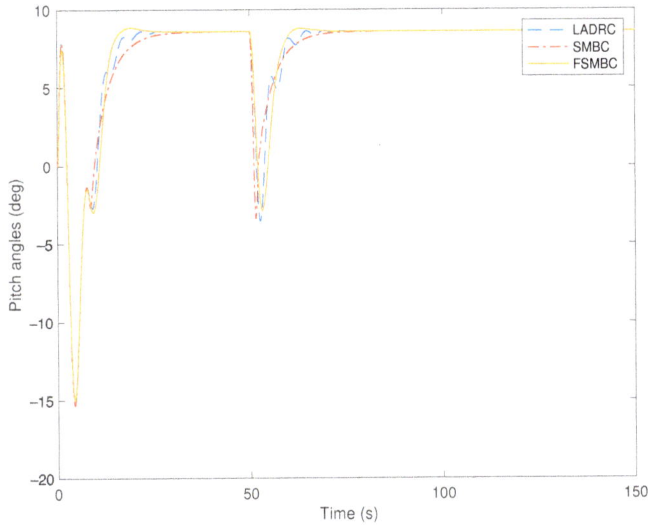

Figure 9. Simulation results of the pitch angle for variable altitude control.

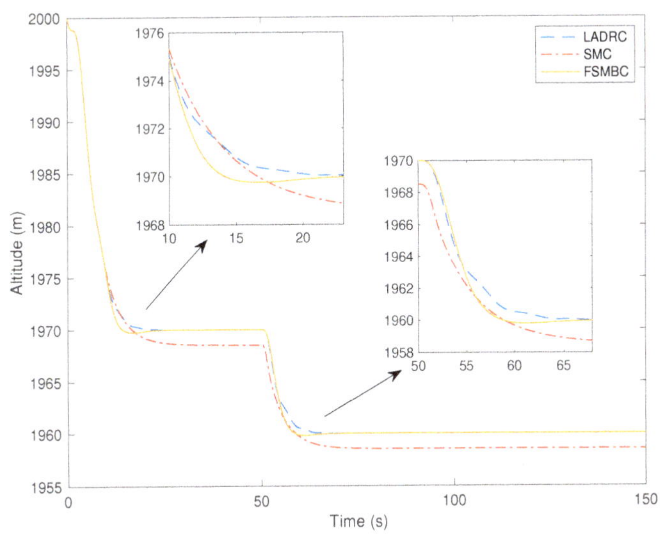

Figure 10. Simulation results of altitude cotrol by LADRC, SMC, and FSMBC for variable altitude control.

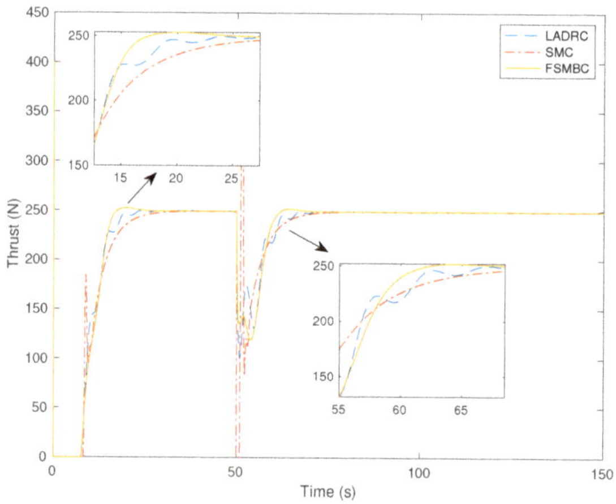

Figure 11. Thrust inputs for LADRC, SMC, and FSMBC for variable altitude control.

Figure 8 shows the observed results of the LESO for variable altitude control. The simulation results of pitch angles are shown in Figure 9. When the desired altitude changed to 1960 m, the altitude of the powered parafoil decreased accordingly and the pitch angle deviated from the steady-state value. When the altitude approached to the desired value, the pitch angle returned to the previous steady-state value. Figure 10 shows that the transient performance is better for the FSMBC than for the LADRC and SMC. The settling time is 8 s for the FSMBC and 11 s for the LADRC and SMC. Figure 11 shows that the thrust input for the SMC vibrates more violently than that of the LADRC and FSMBC during the transient response. The proposed method, that is, FSMBC, achieved good rapidity and smoothness. The performance indexes are shown in Table 3. The transient times of the altitude and the thrust are the same as in Table 2.

Table 3. The performance indexes of the control system for variable altitude control.

	FSMBC	LADRC	SMC
Steady-state error	0.00 m	0.00 m	1.47 m
Settling time of variable altitude	8 s	11 s	11 s

According to the two simulation experiments, due to the lack of ESO, SMC cannot observe and compensate the total disturbance of the system such that there exists the steady-state error. FSMBC adopts the fractional sliding-mode surface such that dynamic characteristics are better than LADRC and SMC, and the control curve is smoother due to the fractional reaching law.

6. Conclusions

This paper addressed the problem of the altitude control of the powered parafoil. The dynamic model of the powered parafoil was derived and converted into the second-order model of the inclination angle. The LESO was devised to estimate the unmodeled dynamics and the exogenous disturbance. Consequently, the estimate was used for compensation in the control law. The fractional sliding-mode surface was employed in the design of the backstepping design to improve the transient performance of the altitude control. The stability criterion of the control system was derived using the Lyapunov method. Simulation results demonstrated the validity and the superiority of the presented method. It was shown that the control performance was better for the FSMBC than for the LADRC and SMC. However, only longitudinal control problem of the powered parafoil was studied in this paper without considering the coupling problem in the horizontal plane. Future work will be focused on the horizontal orientation control of the powered parafoil.

Author Contributions: Conceptualization, E.Z.; methodology, E.Z. and Y.D.; validation and writing—original draft preparation, E.Z.; writing—review, editing, and supervision, Y.D.; funding acquisition, Y.D., W.S., and H.G. All authors have read and agreed to the published version of the manuscript.

Funding: This research was funded by Changzhou Sci&Tech Program under Grant CJ20220068, by Natural Science Foundation of Jiangsu Province under Grant BK20221404, by the Talent Introduction Project of Jiangsu University of Technology under Grant KYY22005, by National Nature Science Foundation under Grant 61803186 and by Anhui University Top-notch Talents Academic Funding Project under Grant gxbjZD2020079.

Institutional Review Board Statement: Not applicable.

Informed Consent Statement: Not applicable.

Data Availability Statement: Not applicable.

Acknowledgments: The authors acknowledge the support from the Robotics and High-end Equipment Innovation Team of the School of Electrical & Information Engineering in Jiangsu University of Technology.

Conflicts of Interest: The authors declare no conflict of interest.

References

1. Chambers, J. Longitudinal Dynamic Modeling and Control of Powered Parachute Aircraft. Ph.D. Thesis, Rochester Institute of Technology, Rochester, NY, USA, 2007.
2. Zhu, E.; Zhao, J.; Li, B.; Gao, H. Trajectory planning for the powered parafoil at insufficient height. In Proceedings of the 33rd Chinese Control and Decision Conference, Kunming, China, 22–24 May 2021.
3. Li, B.; Qi, J.; Lin, T.; Mei, S.; Han, J. Real-Time Data Acquisition and Model Identification for Powered Parafoil UAV. In Proceedings of the 8th International Conference on Intelligent Robotics & Applications, Portsmouth, UK, 24–27 August 2015.
4. Ghoreyshi, M.; Bergeron, K.; Jirasek, A.; Seidel, J.; Lofthouse, R.J.; Cummings, R.M. Computational aerodynamic modelling for flight dynamics simulation of ram-air parachutes. *Aerosp. Sci. Technol.* **2016**, *54*, 286–301. [CrossRef]
5. Tao, J.; Dehmer, M.; Xie, G.; Zhou, Q. A generalized predictive control-based path following method for parafoil systems in wind environments. *IEEE Access* **2019**, *7*, 42586–42595. [CrossRef]
6. Zhu, E.; Gao, H.; Zhao, J. Trajectory planning of the powered parafoil based on multi-phase design. *J. Jiangsu Univ. Technol.* **2020**, *26*, 28–35.
7. Yang, H.; Song, L.; Liu, C.; Huang, J. Study on powered-parafoil longitudinal flight performance with a fast estimation model. *J. Aircr.* **2013**, *50*, 1660–1668.
8. Aoustin, Y.; Martinenko, Y. Control algorithms of the longitude motion of the powered paraglider. In Proceedings of the ASME 11th Biennial Conference on Engineering Systems Design and Analysis, Nantes, France, 2–4 July 2012.
9. Chen, S.; Sun, Q.; Luo, S. Longitudinal Control of Unmanned Powered Parafoil with Precise Control Gain. In Proceedings of the 13rd Chinese Intelligent Systems Conference, Mudanjiang, China, 14–15 October 2017.
10. Chen, Z.; Zhang, H. Altitude control for unmanned powered parafoil based on backstepping method. In Proceedings of the 28th Chinese Control and Decision Conference, Yinchuan, China, 28–30 May 2016.
11. Chen, Z.; Zhang, H.; Qiu, J.; Su, L. Adaptive backstepping altitude control for unmanned powered parafoil. *Control. Eng. China* **2018**, *25*, 554–558.
12. Tan, P.; Sun, Q.; Jiang, Y. Trajectory tracking of powered parafoil based on characteristic model based all-coefficient adaptive control. *J. Cent. South Univ.* **2017**, *24*, 1073–1081. [CrossRef]
13. Zhu, E.; Gao, H. Guidance-based path following control of the powered parafoil. *Control Eng. Appl. Inf.* **2020**, *22*, 42–50.
14. Hu, K-Y.; Wang, X.; Yang, C. Hybrid adaptive dynamic inverse compensation for hypersonic vehicles with inertia uncertainty and disturbance. *Appl. Sci.* **2022**, *12*, 11032. [CrossRef]
15. Wang, C.; Du, Y. Lane-Changing strategy based on a novel sliding mode control approach for connected automated vehicles. *Appl. Sci.* **2022**, *12*, 11000. [CrossRef]
16. Du, Y.; Cao, W.; Wu, M.; She, J.; Fang, M.; Kawata, S. Disturbance rejection and control system design using improved equivalent-input-disturbance approach. *IEEE Trans. Ind. Electron.* **2020**, *67*, 3013–3032. [CrossRef]
17. Du, Y.; Cao, W.; She, J.; Wu, M.; Fang, M.; Kawata, S. Disturbance Rejection and Robustness of Improved Equivalent-Input-Disturbance-Based System. *IEEE Trans. Cybern.* **2022**, *52*, 8537–8546. [CrossRef] [PubMed]
18. Guettal, L.; Chelihi, A.; Ajgou, R.; Touba, M. Robust tracking control for quadrotor with unknown nonlinear dynamics using adaptive neural network based fractional-order backstepping control. *J. Franklin Inst.* **2022**, *359*, 7337–7364. [CrossRef]

19. Pouzesh, M.; Mobayen, S. Event-triggered fractional-order sliding mode control technique for stabilization of disturbed quadrotor unmanned aerial vehicles. *Aerosp. Sci. Technol.* **2022**, *121*, 107337. [CrossRef]
20. Zhu, E.; Sun, Q.; Chen, Z.; Kang, X.; He, Y. Modeling of powered parafoil based on kirchhoff motion equation. *Nonlinear Dyn.* **2015**, *79*, 617–629. [CrossRef]
21. Chen, Z.; Sun, M.; Yang, R. On the stability of linear active disturbance rejection control. *Acta Autom. Sin.* **2013**, *39*, 574–580. [CrossRef]
22. Shao, X.; Wang, L.; Li, J.; Liu, J. High-order ESO based output feedback dynamic surface control for quadrotors under position constraints and uncertainties. *Aerosp. Sci. Technol.* **2019**, *89*, 288–298. [CrossRef]
23. Das, S. *Functional Fractional Calculus*, 2nd ed.; Springer: Berlin, Germany, 2011; pp. 204–206.
24. Monje, C.; Chen, Y.; Vinagre, B.; Xue, D.; Feliu, V. *Fractional-Order Systems and Controls: Fundamentals and Applications*; Springer: London, UK, 2010; pp. 153–155.
25. Roy, P.; Roy, B. Sliding mode control versus fractional-order sliding mode control: Applied to a magnetic levitation system. *J. Control Autom. Electr. Syst.* **2020**, *31*, 597–606. [CrossRef]

applied sciences

MDPI

Article

Model-Assisted Reduced-Order ESO Based Command Filtered Tracking Control of Flexible-Joint Manipulators with Matched and Mismatched Disturbances

Changzhong Pan [1,2,*], Xiangyin Fei [1], Jinsen Xiao [3], Peiyin Xiong [1], Zhijing Li [1] and Hao Huang [1]

1 School of Information and Electrical Engineering, Hunan University of Science and Technology, Xiangtan 411201, China
2 School of Automation, Guangdong University of Petrochemical Technology, Maoming 525000, China
3 Department of Mathematics, Guangdong University of Petrochemical Technology, Maoming 525000, China
* Correspondence: pancz@hnust.edu.cn

Abstract: Flexible-joint manipulators (FJMs) have been widely used in the fields of industry, agriculture, medical service, aerospace, etc. However, the FJMs in practical applications inevitably encounter various uncertainties including matched and mismatched disturbances. In this paper, we consider the high precision tracking control problem of FJMs in the presence of unknown lumped matched and mismatched disturbances. An efficient model-assisted composite control approach is proposed by integrating two reduced-order extended state observers (RESOs), a second-order command filtered backstepping (SCFB) technique and an error compensation dynamic system. Unlike some existing methods, the RESOs constructed with partial known model information are capable of estimating and compensating the matched and mismatched disturbances simultaneously. In addition, by employing the SCFB with an error compensation system, the proposed approach can not only overcome the problem of "explosion of complexity" inherent in backstepping, but also reduce the filtering errors arising from the command filters. The stability of the resulting control system and the convergence of error signals are guaranteed by Lyapunov stability theory. Comparative simulations are conducted for a single-link FJM with both matched and mismatched disturbances, and the results show that the proposed approach achieves a better tracking performance, i.e., compared with conventional backstepping method and adaptive fuzzy command filtered control method, the tracking accuracy is improved by 99.5% and 99.2%, respectively.

Keywords: flexible-joint manipulators; reduced-order extended state observer (RESO); backstepping; command filter; error compensation

Citation: Pan, C.; Fei, X.; Xiao, J.; Xiong, P.; Li, Z.; Huang, H. Model-Assisted Reduced-Order ESO Based Command Filtered Tracking Control of Flexible-Joint Manipulators with Matched and Mismatched Disturbances. *Appl. Sci.* **2022**, *12*, 8511. https://doi.org/10.3390/app12178511

Academic Editor: Alessandro Gasparetto

Received: 28 July 2022
Accepted: 22 August 2022
Published: 25 August 2022

1. Introduction

The last decades have witnessed a tremendous progress in the development of flexible manipulators. Roughly speaking, the flexible manipulators can be divided into flexible-link manipulators (FLMs) [1,2] and flexible-joint manipulators (FJMs) [3–5]. In this paper, we focus on the study of FJMs, whose joints are made up of harmonic reducer, torque sensor, and other elastic components. The FJMs usually exhibit many distinctive features, such as light weight, good flexibility, high human-robot interaction safety, etc. As a result, they have a wide application prospect in the fields of industry, agriculture, medical service, aerospace, and so on [6,7]. However, the flexible joints are easy to produce elastic vibrations during the movements, especially in high-speed operations, which greatly affect the control accuracy of FJMs. In addition, the model of FJMs in practical applications may contain various nonlinear uncertainties including matched and mismatched disturbances. If these uncertainties are ignored in the control design, the performances of the controllers will be deteriorated [8]. Therefore, the study on vibration suppression of FJMs in the presence of uncertainties with a high precision control has attracted great attentions in both control theory and engineering community.

In the past decades, many effective control methods have been proposed for the control of FJMs, e.g., fuzzy control [9,10], singular perturbation control [11–13], feedback linearization control [14,15], backstepping control [16–19], etc. Among them, the backstepping control is a Lyapunov function-based recursive design method, which constructs control Lyapunov functions step by step and designs intermediate virtual control laws recursively until the actual control law is reached. However, it suffers from the drawback of "explosion of complexity", that is, the derivatives of virtual control laws designed in the procedures need to be computed repeatedly. To solve this problem, a dynamic surface control (DSC) was proposed in Ref. [20], where a first-order low-pass filter was introduced in each step to obtain the derivative of the virtual control law instead of taking the derivative directly. Based on the DSC technique, several adaptive control schemes [21–23] were proposed for the single-link FJMs with unknown nonlinearities, and a DSC backstepping-based impedance controller was designed in Ref. [24] for a 5-DOF flexible joint robot. However, these results do not consider the potential errors caused by the filters. Furthermore, the derivatives of virtual control laws in DSC are actually approximated through numerical differentiations, which may amplify the noise and reduce the control accuracy.

To avoid numerical differentiations, a second-order command filtered backstepping (SCFB) method was proposed in Refs. [25,26]. It obtains the derivatives of the virtual control laws through integrations instead of differentiations, which can not only avoid the problem of "explosion of complexity", but also simplify the controller design. With the SCFB technique, an adaptive neural tracking controller was designed in Ref. [27] for uncertain robotic manipulators, and an adaptive fuzzy controller was proposed in Ref. [28] for a two-link robotic manipulator. Furthermore, regarding the filtering errors produced by command filters, two improved SCFB controllers were designed in Refs. [29,30] for FJM systems, where two error compensation mechanisms were constructed to reduce the filtering errors. Unfortunately, most of the above mentioned methods do not consider the practical uncertainties that may exist in the control of FJMs.

To cope with uncertainties, intelligent control methods including neural networks (NNs) and fuzzy logic systems (FLSs), which are well-known for their universal approximation abilities, have been widely utilized for uncertain FJMs. For example, the uncertain model of FJMs was approximated by a Radial Basis Function (RBF) neural network in Ref. [31], on the basis of which an adaptive observer and DSC controller were developed. Similarly, an FLS was employed in Ref. [32] to approximate the unknown functions, and an adaptive fuzzy tracking controller was designed. However, both the NNs and the FLSs require complex online learning mechanisms, which are computationally expensive. Sliding mode control, which is famous for its insensitivity to uncertainties, has also been applied in the control of FJMs [33–36], but the phenomenon of chattering cannot be avoided.

Observer-based technique is an alternative to deal with uncertainties, which estimates the uncertainties by designing a state/disturbance observer. In Ref. [37], a high gain observer-based robust output feedback control approach was proposed for a single-link FJM with matched disturbances and parametric uncertainties. In Ref. [38], a nonlinear disturbance observer (NDO) based DSC approach was proposed for the FJM with input saturation and unknown nonlinear disturbance, where the NDO was applied to estimate the unknown external disturbance and compensate the saturation constrain. However, only matched disturbances were considered in these control methods. Extended state observer (ESO), which regards internal and external disturbances as an extended system state variable, is another effective and practical disturbance estimation and attenuation approach [39]. As for FJMs in the presence of disturbances, an ESO-based feedback linearization control method was proposed in Ref. [40], and a cascaded-ESO based sliding-model control strategy was proposed in Ref. [41]. However, the conventional ESO is only applicable for integral chain systems, which should satisfy the so-called matching conditions. Although the ESOs in Refs. [40,41] can transform a mismatched disturbance into a matched one, they require a series of complex coordinate transformations, which make the control algorithms computationally complicated. In addition, the order of the constructed ESOs is greater than the system. For high-order systems like the FJM, it may bring about some negative effects of the high gain action, such as noise amplification and the peaking phenomenon [41].

Unlike conventional ESO, the reduced-order ESO (RESO) [42], which makes full use of the measurable system state information, can attenuate the peaking phenomenon and yield a better estimation performance. More importantly, the RESO can be applied to non-integral chain systems with matched and mismatched disturbances. Due to these advantages, the RESO has been widely used for the control of various engineering systems, such as missiles [43], gear-shifting actuators [44], underwater vehicles [45], DC-DC buck converters [46], all-clamped plates [47], etc. However, the application of RESO on FJMs has not been reported.

Based on the above literature review and analysis, it is noted that there are still many crucial problems worthy of being further investigated in the control of FJMs, which can be summarized as follows.

(1) The FJMs in practical applications inevitably encounter various uncertainties including matched and mismatched disturbances. Unfortunately, the current researches focus on the matched disturbances, while the mismatched ones are not considered. Although the conventional ESO can transform a mismatched disturbance into a matched one, it requires a series of complex coordinate transformations, which make the control algorithms computationally complicated;
(2) The backstepping technique employed for the control design of FJMs suffers from the drawback of "explosion of complexity". Although the DSC or SCFCB can deal with the computation problem, the potential errors caused by the introduction of filters are not considered, which may greatly reduce the tracking accuracy.

Motivated by the above considerations, this paper aims to propose an efficient model-assisted composite control approach for the high precision tracking control of FJMs in the presence of lumped matched and mismatched disturbances by integrating the techniques of RESO and the SCFB. More specifically, the uncertain model of a single-link FJM is first given, where all the uncertainties affecting the system including friction/damping terms and external disturbances are lumped as matched and mismatched disturbances. Then, two model-assisted RESOs are constructed to estimate the matched and mismatched disturbances in real time. On the basis of the estimation values from the RESOs, a feedback controller is derived by using the recursive backstepping methodology, where three second-order command filters (SCFs) are incorporated to overcome the problem of "explosion of complexity". In addition, an error compensation dynamic system is designed to reduce the filtering errors caused by the SCFs. By utilizing Lyapunov stability theory, it rigorously proves that all the error signals in the closed-loop control system are uniformly ultimately bounded, and converge to a small neighbourhood of the origin. Numerical simulations with comparisons to existing methods are finally presented to verify the effectiveness and efficiency of the proposed approach.

The novel features and main contributions of this paper are highlighted as follows.

(1) The RESOs constructed with partial known model information are capable of estimating and compensating the matched and mismatched disturbances simultaneously. This is much different from the existing ESO-based methods where complex coordinate transformations are required to convert a mismatched disturbance into a matched one. The developed control algorithm is thus robust and efficient;
(2) The inherent complexity problem of backstepping is addressed by employing the SCFB control, where the derivatives of the virtual control laws are obtained through integrations instead of differentiations. The transient control performance of the controller is thus improved;
(3) The potential filtering errors caused by the command filters are taken into account, and they are reduced by the error compensation dynamic system, which improves the steady-state tracking control accuracy.

The remainder of this paper is organized as follows. The dynamic model of a single-link FJM with disturbances and the corresponding control problem are presented in Section 2. The detailed control design including two RESOs, a backstepping controller, and an error compensation system is given in Section 3. The stability of the resulting control system is analysed in Section 4. Simulation results are presented in Section 5 to verify the effectiveness and efficiency of the proposed approach. The concluding remarks are finally discussed in Section 6.

2. Problem Formulation

This paper studies the control problem of an FJM with one flexible joint and one rigid link, whose physical model is shown in Figure 1.

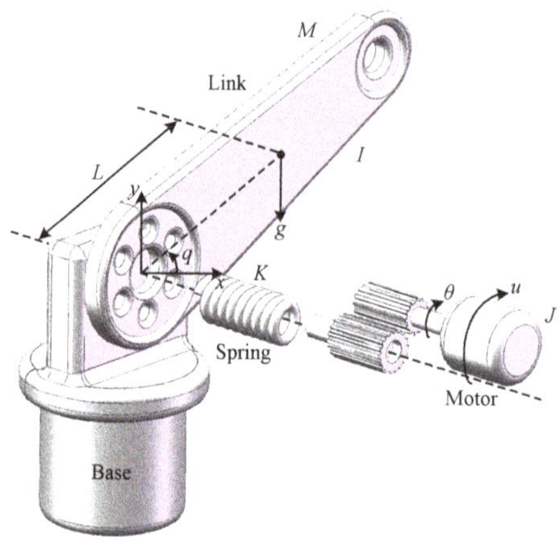

Figure 1. Physical model of a single-link FJM.

The parameters and variables in Figure 1 are given as follows: q and θ represent the angles of the link and the motor shaft, respectively; u is the control torque generated by the motor; g is the acceleration of gravity; M, K and L are the mass of the link, the spring stiffness of the flexible joint, the distance between the flexible joint, and the mass centre of the link, respectively; I and J are the rotational inertia of the rigid link and the motor, respectively.

According to the Euler–Lagrangian equation, the mathematical model of the single-link FJM is obtained as [16,48]:

$$\begin{cases} I\ddot{q} + K(q-\theta) + MgL\sin q = w_1 \\ J\ddot{\theta} - K(q-\theta) = u + w_2 \end{cases} \tag{1}$$

where w_1 and w_2 are the lumped disturbances including system friction/natural damping terms and unknown external disturbances.

Define $x = [x_1, x_2, x_3, x_4]^\top = [q, \theta, \dot{q}, \dot{\theta}]^\top$ as the state variable, and y as the output of the system, then the state-space equation of model (1) is written as:

$$\begin{cases} \dot{x}_1 = x_2 \\ \dot{x}_2 = x_3 + g_1(x) + d_1 \\ \dot{x}_3 = x_4 \\ \dot{x}_4 = \frac{1}{J}u + g_2(x) + d_2 \\ y = x_1 \end{cases} \tag{2}$$

where

$$
\begin{aligned}
g_1(x) &= -x_3 - \frac{K}{I}(x_1 - x_3) - \frac{MgL}{I}\sin x_1 \\
g_2(x) &= \frac{K}{J}(x_1 - x_3) \\
d_1 &= \frac{1}{I}w_1 \\
d_2 &= \frac{1}{J}w_2
\end{aligned} \tag{3}
$$

Note that the terms $g_1(x)$ and $g_2(x)$ are straightforwardly derived from (1), and they are treated as nominal functions of the FJM model, which will be used for the control design. The main feature of the FJM system expressed as (2) is that it contains two terms d_1 and d_2, which are the mismatched and matched lumped disturbances, respectively. Generally speaking, the mismatched lumped disturbance d_1 is determined by the friction term and external disturbance, while d_2 represents the lumped disturbance caused by the natural damping and the disturbance generated by the control torque. In this paper, the mismatched and matched disturbances are simultaneously estimated by employing two RESOs, and the estimation values are fed back to the controller to compensate for their effects.

Considering the physical limitations on the FJM in practical applications, some assumptions are given as follows.

Assumption 1 ([30]). *The desired reference trajectory y_d and its first-order time derivative $\dot{y}_d$ are available.*

Assumption 2 ([43]). *The lumped disturbances d_1, d_2, and their derivatives are all bounded, i.e., there exist positive constants $\bar{d}_1$, $\bar{d}_{1d}$, $\bar{d}_2$ and $\bar{d}_{2d}$ that satisfy $|d_1| \le \bar{d}_1$, $|\dot{d}_1| \le \bar{d}_{1d}$, $|d_2| \le \bar{d}_2$, $|\dot{d}_2| \le \bar{d}_{2d}$, $\forall t \in [0, +\infty)$.*

To facilitate the control design and stability analysis, the following lemmas are needed in the subsequent context.

Lemma 1 ([25]). *Consider the second-order command filters (SCFs) defined as*

$$
\begin{cases} \dot{z}_i = z_{id} \\ \dot{z}_{id} = -2\zeta\omega_n z_{id} - \omega_n^2(z_i - \alpha_i) \end{cases} \quad (i = 1,2,3) \tag{4}
$$

where α_i are the inputs and $\alpha_i^c = z_i$, $\dot{\alpha}_i^c = z_{id}$ are the outputs of the SCFs; ζ and ω_n are the damping ratio and bandwidth, respectively. Set the initial conditions as $z_i(0) = \alpha_i(0)$ and $z_{id}(0) = 0$. $\forall$ $t \ge 0$, if the inputs satisfy $|\dot{\alpha}_i| \le \varsigma_1$, $|\ddot{\alpha}_i| \le \varsigma_2$, where $\varsigma_1 > 0$, $\varsigma_2 > 0$, then there exist $\zeta \in (0,1]$, $\omega_n > 0$ and $\mu > 0$, such that $|z_i - \alpha_i| \le \mu$, $|z_{id} - \dot{\alpha}_i| \le \mu$, and $|\dot{z}_i|$, $|\ddot{z}_i|$, $|\dddot{z}_i|$ are bounded. Theoretically, the filtering errors of the SCFs can be made arbitrarily small by increasing ω_n.

The structure of an SCF is shown in Figure 2, from which it is clearly seen that the derivative of α_i is obtained through an integrator rather than a differentiator. This can reduce the measurement noise caused by differential operation and improve the control accuracy.

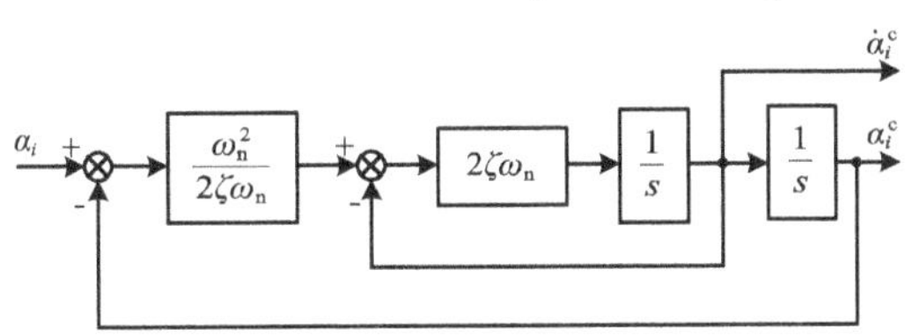

Figure 2. Structure of a second-order command filter.

Lemma 2 ([19,49]). *For $V : [0, +\infty) \in \mathbb{R}$, the solutions of inequality equations of $\dot{V}(t) \leq -aV(t) + f$ are*

$$V(t) \leq e^{-a(t-t_0)} V(t_0) + \int_{t_0}^{t} e^{-a(t-h)} f(h) dh, \ \forall t \geq t_0 \geq 0 \tag{5}$$

where a is any constant.

The control problem of this paper is formulated as following. Consider the single-link FJM (2) in the presence of lumped mismatched and matched disturbances. Design a proper controller by integrating the techniques of RESO and SCFB such that the output y tracks a desired trajectory y_d quickly and precisely.

3. RESO-Based Backstepping Control Design

To achieve the above control objective, in this section, two model-assisted RESOs are first designed to estimate the lumped matched and mismatched disturbances, and then a feedback controller with an error compensation mechanism is developed by employing the SCFB technique.

3.1. Reduced-Order ESO (RESO)

In order to estimate the lumped disturbances d_1 and d_2 in system (2), according to the design principle of RESO [42], two specified RESOs are given as

$$\begin{cases} \hat{d}_1 = p_1 + \beta_1 x_2 \\ \dot{p}_1 = -\beta_1 p_1 - \beta_1^2 x_2 - \beta_1 x_3 \end{cases} \tag{6}$$

$$\begin{cases} \hat{d}_2 = p_2 + \beta_2 x_4 \\ \dot{p}_2 = -\beta_2 p_2 - \beta_2^2 x_4 - \beta_2 J^{-1} u \end{cases} \tag{7}$$

where $\hat{d}_1$ stands for the estimate of the sum term of $g_1(x)$ and d_1; $\hat{d}_2$ stands for the estimate of the sum term of $g_2(x)$ and d_2; p_i and $\beta_i > 0 (i = 1, 2)$ are the auxiliary states and the observer gains, respectively.

Note that the internal dynamics $g_1(x)$ and $g_2(x)$ are available, which can be directly used as given model information for the observers. Hence, to reduce the estimation burden, the model-assisted RESOs are designed as

$$\begin{cases} \hat{d}_1 = p_1 + \beta_1 x_2 \\ \dot{p}_1 = -\beta_1 p_1 - \beta_1^2 x_2 - \beta_1 [x_3 + g_1(x)] \end{cases} \tag{8}$$

$$\begin{cases} \hat{d}_2 = p_2 + \beta_2 x_4 \\ \dot{p}_2 = -\beta_2 p_2 - \beta_2^2 x_4 - \beta_2 [J^{-1} u + g_2(x)] \end{cases} \tag{9}$$

where $\hat{d}_1$ and $\hat{d}_2$ stand for the estimates of the lumped uncertainties d_1 and d_2, respectively. Define the estimation errors as

$$\tilde{d}_1 = \hat{d}_1 - d_1 \tag{10}$$

$$\tilde{d}_2 = \hat{d}_2 - d_2 \tag{11}$$

Then, based on (8) and (10), the time derivative of $\tilde{d}_1$ can be obtained as

$$\begin{aligned} \dot{\tilde{d}}_1 &= \dot{\hat{d}}_1 - \dot{d}_1 = \dot{p}_1 + \beta_1 \dot{x}_2 - \dot{d}_1 \\ &= -\beta_1 p_1 - \beta_1^2 x_2 - \beta_1 [x_3 + g_1(x)] + \beta_1 [x_3 + g_1(x) + d_1] - \dot{d}_1 \\ &= -\beta_1 (p_1 + \beta_1 x_2) + \beta_1 d_1 - \dot{d}_1 \\ &= -\beta_1 \tilde{d}_1 - \dot{d}_1 \end{aligned} \tag{12}$$

According to the same derivation process, $\dot{\tilde{d}}_2$ can also be obtained as

$$\begin{aligned}\dot{\tilde{d}}_2 &= \dot{\hat{d}}_2 - \dot{d}_2 = \dot{p}_2 + \beta_2\dot{x}_4 - \dot{d}_2 \\ &= -\beta_2 p_2 - \beta_2^2 x_4 - \beta_2[J^{-1}u + g_2(x)] + \beta_2[J^{-1}u + g_2(x) + d_2] - \dot{d}_2 \\ &= -\beta_2(p_2 + \beta_2 x_4) + \beta_2 d_2 - \dot{d}_2 \\ &= -\beta_2\tilde{d}_2 - \dot{d}_2\end{aligned} \tag{13}$$

Combining (12) with (13), the disturbance estimation error dynamics for system (2) can be written in the following compact form:

$$\dot{\boldsymbol{e}}_o = \boldsymbol{A}_o\boldsymbol{e}_o + \dot{\boldsymbol{d}} \tag{14}$$

where $\boldsymbol{e}_o = \left[\tilde{d}_1, \tilde{d}_2\right]^{\mathrm{T}}$, $\dot{\boldsymbol{d}} = \left[-\dot{d}_1, -\dot{d}_2\right]^{\mathrm{T}}$ and $\boldsymbol{A}_o = \mathrm{diag}\{-\beta_1, -\beta_2\}$. Note that the gains in (8) and (9) are chosen as $\beta_i > 0 (i = 1, 2)$, which ensure that $\boldsymbol{A}_o$ is a Hurwitz matrix.

3.2. Second-Order Command Filtered Backstepping (SCFB) Controller

Based on the RESOs, a feedback tracking controller is designed by using the SCFB technique, and an error compensation system is proposed to reduce the filtering errors.

Firstly, we define the tracking errors as

$$\begin{cases} e_1 = y - y_{\mathrm{d}} \\ e_2 = x_2 - z_1 \\ e_3 = x_3 - z_2 \\ e_4 = x_4 - z_3 \end{cases} \tag{15}$$

where $z_i (i = 1, 2, 3)$ are the outputs of the SCFs defined in (4). In order to reduce the filtering errors produced by the SCFs, an error compensation dynamic system is designed as:

$$\begin{cases} \dot{\xi}_1 = -k_1\xi_1 + \xi_2 + (z_1 - \alpha_1) \\ \dot{\xi}_2 = -\frac{1}{J}k_2\xi_2 + \frac{K}{J}[\xi_3 + (z_2 - \alpha_2)] \\ \dot{\xi}_3 = -k_3\xi_3 + \xi_4 + (z_3 - \alpha_3) \\ \dot{\xi}_4 = -k_4\xi_4 \end{cases} \tag{16}$$

where $k_i > 0$ $(i = 1, 2, 3, 4)$ are design parameters, and the initial values of ξ_i $(i = 1, 2, 3, 4)$ are $\xi_i(0) = 0$. Then the compensated tracking errors are defined as

$$\begin{cases} v_1 = e_1 - \xi_1 \\ v_2 = e_2 - \xi_2 \\ v_3 = e_3 - \xi_3 \\ v_4 = e_4 - \xi_4 \end{cases} \tag{17}$$

Next, we derive the control laws to stabilize the tracking errors v_i by using the recursive backstepping methodology. The whole design procedure is divided into the following four steps.

Step 1: To stabilize v_1, the first Lyapunov function candidate is chosen as:

$$V_1 = \frac{1}{2}v_1^2 \tag{18}$$

Based on (2), (15) and (17), the time derivative of v_1 is computed by

$$\begin{aligned}\dot{v}_1 &= \dot{e}_1 - \dot{\xi}_1 = \dot{x}_1 - \dot{y}_{\mathrm{d}} - \dot{\xi}_1 \\ &= x_2 - \dot{y}_{\mathrm{d}} - \dot{\xi}_1 = v_2 + z_1 - \dot{y}_{\mathrm{d}} + \xi_2 - \dot{\xi}_1\end{aligned} \tag{19}$$

Taking the time derivative of V_1 along (19) and using the first equation of (16) yields

$$\begin{aligned}\dot{V}_1 &= v_1\dot{v}_1 = v_1(v_2 + z_1 - \dot{y}_\mathrm{d} + \xi_2 - \dot{\xi}_1) \\ &= v_1(v_2 + \alpha_1 - \dot{y}_\mathrm{d} + k_1\xi_1) \\ &= v_1(v_2 + \alpha_1 - \dot{y}_\mathrm{d} + k_1 e_1 - k_1 v_1)\end{aligned} \tag{20}$$

To make V_1 negative, the virtual control law α_1 is chosen as

$$\alpha_1 = -k_1 e_1 + \dot{y}_\mathrm{d} \tag{21}$$

where k_1 is a positive design parameter. Substituting (21) into (20) yields

$$\dot{V}_1 = -k_1 v_1^2 + v_1 v_2 \tag{22}$$

Obviously, if $v_2 = 0$, then $\dot{V}_1 \leq 0$.

Step 2: Similarly, to stabilize v_2, the second Lyapunov function candidate is chosen as:

$$V_2 = V_1 + \frac{1}{2} I v_2^2 \tag{23}$$

The time derivative of v_2 is obtained as

$$\begin{aligned}\dot{v}_2 &= \dot{e}_2 - \dot{\xi}_2 = \dot{x}_2 - \dot{z}_1 - \dot{\xi}_2 \\ &= -\frac{K}{I}(x_1 - x_3) - \frac{1}{I} MgL\sin x_1 + d_1 - z_{1d} - \dot{\xi}_2 \\ &= \frac{K}{I}(v_3 + z_2 + \xi_3) - \frac{K}{I}x_1 - \frac{1}{I} MgL\sin x_1 + d_1 - z_{1\mathrm{d}} - \dot{\xi}_2\end{aligned} \tag{24}$$

Taking the time derivative of V_2 along (24) and using the second equation of (16) yields

$$\begin{aligned}\dot{V}_2 &= \dot{V}_1 + I v_2 \dot{v}_2 \\ &= -k_1 v_1^2 + v_1 v_2 + v_2\left[K(v_3 + z_2 + \xi_3) - Kx_1 - MgL\sin x_1 + Id_1 - Iz_{1\mathrm{d}} - I\dot{\xi}_2\right] \\ &= -k_1 v_1^2 + v_2(v_1 + Kv_3 + K\alpha_2 - Kx_1 - MgL\sin x_1 + Id_1 - Iz_{1d} + k_2 e_2 - k_2 v_2)\end{aligned} \tag{25}$$

To make V_2 negative, the virtual control law α_2 is chosen as

$$\alpha_2 = \frac{1}{K}\left(-k_2 e_2 + Kx_1 + MgL\sin x_1 + Iz_{1\mathrm{d}} - v_1 - I\hat{d}_1\right) \tag{26}$$

where k_2 is a positive design parameter; $\hat{d}_1$ is the estimate of d_1 from the RESO (8). Subsisting (26) into (25) yields

$$\dot{V}_2 = -k_1 v_1^2 - k_2 v_2^2 + K v_2 v_3 - I v_2 \tilde{d}_1 \tag{27}$$

If $v_3 = 0$ and $\tilde{d}_1 = 0$, then $\dot{V}_2 \leq 0$.

Step 3: To stabilize v_3, the third Lyapunov function candidate is chosen as:

$$V_3 = V_2 + \frac{1}{2} v_3^2 \tag{28}$$

The time derivative of v_3 is computed by

$$\begin{aligned}\dot{v}_3 &= \dot{e}_3 - \dot{\xi}_3 = \dot{x}_3 - \dot{z}_2 - \dot{\xi}_3 \\ &= x_4 - z_{2\mathrm{d}} - \dot{\xi}_3 = v_4 + z_3 + \xi_4 - z_{2\mathrm{d}} - \dot{\xi}_3\end{aligned} \tag{29}$$

Taking the time derivative of V_3 along (29) and using the third equation of (16) yields

$$\begin{aligned} \dot{V}_3 &= \dot{V}_2 + v_3\dot{v}_3 \\ &= -k_1v_1^2 - k_2v_2^2 + Kv_2v_3 - Iv_2\tilde{d}_1 + v_3(v_4 + z_3 + \xi_4 - z_{2d} - \dot{\xi}_3) \\ &= -k_1v_1^2 - k_2v_2^2 - Iv_2\tilde{d}_1 + v_3(Kv_2 + v_4 + \alpha_3 - z_{2d} + k_3e_3 - k_3v_3) \end{aligned} \tag{30}$$

To make V_3 negative, the virtual control law α_3 is chosen as

$$\alpha_3 = -k_3e_3 + z_{2d} - Kv_2 \tag{31}$$

where k_3 is a positive design parameter. Substituting (31) into (30) yields

$$\dot{V}_3 = -k_1v_1^2 - k_2v_2^2 - k_3v_3^2 + v_3v_4 - Iv_2\tilde{d}_1 \tag{32}$$

If $v_4 = 0$ and $\tilde{d}_1 = 0$, then $\dot{V}_3 \leq 0$.

Step 4: To stabilize v_4, the final Lyapunov function candidate is chosen as:

$$V_4 = V_3 + \frac{1}{2}Jv_4^2 \tag{33}$$

The time derivative of v_4 is

$$\dot{v}_4 = \dot{e}_4 - \dot{\xi}_4 = \dot{x}_4 - \dot{z}_3 - \dot{\xi}_4 = \frac{u}{J} + \frac{K}{J}(x_1 - x_3) + d_2 - z_{3d} - \dot{\xi}_4 \tag{34}$$

Taking the time derivative of V_4 along (34) and using the fourth equation of (16) yields

$$\begin{aligned} \dot{V}_4 &= \dot{V}_3 + Jv_4\dot{v}_4 \\ &= -k_1v_1^2 - k_2v_2^2 - k_3v_3^2 + v_3v_4 - Iv_2\tilde{d}_1 \\ &\quad + v_4[u + K(x_1 - x_3) + Jd_2 - Jz_{3d} - J\dot{\xi}_4] \\ &= -k_1v_1^2 - k_2v_2^2 - k_3v_3^2 - Iv_2\tilde{d}_1 \\ &\quad + v_4[v_3 + u + K(x_1 - x_3) + Jd_2 - Jz_{3d} + k_4e_4 - k_4v_4] \end{aligned} \tag{35}$$

To make V_4 negative, the actual control law u is designed as

$$u = -k_4e_4 - K(x_1 - x_3) + Jz_{3d} - v_3 - J\hat{d}_2 \tag{36}$$

where k_4 is a positive design parameter, and $\hat{d}_2$ is the estimate of d_2 from the RESO (9). Substituting (36) into (35) yields

$$\dot{V}_4 = -k_1v_1^2 - k_2v_2^2 - k_3v_3^2 - k_4v_4^2 - Iv_2\tilde{d}_1 - Jv_4\tilde{d}_2 \tag{37}$$

If $\tilde{d}_1 = 0$ and $\tilde{d}_2 = 0$, then $\dot{V}_4 \leq 0$.

Summarizing the above design steps, the proposed SCFB controller is composed of the following control laws.

$$\begin{cases} \alpha_1 = -k_1e_1 + \dot{y}_d \\ \alpha_2 = \dfrac{1}{K}\left(-k_2e_2 + Kx_1 + MgL\sin x_1 + Iz_{1d} - v_1 - I\hat{d}_1\right) \\ \alpha_3 = -k_3e_3 + z_{2d} - Kv_2 \\ u = -k_4e_4 - K(x_1 - x_3) + Jz_{3d} - v_3 - J\hat{d}_2 \end{cases} \tag{38}$$

where $z_{id}(i = 1, 2, 3)$ come from the SCFs (4), and $\hat{d}_1, \hat{d}_2$ are from the RESOs (8) and (9). The block diagram of the proposed RESO based SCFB composite control system for the single-link FJM is shown in Figure 3.

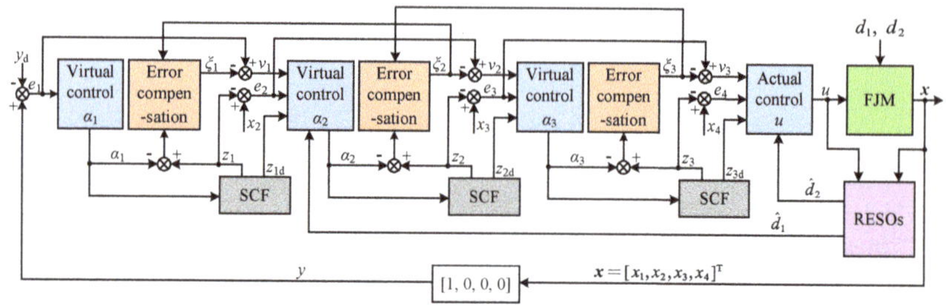

Figure 3. Block diagram of the composite control system for the single–link FJM control system.

4. Stability Analysis

In this section, the convergence of the error signals and the stability of the resulting control system are analyzed using the Lyapunov stability theory.

Theorem 1. *Consider the proposed RESOs (8) and (9) for the single-link FJM described by (2) with matched and mismatched disturbances. If Assumption 2 is satisfied, then the error norm of the RESOs is bounded by*

$$\|e_o\| \leq \frac{\max(\bar{d}_i)}{\min(|\beta_i|)}(i=1,2) \tag{39}$$

Proof. Consider the Lyapunov function candidate V_o

$$V_o = \frac{1}{2}e_o^\top P e_o \tag{40}$$

where P is a positive definite matrix. Since A_o is a Hurwitz matrix, there exists a positive definite matrix Q such that

$$A_o^\top P + PA_o = -Q \tag{41}$$

Taking the time derivative of V_o along (14) and (41) yields

$$\begin{aligned} \dot{V}_o &= \frac{1}{2}\dot{e}_o^T P e_o + \frac{1}{2}e_o^T P \dot{e}_o \\ &= \frac{1}{2}e_o^T\left(A_o^T P + PA_o\right)e_o + e_o^T P\dot{d} \\ &= -\frac{1}{2}e_o^T Q e_o + e_o^T P\dot{d} \\ &\leq -\frac{1}{2}\|e_o\|^2 Q + \|e_o\|\|P\|\max(\bar{d}_i) \\ &\leq -\|e_o\|[\|e_o\|\|P\|\min(|\beta_i|) - \|P\|\max(\bar{d}_i)] \end{aligned} \tag{42}$$

where $\min(|\beta_i|)(i=1,2)$ represents the smallest eigenvalue of A_o. Therefore, within finite time, the norm of the estimation errors is bounded by

$$\|e_o\| \leq \frac{\max(\bar{d}_i)}{\min(|\beta_i|)}(i=1,2) \tag{43}$$

This completes the proof of Theorem 1. □

Theorem 2. *Consider the single-link FJM described by (2) with matched and mismatched disturbances. Suppose that Assumptions 1 and 2 are satisfied. The RESOs (8) and (9), the error compensation system (16), and the SCFB controller (38) guarantee that all the signals in the control system are uniformly ultimately bounded, and the tracking errors converge to a small neighborhood around zero.*

Proof. Based on the design procedure in Section 3.2, the final Lyapunov function candidate for the feedback control is

$$V_f = \frac{1}{2}v_1^2 + \frac{I}{2}v_2^2 + \frac{1}{2}v_3^2 + \frac{J}{2}v_4^2 \tag{44}$$

According to (37), the time derivative of V_f can be described as

$$\dot{V}_f = -k_1 v_1^2 - k_2 v_2^2 - k_3 v_3^2 - k_4 v_4^2 - I v_2 \tilde{d}_1 - J v_4 \tilde{d}_2 \tag{45}$$

Using the Young's inequality, we have

$$\begin{aligned} \left|I v_2 \tilde{d}_1\right| &\leq \frac{I}{2}v_2^2 + \frac{I}{2}\tilde{d}_1^2 \\ \left|J v_4 \tilde{d}_2\right| &\leq \frac{J}{2}v_4^2 + \frac{J}{2}\tilde{d}_2^2 \end{aligned} \tag{46}$$

Substituting (46) into (45) yields

$$\dot{V}_f \leq -k_1 v_1^2 - \left(k_2 - \frac{I}{2}\right)v_2^2 - k_3 v_3^2 - \left(k_4 - \frac{J}{2}\right)v_4^2 - \frac{I}{2}\tilde{d}_1^2 - \frac{J}{2}\tilde{d}_2^2 \tag{47}$$

Rewriting inequality (47) in a compact form, we have

$$\dot{V}_f \leq -\alpha_0 V_f + D \tag{48}$$

where

$$\begin{aligned} \alpha_0 &= \min\left(2k_1, 2k_2 - I, 2k_3, 2k_4 - J\right) \\ D &= -\frac{I}{2}\tilde{d}_1^2 - \frac{J}{2}\tilde{d}_2^2 \end{aligned} \tag{49}$$

Selecting the design parameters $k_1 > 0, k_2 > I/2, k_3 > 0, k_4 > J/2$ to ensure $\alpha_0 > 0$. Then, according to Lemma 2, the solution of (48) is

$$V_f(t) \leq \frac{D}{\alpha_0} + \left[V_f(0) - \frac{D}{\alpha_0}\right]e^{-\alpha_0 t} \tag{50}$$

which means that $V_f(t)$ converges exponentially to the upper bound of $\frac{D}{\alpha_0}$, i.e., as $t \to \infty$, $V_f(t) \leq \frac{D}{\alpha_0}$. As a result, the compensated errors $v_i(i = 1,2,3,4)$ are bounded.

Furthermore, it has been proved in Ref. [50] that the compensation signals $\xi_i(i = 1,2,3,4)$ in (16) are bounded. Since $e_i = v_i + \xi_i$, it is clearly known that the tracking errors e_i are also bounded. According to Lemma 1 and Theorem 1, the filter outputs $z_i, z_{id}(i = 1,2,3)$ and the estimations $\hat{d}_1, \hat{d}_2$ are bounded. From (38), it is evident that the virtual control laws $\alpha_1(\dot{x}_d, e_1)$, $\alpha_2\left(x_1, e_2, \hat{d}_1, z_{1d}, v_1\right)$, $\alpha_3(e_3, z_{2d}, v_2)$, and the actual law $u\left(x_1, x_3, e_4, \hat{d}_2, z_{3d}, v_3\right)$ are also bounded because of the boundedness of their independent variables.

Therefore, all the signals in the closed-loop control system are uniformly ultimately bounded. In addition, by properly choosing the design parameters, such as large k_i(i=1,2,3,4) and $\beta_i(i = 1,2)$, the tracking errors $e_i(i = 1,2,3,4)$ can converge to a small neighbourhood around zero. This completes the proof of Theorem 2. □

5. Numerical Simulations

To evaluate the effectiveness and efficiency of the proposed control method, two simulation tests are conducted in the MATLAB/SIMULINK platform. Specifically, the first simulation is presented to examine the tracking control performance of the proposed approach for the single-link FJM with matched and mismatched disturbances as well as measurement noises. In the second simulation, a comparison study between the proposed method and the existing approaches are presented to show the superior performance of the proposed controller.

The physical parameters of the single-link FJM in (1) are given as: $M = 0.25$ kg, $g = 9.8\ \mathrm{m/s^2}$, $L = 0.45$ m, $K = 5\ \mathrm{N \cdot m/rad}$, $I = 0.05\ \mathrm{m/s^2}$, $J = 0.0005\ \mathrm{m/s^2}$. The gains of the controller (38) are chosen as $k_1 = 7$, $k_2 = 25$, $k_3 = 8$, $k_4 = 8$. Meanwhile, the parameters of the SCFs (4) are chosen as $\omega_n = 1100$, $\zeta = 0.8$, and the parameters of the proposed RESOs in (8) and (9) are: $\beta_1 = 50$, $\beta_2 = 50$.

The initial condition of the FJM is set as $x(0) = [0.5, 0, 0, 0]^{\top}$. The initial states of the error compensation system and the SCFs are all zeros. The desired reference trajectory is chosen as $y_{\mathrm{d}} = 0.5(\sin t + \sin 0.5t)$, and the lumped mismatched and matched disturbances added for the two simulations are given as:

$$\begin{cases} w_1 = 0.005 \cos \dot{q} + 0.3 \sin(2\pi t) \\ w_2 = 0.1\dot{\theta} + 0.3 \cos(2\pi t) \end{cases} \tag{51}$$

5.1. Simulation Results with Disturbances and Noises

To imitate the measurement noises of encoders equipped in the FJM, Gaussian white noise with zero mean and standard deviation of 0.1, which can be generated by the "randn" function in MATLAB, is added for all the state measurements. The simulation results are depicted in Figures 4–9.

Figure 4 displays the trajectories of the system output y and the given reference signal y_{d}. Figure 5 shows the curve of tracking error. From these figures, we can see that the proposed approach achieves a satisfactory tracking control performance despite the presence of mismatched and matched disturbances as well as noises affecting measurements. The estimates of the mismatched and matched disturbances are respectively illustrated in Figures 6 and 7, from which we can see that the designed RESOs can estimate the disturbances quickly and precisely.

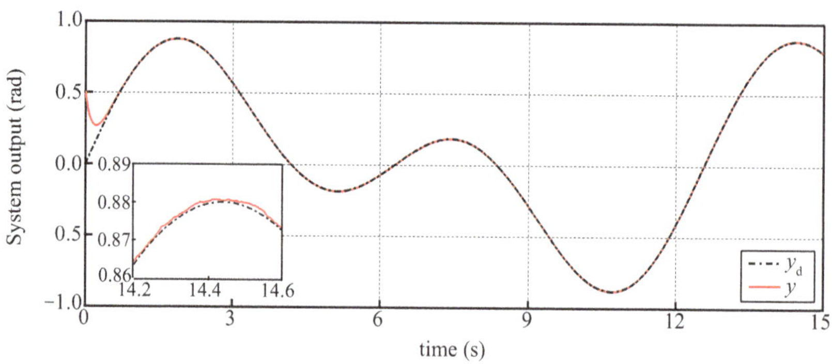

Figure 4. Tracking performance of the system output y under the proposed approach with noises.

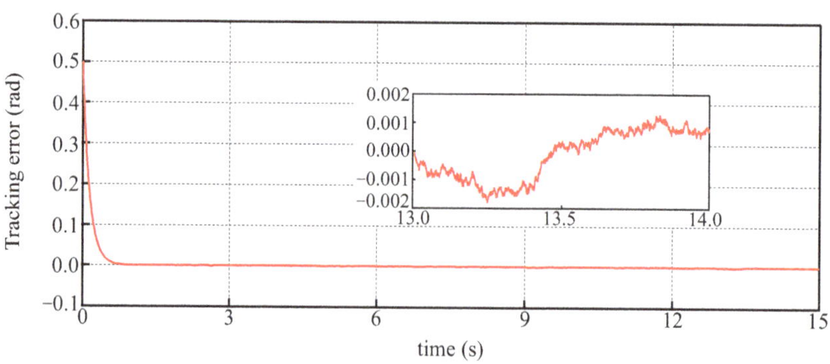

Figure 5. Tracking error e_1 under the proposed approach with noises.

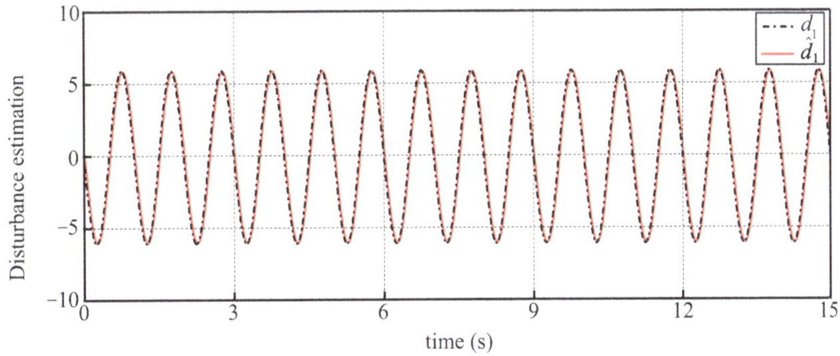

Figure 6. Estimation performance of d_1 via RESO.

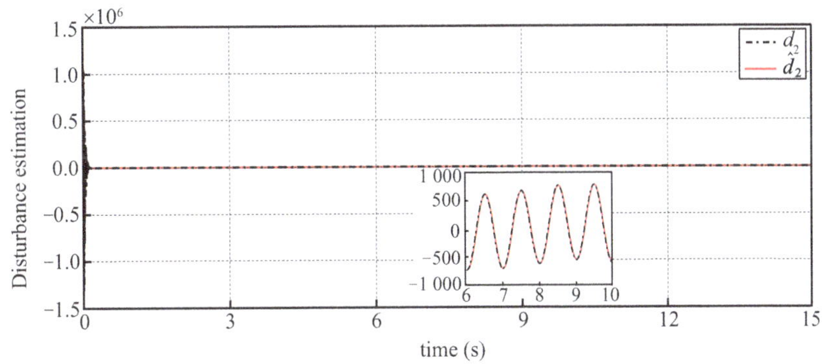

Figure 7. Estimation performance of d_2 via RESO.

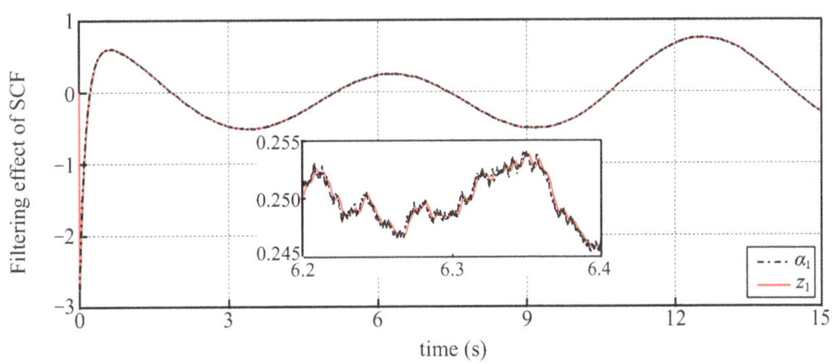

Figure 8. Filtering performance of the SCF with α_1 being the input and z_1 being the output.

Figure 8 depicts the intermediate virtual control law α_1 and the corresponding filtered control signal z_1 from the SCF. It is clearly seen that the virtual control law α_1 is corrupted by the white noises, but the filtered signal z_1 through the SCF is smooth. In addition, Figure 9 shows the comparison results between the direct time derivative of α_1 and the output z_{1d} of the SCF in the presence of white noises. It is noticed that the chattering amplitude of $\dot{\alpha}_1$ is much bigger than that of z_{1d}, which indicates that the noises in $\dot{\alpha}_1$ are amplified. The reason is that the derivative of α_1 is approximated by the SCF through integration rather than differentiation (as shown in Figure 2). These results demonstrate that the SCF employed for FJMs with measurement noises can not only filter the noises for the control signal to some extent, but also avoid the amplification of the noises in the recursive backstepping design. Similar results are obtained for α_i $(i = 2,3)$ and z_{id}, which are omitted for the space saving.

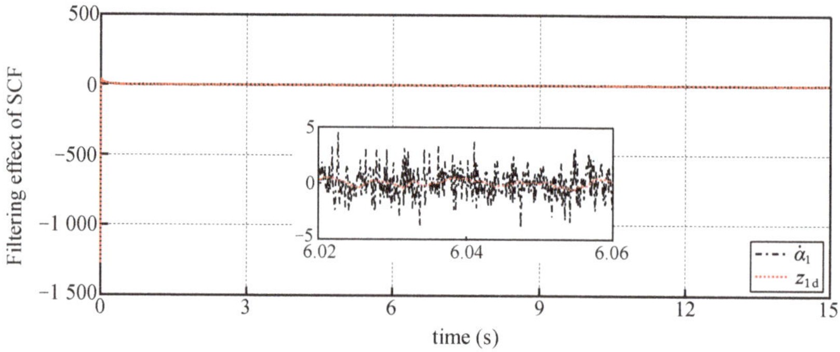

Figure 9. Comparison results between $\dot{\alpha}_1$ and the output of the SCF z_{1d}.

5.2. Comparison Results with CBC and AFCFC Methods

In order to further show the superior performance of the proposed RESO based command filtered controller (RBCFC), the following two controllers are selected to make a comparison study.

(1) **Conventional backstepping controller (CBC) proposed in [16].** The structure of the controller is given as:

$$\begin{cases} \alpha_1 = -k_1 e_1 + \dot{y}_d \\ \alpha_2 = \dfrac{1}{K}(-k_2 e_2 - e_1 + MgL\sin x_1 + Kx_1 + I\dot{\alpha}_1) \\ \alpha_3 = -k_3 e_3 - Ke_2 + \dot{\alpha}_2 \\ u = -k_4 e_4 - e_3 - K(x_1 - x_3) + J\dot{\alpha}_3 \end{cases} \tag{52}$$

(2) **Adaptive fuzzy command filtered controller (AFCFC) proposed in [30].** The structure of the controller is given as:

$$\begin{cases} \alpha_1 = -k_1 e_1 + \dot{y}_d \\ \alpha_2 = \dfrac{1}{M^{-1}K}(-k_2 e_2 - \dfrac{1}{2}v_2 - e_1 + z_{1d} - \dfrac{v_2\hat{\theta}\psi_1^\top\psi_1}{2l_1^2}) \\ \alpha_3 = -k_3 e_3 - M^{-1}Ke_2 + z_{2d} \\ u = B(e_4 + z_3) + K(e_3 + z_2 - e_1 - y_d) + J(-k_4 e_4 - e_3 + z_{3d}) \\ \dot{\hat{\theta}} = \dfrac{r v_2^\top v_2 \psi_1^\top \psi_1}{2l_1^2} - \bar{r}\hat{\theta} \end{cases} \tag{53}$$

where $z_{id}(i = 1,2,3)$ are the outputs of SCFs, $l_1, r, \bar{r}$ are positive design parameters, $\hat{\theta}$ is the estimation of the adaptive parameter θ, ψ_1 is the vector of fuzzy basis functions. More details about these control parameters are referred to in Ref. [30].

Remark 1. *It is noticed that the structure of the CBC is the simplest among the three controllers, but it suffers from the drawback of computational complexity, and does not consider the influence of disturbances. Both the proposed RBCFC and AFCFC can address these problems, but the AFCFC employs a fuzzy system with an online adaptive learning law to approximate the unknown disturbances, while the proposed RBCFC utilizes two simple RESOs to estimate the disturbances.*

Since the three controllers are designed using the same backstepping methodology, their gains are chosen as the same for the sake of fair comparison, i.e., $k_1 = 7$, $k_2 = 25$, $k_3 = 8$, $k_4 = 8$. Meanwhile, the same parameters of the SCFs are chosen for the proposed controller and the AFCFC, i.e., $\omega_n = 1100$, $\zeta = 0.8$. Other parameters for the AFCFC are chosen as in Ref. [30]. No measurement noises are considered in this circumstance for a clear and fair comparison.

The simulation results of the single-link FJM under the three controllers are depicted in Figures 10–13, which record the curves of the tracking performance of the system output y, the tracking error e_1, the state of the motor angle θ, and the control torque u, respectively.

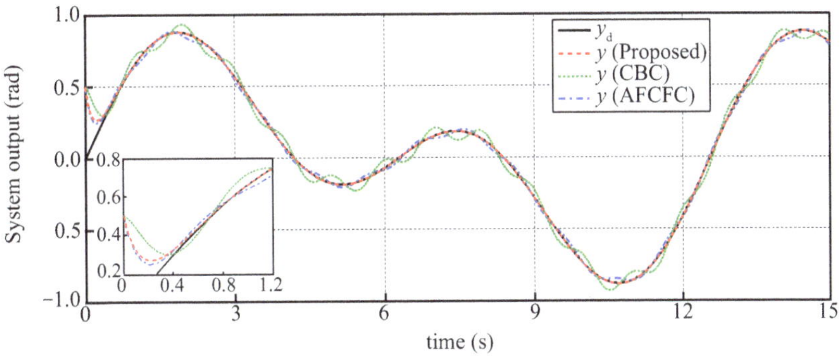

Figure 10. Tracking performance of the system output y under the three controllers.

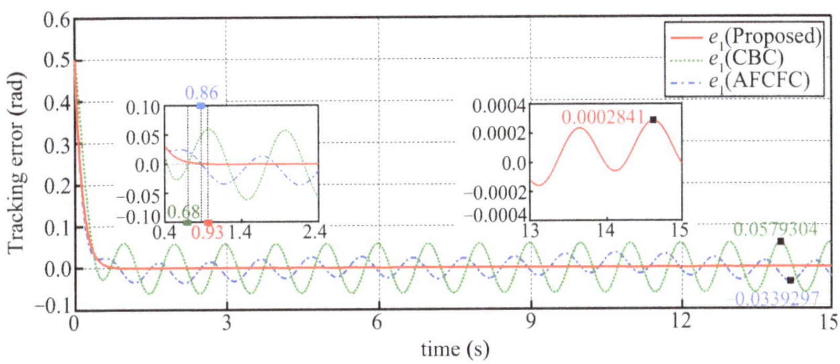

Figure 11. Tracking error e_1 under the three controllers.

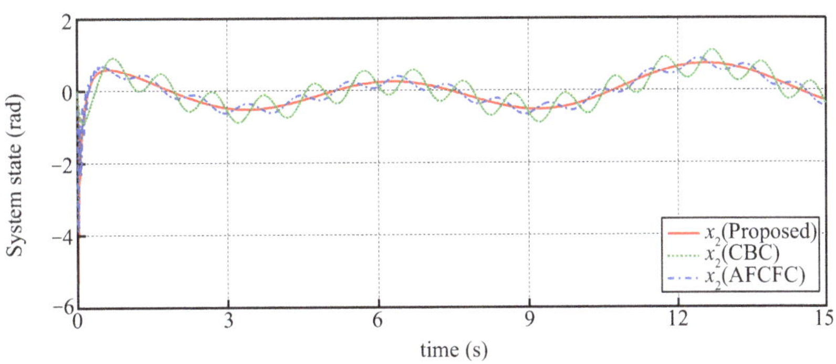

Figure 12. System state x_2 (the motor angle θ) under the three controllers.

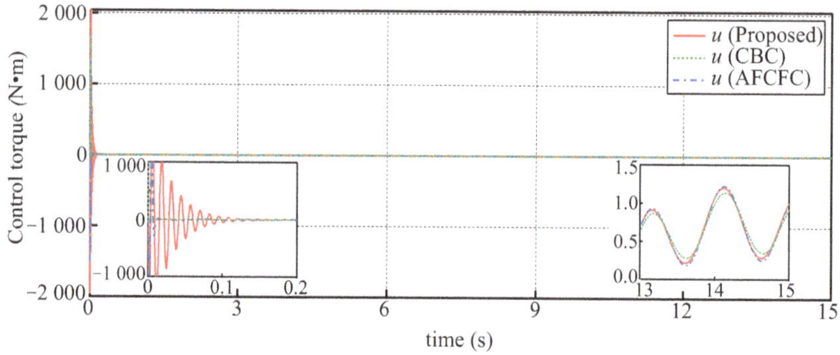

Figure 13. Control torque u under the three controllers.

From Figure 10, it is seen that both the proposed controller and the AFCFC can attenuate the influence of lumped disturbances and achieve satisfactory control performance. However, the system output with the CBC is seriously affected by the disturbances, and fluctuates around the reference trajectory. This is mainly because the CBC does not contain a compensation term for the disturbances. The same results can be further verified by Figures 11 and 12, which respectively show that the tracking error e_1 and the motor angle θ under the three controllers converge to a small neighborhood around zero. It is clearly seen that the convergence radius of the proposed controller is the smallest among the three controllers.

In order to quantitatively analyze the control performance of the three controllers, we define three performance indexes including the settling time t_s, the mean squared error in the transient stage between 0.4 s and 2.4 s, i.e.,

$$e_{1\mathrm{MSE}} = \frac{1}{N}\sum_{i=1}^{N}[e(i)]^2 \tag{54}$$

and the maximum tracking error in the steady stage $|e_{1\infty\,\max}|$. The details of the quantified performance indexes of the three controllers are given in Table 1.

Table 1. Performance indexes under the three controllers.

Methods	t_s (s)	$e_{1\mathrm{MSE}}$ (rad)	$\lvert e_{1\infty\,\max}\rvert$ (rad)
RBCFC [Proposed]	0.93	3.09×10^{-5}	2.84×10^{-4}
CBC [16]	0.68	1.51×10^{-3}	5.79×10^{-2}
AFCFC [30]	0.86	4.84×10^{-4}	3.39×10^{-2}

As seen from Table 1, the proposed RBCFC exhibits better performance than the other two controllers in aspects of transient and steady tracking errors. Although the settling time of the proposed controller is a little longer than those of other controllers, the mean squared error of the proposed controller in the transient stage is reduced almost down to 3.09×10^{-5}, which is quite smaller than that (i.e., 1.51×10^{-3}) of CBC, as well as that (i.e., 4.84×10^{-4}) of AFCFC. In addition, the maximum tracking error of the proposed controller in the steady stage is about 2.84×10^{-4}, while those of CBC and AFCFC are 5.79×10^{-2} and 3.39×10^{-2}, respectively. Compared with the CBC and the AFCFC, the tracking accuracy of the proposed controller is improved by 99.5% and 99.2%, respectively.

From Figure 13, it is seen that the control torque u of the proposed controller in the initial stage between 0 s and 0.2 s shows large fluctuations. The reason for this may be that the proposed controller in the system adjustment stage is more susceptible to the unknown disturbances. By introducing the RESOs, however, the disturbances are quickly estimated and compensated for in the feedback control.

According to the above comparative simulation results, it can be concluded that compared with other methods, the proposed controller can estimate and compensate the unknown matched and mismatched disturbances effectively, and achieve a better transient and steady tracking performance.

6. Conclusions

This paper has successfully proposed an efficient model-assisted composite control approach for the high precision tracking control of FJMs in the presence of lumped matched and mismatched disturbances. Two RESOs are constructed with partial known model information of FJMs to estimate and compensate the disturbances, three second-order command filters are incorporated into the backstepping control design to avoid the problem of "explosion of complexity", and an error compensation dynamic system is designed to reduce the filtering errors. The stability of the resulting control system is rigorously proven via Lyapunov stability theory, and the tracking errors are guaranteed to be uniformly ultimately bounded. The numerical simulation results prove that the proposed RESOs deliver accurate estimates of both the matched and mismatched disturbances. In addition, compared with conventional backstepping method and adaptive fuzzy command filtered control method, the proposed approach achieves a better tracking performance, i.e., the tracking accuracy is improved by 99.5% and 99.2%, respectively.

It is worth pointing out that the proposed approach requires the knowledge of all the state variables, which may be unfeasible or inconvenient in practice. It is interesting to study the output feedback control problem of FJMs with unmeasured state variables by integrating the state estimation technique, e.g., Kalman state observers presented in Refs. [1,2]. In addition, it is necessary and significant to evaluate the control performance of the proposed approach under varying load mass and extend the proposed approach to a more general 6-DOF FJM. These issues will be investigated in our future works.

Author Contributions: Conceptualization, C.P. and X.F.; methodology, X.F.; software, X.F. and Z.L.; validation, C.P., X.F. and H.H.; formal analysis, Z.L.; investigation, C.P. and X.F.; resources, C.P., X.F. and H.H.; data curation, X.F.; writing—original draft preparation, X.F.; writing—review and editing, C.P. and J.X.; visualization, P.X. and Z.L.; supervision, C.P.; project administration, C.P. and J.X.; funding acquisition, C.P. and J.X. All authors have read and agreed to the published version of the manuscript.

Funding: This work is supported in part by: (i) National Natural Science Foundation of China (Grant No. 62173138); (ii) Guangdong Basic and Applied Basic Research Foundation (Grant Nos. 2020A1515011082 and 2019A1515010955); (iii) Scientific Research Fund of Hunan Provincial Education Department (Grant Nos. 20A186 and 21C0329); (iv) Hunan Provincial Natural Science Foundation of China (Grant No. 2022JJ30263).

Institutional Review Board Statement: Not applicable.

Informed Consent Statement: Not applicable.

Data Availability Statement: Not applicable.

Conflicts of Interest: The authors declare no conflict of interest.

References

1. Barjuei, E.S.; Gasparetto, A. Predictive control of spatial flexible mechanisms. *Int. J. Mech. Control* **2015**, *16*, 85–96.
2. Palomba, I.; Richiedei, D.; Trevisani, A. Reduced-order observers for nonlinear state estimation in flexible multibody systems. *Shock Vib.* **2018**, *2018*, 6538737. [CrossRef]
3. Zhang, A.; Lin, Z.; Wang, B.; Han Z. Nonlinear model predictive control of single-link flexible-joint robot using recurrent neural network and differential evolution optimization. *Electronics* **2021**, *10*, 2426. [CrossRef]
4. Spyrakos-Papastavridis, E.; Da, J.S. Minimally model-based trajectory tracking and variable impedance control of flexible-joint robots. *IEEE Trans. Ind. Electron.* **2020**, *68*, 6031–6041. [CrossRef]
5. Li, Z.; Yang, C.; Su, C.Y.; Deng, A.; Fuchun, S.; Sun, F.; Zhang, W. Decentralized fuzzy control of multiple cooperating robotic manipulators with impedance interaction. *IEEE Trans. Fuzzy Syst.* **2014**, *23*, 1044–1056. [CrossRef]

6. Chen, T.; Shan, J. Distributed control of multiple flexible manipulators with unknown disturbances and dead-zone input. *IEEE Trans. Ind. Electron.* **2019**, *67*, 9937–9947. [CrossRef]
7. Yang, C.; Jiang, Y.; He, W.; Na, J.; Li, Z.; Xu, B. Adaptive parameter estimation and control design for robot manipulators with finite-time convergence. *IEEE Trans. Ind. Electron.* **2018**, *65*, 8112–8123. [CrossRef]
8. Ibrahim, K.; Sharkawy, A.B. A hybrid PID control scheme for flexible joint manipulators and a comparison with sliding mode control. *Ain Shams Eng. J.* **2018**, *9*, 3451–3457. [CrossRef]
9. Kelekci, E.; Kizir, S. Trajectory and vibration control of a flexible joint manipulator using interval type-2 fuzzy logic. *ISA Trans.* **2019**, *94*, 218–233. [CrossRef]
10. Ak, A. Sliding mode controller design using fuzzy sliding surface for flexible joint manipulator. *IETE J. Res.* **2021**, *68*, 760–767. [CrossRef]
11. Sun, T.; Liang, D.; Song, Y. Singular-perturbation-based nonlinear hybrid control of redundant parallel robot. *IEEE Trans. Ind. Electron.* **2017**, *65*, 3326–3336. [CrossRef]
12. Chen, H.; Dong, X.; Yang, Y.; Liu, J. Backstepping sliding mode control of uncertainty flexible joint manipulator with actuator saturation. In Proceedings of the 2020 International Symposium on Automation, Information and Computing, Beijing, China, 2–4 December 2020; Volume 1828.
13. Hooshmand, H.; Fateh, M.M. Voltage control of flexible-joint robot manipulators using singular perturbation technique for model order reduction. *J. Electr. Comput. Eng. Innov.* **2022**, *10*, 123–142.
14. Korayem, M.H.; Nekoo, S.R.; Kazemi, S. Finite-time feedback linearization (FTFL) controller considering optimal gains on mobile mechanical manipulators. *J. Intell. Robot. Syst.* **2019**, *94*, 727–744. [CrossRef]
15. Abdul-Adheem, W.R.; Ibraheem, I.K.; Humaidi, A.J.; Azar, A.T. Model-free active input–output feedback linearization of a single-link flexible joint manipulator: An improved active disturbance rejection control approach. *Meas. Control* **2021**, *54*, 856–871. [CrossRef]
16. Oh, J.H.; Lee, J.S. Control of flexible joint robot system by backstepping design approach. *Intell. Autom. Soft Comput.* **1999**, *5*, 267–278. [CrossRef]
17. Huang, A.C.; Chen, Y.C. Adaptive sliding control for single-link flexible-joint robot with mismatched uncertainties. *IEEE Trans. Control Syst. Technol.* **2004**, *12*, 770–775. [CrossRef]
18. Dian, S.; Hu, Y.; Zhao, T.; Han, J. Adaptive backstepping control for flexible-joint manipulator using interval type-2 fuzzy neural network approximator. *Nonlinear Dyn.* **2019**, *97*, 1567–1580. [CrossRef]
19. Wang, L.; Shi, Q.; Liu, J.; Zhang, D. Backstepping control of flexible joint manipulator based on hyperbolic tangent function with control input and rate constraints. *Asian J. Control* **2020**, *22*, 1268–1279. [CrossRef]
20. Swaroop, D.; Hedrick, J.K.; Yip, P.P.; Gerdes, J.C. Dynamic surface control for a class of nonlinear systems. *IEEE Trans. Aautomatic Control* **2000**, *45*, 1893–1899. [CrossRef]
21. Ling, S.; Wang, H.; Liu, P.X. Adaptive fuzzy dynamic surface control of flexible-joint robot systems with input saturation. *IEEE/CAA J. Autom. Sin.* **2019**, *6*, 97–107. [CrossRef]
22. Yoo, S.J.; Park, J.B.; Choi, Y.H. Adaptive output feedback control of flexible-joint robots using neural networks: Dynamic surface design approach. *IEEE Trans. Neural Netw.* **2008**, *19*, 1712–1726. [PubMed]
23. Li, Y.; Tong, S.; Li, T. Fuzzy adaptive dynamic surface control for a single-link flexible-joint robot. *Nonlinear Dyn.* **2012**, *70*, 2035–2048. [CrossRef]
24. Xiong, G.; Xie, Z.; Huang, J.; Liu, H.; Jiang, Z.; Sun, K. Dynamic surface control-backstepping based impedance control for 5-DOF flexible joint robots. *J. Cent. South Univ. Technol.* **2010**, *17*, 807–815. [CrossRef]
25. Farrell, J.A.; Polycarpou, M.; Sharma, M.; Dong, W. Command filtered backstepping. *IEEE Trans. Autom. Control* **2009**, *54*, 1391–1395. [CrossRef]
26. Dong, W.; Farrell, J.A.; Polycarpou, M.M.; Djapic, V.; Sharma, M. Command filtered adaptive backstepping. *IEEE Trans. Control Syst. Technol.* **2012**, *20*, 566–580. [CrossRef]
27. Wang, L.; Yang, C. Neural network adaptive command filtered control of robotic manipulators with input saturation. *Int. J. Adv. Robot. Syst.* **2019**, *16*, 1729881419894779. [CrossRef]
28. Deng, Y. Adaptive finite-time fuzzy command filtered controller design for uncertain robotic manipulators. *Int. J. Adv. Robot. Syst.* **2019**, *16*, 1729881419828148. [CrossRef]
29. Lin, M.; Lei, Z.; Zhang, Y.; Li, P.; Wang, X. Command filter backstepping control with error compensation for flexible-joint manipulator. In Proceedings of the 2020 Chinese Control And Decision Conference, Hefei, China, 22–24 August 2020; pp. 3696–3701.
30. Ling, S.; Wang, H.; Liu, P.X. Adaptive fuzzy tracking control of flexible-joint robots based on command filtering. *IEEE Trans. Ind. Electron.* **2019**, *67*, 4046–4055. [CrossRef]
31. Liu, X.; Yang, C.; Chen, Z.; Wang, M.; Su, C. Neuro-adaptive observer based control of flexible joint robot. *Neurocomputing* **2018**, *275*, 73–82. [CrossRef]
32. Sun, W.; Su, S.F.; Xia, J.; Nguyen, V. Adaptive fuzzy tracking control of flexible-joint robots with full-state constraints. *IEEE Trans. Syst. Man Cybern. Syst.* **2018**, *49*, 2201–2209. [CrossRef]
33. Soltanpour, M.R.; Moattari, M. Voltage based sliding mode control of flexible joint robot manipulators in presence of uncertainties. *Robot. Auton. Syst.* **2019**, *118*, 204–219.

34. Cheng, X.; Liu, H.; Lu, W. Chattering-suppressed sliding mode control for flexible-joint robot manipulators. *Actuators Multidiscip. Digit. Publ. Inst.* **2021**, *10*, 288. [CrossRef]
35. Alam, W.; Ahmad, S.; Mehmood, A.; Iqbal, J. Robust sliding mode control for flexible joint robotic manipulator via disturbance observer. *Interdiscip. Descr. Complex Syst. INDECS* **2019**, *17*, 85–97. [CrossRef]
36. Zaare, S.; Soltanpour, M.R. Continuous fuzzy nonsingular terminal sliding mode control of flexible joints robot manipulators based on nonlinear finite time observer in the presence of matched and mismatched uncertainties. *J. Frankl. Inst.* **2020**, *357*, 6539–6570. [CrossRef]
37. Ullah, H.; Malik, F.M.; Raza, A.; Mazhar, N.; Khan, R.; Saeed, A.; Ahmad, I. Robust output feedback control of single-link flexible-joint robot manipulator with matched disturbances using high gain observer. *Sensors* **2021**, *21*, 3252. [CrossRef]
38. Hu, Y.; Dian, S.; Guo, R.; Li, S.; Zhao, T. Observer-based dynamic surface control for flexible-joint manipulator system with input saturation and unknown disturbance using type-2 fuzzy neural network. *Neurocomputing* **2021**, *436*, 162–173. [CrossRef]
39. Han, J.Q. From PID to active disturbance rejection control. *IEEE Trans. Ind. Electron.* **2009**, *56*, 900–906. [CrossRef]
40. Talole, S.E.; Kolhe, J.P.; Phadke, S.B. Extended-state-observer-based control of flexible-joint system with experimental validation. *IEEE Trans. Ind. Electron.* **2009**, *57*, 1411–1419. [CrossRef]
41. Rsetam, K.; Cao, Z.; Man, Z. Cascaded-extended-state-observer-based sliding-mode control for underactuated flexible joint robot. *IEEE Trans. Ind. Electron.* **2019**, *67*, 10822–10832. [CrossRef]
42. Huang, Y.; Xue, W. Active disturbance rejection control: Methodology and theoretical analysis. *ISA Trans.* **2014**, *53*, 963–976. [CrossRef]
43. Shao, X.; Wang, H. Back-stepping active disturbance rejection control design for integrated missile guidance and control system via reduced-order ESO. *ISA Trans.* **2015**, *57*, 10–22.
44. Zhou, Y.; Chang, S. A model-assisted reduced-order ESO based cascade controller for sensorless control of independent gear-shifting actuators. *Appl. Sci.* **2017**, *7*, 1283. [CrossRef]
45. Xie, T.; Li, Y.; Jiang, Y.; An, L.; Wu, H. Backstepping active disturbance rejection control for trajectory tracking of underactuated autonomous underwater vehicles with position error constraint. *Int. J. Adv. Robot. Syst.* **2020**, *17*, 1–12. [CrossRef]
46. Wang, J.; Rong, J.; Yu, L. Reduced-order extended state observer based event-triggered sliding mode control for DC-DC buck converter system with parameter perturbation. *Asian J. Control* **2021**, *23*, 1591–1601. [CrossRef]
47. Zhai, J.; Li, S.; Xu, Z.; Zhang, L.; Li, J. Reduced-order extended state observer-based sliding mode control for all-clamped plate using an inertial actuator. *Energies* **2022**, *15*, 1780. [CrossRef]
48. Chien, M.C.; Huang, A.C. Adaptive control for flexible-joint electrically driven robot with time-varying uncertainties. *IEEE Trans. Ind. Electron.* **2007**, *54*, 1032–1038. [CrossRef]
49. Narendra, K.S.; Annaswamy, A.M. *Robust Adaptive Control. Adaptive and Learning Systems*; Springer: Boston, MA, USA, 1986; pp. 3–31.
50. Yu, J.; Shi, P.; Dong, W.; Yu, H. Observer and command-filter-based adaptive fuzzy output feedback control of uncertain nonlinear systems. *IEEE Trans. Ind. Electron.* **2015**, *62*, 5962–5970. [CrossRef]

MDPI
St. Alban-Anlage 66
4052 Basel
Switzerland
www.mdpi.com

Applied Sciences Editorial Office
E-mail: applsci@mdpi.com
www.mdpi.com/journal/applsci

www.ingramcontent.com/pod-product-compliance
Lightning Source LLC
Chambersburg PA
CBHW040813250926
51850CB00007BA/1159
9783036597560